D0991056

The Respiratory Physiology of Animals

The Respiratory Physiology of Animals

JAMES N. CAMERON

Marine Science Institute
The University of Texas

New York Oxford
OXFORD UNIVERSITY PRESS
1989

Oxford University Press

Oxford New York Toronto
Delhi Bombay Calcutta Madras Karachi
Petaling Jaya Singapore Hong Kong Tokyo
Nairobi Dar es Salaam Cape Town
Melbourne Auckland

and associated companies in
Berlin Ibadan

Published by Oxford University Press, Inc.
200 Madison Avenue, New York, New York 10016

Library of Congress Cataloging-in-Publication Data
Cameron, James N.
The respiratory physiology of animals / James N. Cameron.
p. cm. Bibliography: p. Includes index.
ISBN 0-19-506019-9
1. Respiration. 2. Animals—Physiology.
3. Physiology, Comparative.
I. Title.
QP121.C26 1989
591.1′2—dc20 89-8795 CIP

2 4 6 8 9 7 5 3 1

Printed in the United States of America
on acid-free paper

Preface

All animals, from single-celled amebae to the fleetest falcon, share the common problems of obtaining oxygen fuel for metabolic energy and eliminating the carbon dioxide waste. The conservatism of biochemical evolution means also that related problems such as maintaining a proper internal pH are shared throughout the animal kingdom. My objective in writing this book was to treat the common processes of respiration, circulation, and acid-base regulation throughout the spectrum of animal complexity.

The book is in two parts: The first eight chapters deal with physiological principles in general topics, beginning with physicochemical principles, including types of exchangers, blood pigments, circulatory systems, and acid-base regulation, and ending with a discussion of respiratory control systems. The second part is a series of case studies, in which these principles are illustrated with specific data for representatives of different phyla and (perhaps) different levels of evolutionary complexity. Examples of invertebrates, aquatic breathers, partially air-breathing animals, and completely terrestrial animals both ectotherm and homeotherm are discussed. Research in respiratory physiology has been somewhat concentrated on species that are readily available and tractable in the laboratory, which has dictated to a certain extent the choice of case studies. The rainbow trout, for example, has become a sort of "white rat" for fish physiologists. For each of the chapters, the objectives have been to discuss the important ideas and to provide entry to recent literature for those wishing exhaustive detail.

A debt to Professor Pierre Dejours is to be acknowledged: his excellent treatise published in 1975 was the first successful attempt to discuss respiratory physiology from a broadly comparative standpoint since Krogh's writings of some thirty years earlier. His book, however, approached the subject with somewhat of a mammalian or clinical bias, and has, of course, become obsolete by the present standards of rapid scientific advancement. Assuming that Prof. Dejours had access to literature through 1974, more than half of the references cited in this work have been published since his *Principles of Comparative Respiratory Physiology* was written.

In addition to the text material, a supplemental computer disk is provided. This disk contains a number of programs that will do measurement unit conversions, calculate temperature-dependent parameters such as gas solubility, perform several routine analyses such as Hill plots and Davenport diagrams, and illustrate the use of computer simulation modelling. The programs are provided in IBM-compatible format (MS-DOS) in the hope that nearly everyone has access to such a machine.

In the course of completing my task, I have been continually impressed by the common features of respiratory systems in animals of totally different appearance, origin, and structure. Clinical hypothermia and the temperature responses of clams would seem to have little in common, for example, but viewed from the standpoint of temperature and pH regulation across many phyla the common pattern is plain. An understanding of how certain problems are solved in a variety of animals lends perspective to study of any one of them.

J. N. C.
Port Aransas, Texas
April, 1989

Acknowledgements

My debts to many people must be acknowledged. The support and valuable criticism provided throughout the project by William Curtis, my editor at Oxford, is gratefully acknowledged. Several colleagues kindly agreed to read and comment upon various chapter drafts; I am particularly indebted to D. R. Jones, C. P. Mangum, W. K. Milsom, N. J. Smatresk, and C. M. Wood, although any persistent deficiencies must be blamed on the author. I acknowledge with gratitude the excellent photographs supplied by J. B. Aldridge, H.-R. Duncker, B. Gannon, C. L. Kitting, D. P. Toews, S. F. Perry, and E. Weibel.

The author gratefully acknowledges permission to reprint material from many publications, including:

The American Journal of Physiology, American Physiological Society
American Zoologist, American Society of Zoologists
An Introduction to General and Comparative Physiology, W. B. Saunders
Avian Biology, Academic Press
Biological Bulletin, Marine Biological Laboratory
The Biology of Crustacea, Academic Press
Canadian Journal of Zoology, Environment Canada
Comparative Biochemistry and Physiology, Pergamon Press
Epithelial Transport in Lower Vertebrates, Cambridge University Press
Federation Proceedings, FASEB Journal, FASEB
Fish Physiology, Academic Press
Handbook of Physiology, American Physiological Society
The Heart, McGraw-Hill Book Company
How to Understand Acid-Base, Elsevier Science Publishers
Insect Physiology, Chapman and Hall
The Invertebrates, McGraw-Hill Book Company
Journal of Comparative Physiology, Springer-Verlag
Journal of Experimental Biology, Company of Biologists, Ltd.
Journal of Experimental Marine Biology and Ecology, Elsevier Science Publishers
The Journal of Experimental Zoology, Alan R. Liss
The Journal of Morphology, Alan R. Liss
Marine Biology Letters, Elsevier Science Publishing
The Mollusca, Academic Press
The Pathway for Oxygen, Harvard University Press
Perspectives in Experimental Biology, Pergamon Press
Physiology and Biophysics, W. B. Saunders Co.
The Physiology of Crustacea, Academic Press

Respiration Physiology, Elsevier Science Publishing

Support for my work over the years has been generously provided by the National Science Foundation, and I owe a number of colleagues for their support, guidance, and stimulation. For infinite patience and support I thank my wife and best friend, Sharon.

Contents

Part I: General Principles

Part II: Case Studies of Animals

I
General Principles

Chapter 1

Basic Gas and Solution Concepts

The metabolic activities of life involve the exchange of respiratory gases, and because the basic life processes take place in aqueous solution, a study of respiration can hardly proceed without a thorough understanding of the physical properties of both gases and liquids. Equally important is an understanding of physicochemical processes such as diffusion, solution equilibria, and surface tension. As an introduction to the subject of respiratory physiology, a brief description of some important concepts is presented in the following sections, with references to more detailed texts or to treatments of specialized topics.

DEFINITIONS

Fluid (from Latin *fluere*, to flow) means a state of matter in which the molecules are relatively free to change their positions with respect to each other in a continuous fashion. The ease with which positional change can occur varies widely from one fluid to another, depending on the nature of the substance and the kinetic energy of the molecules.

A *gas* (probably from Greek *chaos*) is a fluid in which the molecules are practically unrestricted by cohesive forces, and so a gas has no definite shape and no intrinsic volume. Both shape and volume are imparted by the container of the gas; without a container a gas will tend to expand infinitely.

A *liquid* (from Latin *liquidus*) is also a fluid, but the important distinction is that in a liquid the weak cohesive forces between molecules become important, so the molecules are still relatively free to change position with respect to each other but are restricted to an area near other molecules. A liquid thus has an intrinsic, relatively fixed volume independent of its container. Without a container a liquid will remain coherent, and will assume a minimum potential energy configuration. The weak intermolecular cohesive forces, or *Van der Waal's forces*, vary with the nature of the fluid and with temperature. At high temperatures the cohesive forces may be overcome by kinetic energy of the molecules, changing the liquid to a gas. Thus many familiar substances are gaseous at high temperatures and liquid at low temperatures. In water, the Van der Waal's forces are relatively strong, so that the gas/liquid transition occurs at the relatively high temperature of 100°C (373.2°K). For other fluids of physiological importance, particularly oxygen and carbon dioxide, this transition occurs far below the physiological temperature range, the Van der Waal's forces of these materials being much weaker. Another consequence of the cohesive forces of liquids is that they may be considered incompressible within the physiological pressure range.

Density is defined as the mass per unit volume of a particular substance. Densities of common gases vary by more than an order of magnitude, but even the densest of them is almost 1000 times less dense than water (Table 1.1). Density of gases is pressure-sensitive, of course, so the stated values are always referred to standard temperature and pressure (STP = 0°C and 1 atm pressure). Density is an important parameter in calculations of drag, resistance to flow, and the development of turbulence at high velocity. The usual symbol for density is ρ.

Pressure (from Latin *pressura*, to squeeze) is defined as a force exerted over an area, which for a gas would be the walls of the container. The pressure depends on the number of molecules contained in the volume and their kinetic (thermal) energy; these two factors determine the frequency of collisions between the gas molecules and the container wall per unit time, and thus the pressure. Increased temperature, which imparts a greater velocity to each gas molecule, will increase the number of collisions and the energy of each collision, thus increasing the pressure. Similarly, if the volume is reduced by compression, there will be a greater number of molecules per unit volume adjacent to the wall, and again the probability of collision with the walls is increased. The relation between temperature, pressure, and volume is expressed by the *ideal gas law*, a combination of laws discovered by Boyle and Charles:

$$PV = nRT \tag{Eq.1.1}$$

where P is the pressure, V the volume, n the number of moles of gas, R is the gas constant and T is the temperature in °K. The parameter R is a thermodynamic constant whose numerical value depends on the units employed for the other variables in the equation; with volume in liters, pressure in millimeters of mercury (mm Hg) and temperature in °K, the value for R is 62.36. By applying the appropriate unit conversion factors (see Program UNITS), the appropriate value for R may be calculated. The ideal molar volume for a gas is 22.414 L, but the gases of physiological interest exhibit nonideal behavior, with molar volumes between 22.09 and 22.43 (Table 1.1).

Several units of pressure are in general use. In the strict physical sense, pressure is defined as force applied to, or distributed over, a surface and is measured as force per unit area. In the cgs system, the proper units of pressure are *baryes*, where 1 barye = 1 dyne cm^{-2}, or the more familiar *bar*, equal to 10^6 baryes. These units are rarely used by physiologists; more commonly pressure is expressed either in the empirical unit, mm Hg, or the newer unit, the kilopascal (kP). The "torr" is also used as an equivalent term to the mm Hg; it is sometimes capitalized and is named after Torricelli, a seventeenth century student of Galileo who was the inventor of the mercury barometer. The torr is defined as a pressure (i.e. force) sufficient to balance the force exerted by a column of mercury 1 mm high. Although the torr is not a standard cgs or SI unit, it is an intuitively comfortable one through familiarity with barometers, mercury manometers, etc., so its use continues. Occasionally, when very small pressures are being measured, investigators find it convenient to measure them against a column of water or saline solution. At room temperature these pressures can be interconverted by the using the ratio of the density of mercury to that of water:

$$P\ (\text{torr}) \times 13.6 = P\ (\text{mm } H_2O) \tag{Eq. 1.2}$$

When sea water is used, the correction is 13.1, and for intermediate saline concentrations (densities), it is something in between.

The SI system has adopted the Pascal (abbreviated Pa; 1 torr = 133.322 Pa) as the unit of pressure, but it does not appear to have gained widespread use. Other units of pressure measurement and conversion factors are incorporated into the Program UNITS (see accompanying disk).

PROPERTIES OF GASES

A simple kinetic theory to describe the behavior of gases treats each molecule of the gas as a very small, elastic particle, each particle in continuous motion with random direction (point model). Although some characteristics of gas behavior may be accounted for reasonably well by such a model, a complete modern kinetic theory is much more complex and takes such matters as intermolecular attraction or repulsion into account.

A mole of a gas consists of 6.023×10^{23} molecules (Avogadro's number),so 1 cm^3 of a gas will contain 2.687×10^{19} molecules. It follows then that the average intermolecular distance is about 3.3×10^{-7} cm, or 33 Å (1 Å = 10^{-10} m). By comparison, the molecular diameters of common gases range from about 2 to 5 Å or about one-tenth of the intermolecular separation. The molecular separation becomes less as pressure rises, but even at atmospheric pressure and ordinary temperatures the intermolecular forces will cause significant deviations from the point models of gases. The deviations cause departures from predictions based upon simple gas theory, but for most respiratory gases at ordinary temperatures and pressures these refinements can be ignored.

Compressibility

A primary characteristic of gases, implied in the definition of the gaseous state (above), is that they are easily compressible. The relation between pressure, volume, and temperature for an ideal gas is given by the gas law (Eq. 1.1). One should keep in mind, however, that if the pressure of a real gas in a closed container is increased, its temperature will rise in proportion to the work done (energy expended) in compressing it. This is called *adiabatic* heatin, and reflects the rise in internal energy of the gas imparted by the compression. Conversely, if a gas is allowed to expand, thereby reducing its pressure, its temperature will fall, i.e. it will cool adiabatically. This phenomenon is important in understanding changes in the atmosphere; when warm, moist air rises, it cools adiabatically causing condensation of moisture and cloud formation. The principle is also applied in compression/expansion cycle machines such as air conditioners used for heating and cooling. Temperature changes induced by physiological pressure changes, as in ventilation, are probably always negligible.

Viscosity

The viscosity of a material (symbolized by η) is a measure of its resistance to *shear*, and shear is in turn defined as drag exerted on a moving particle or layer of a fluid by adjacent particles or layers. Drag is simply a force opposed to the direction of movement. These concepts of drag, shear, and viscosity are perhaps more familiar in the context of liquid flows (see below) but are also important in the mechanics of gases. The viscosities of some common gases are given in Table 1.1.

Table 1.1: Properties of common gases and water at standard temperature and pressure (STP = 0°C and 1 atm).

Gas	Formula	Mole Wt, g	Mole Vol., L	Boiling Point	Density	Viscosity	Thermal Conductivity	Specific Heat	Kinematic Viscosity
Oxygen	O_2	31.9988	22.39	-218.4	1.429	192.6	58.5	0.157	0.135
Nitrogen	N_2	28.0134	22.40	-195.8	1.251	167.4	58.0	0.177	0.134
Carbon dioxide	CO_2	44.0098	22.26	-41.6	1.832	138.0	34.0	0.154	0.0753
Carbon monoxide	CO_2	8.0104	22.40	-190.0	1.250	166.5	55.9	0.178	0.133
Helium	He	4.0026	22.43	-268.9	0.1785	188.7	352.0	0.745	1.057
Hydrogen	H_2	2.0158	22.43	-252.8	0.0899	85.0	416.0	2.411	0.946
Argon	Ar	39.948	22.39	-185.7	1.784	210.4	39.7	0.0745	0.118
Neon	Ne	20.179	22.42	-245.9	0.900	298.1	108.7	0.148	0.331
Ammonia	NH_3	17.0304	22.09	-33.3	0.771	94.4	52.2	0.391	0.122
Water	H_2O	18.01534	0.18	100.0	999.8	17,980	1430	0.9998	0.018

Density in grams per liter. Viscosity in micropoises. Thermal conductivity at 20°C in μcal cm^{-1} sec^{-1} $°K^{-1}$. Specific heat in cal g^{-1} $°C^{-1}$. Kinematic viscosity in dyne sec cm g^{-1} (=stokes).

Thermal Conductivity

If a temperature gradient exists in a gas, heat will be transferred from the hotter to the cooler regions at a rate dependent primarily upon the density, kinetic energy, and molecular characteristics of the gas. Thermal conductivities of common gases vary by more than an order of magnitude and are given in Table 1.1.

Partial Pressure

An important extension of the gas laws is Dalton's *law of partial pressures*, which states that the total pressure in a container of a mixture of gases will be the sum of the pressures of each of the component gases, or:

$$P_{Total} = P_A + P_B + \ldots + P_N \qquad \text{(Eq.1.3)}$$

where P_A, P_B, etc. are the individual components of the gas. It is an important physical law for physiology because it states that each gas exerts a pressure independent of any other gases present, and that there is a direct proportionality between the composition of a mixture of gases and the pressure exerted by each. An example of the utility of this principle is provided by considering a sample of air at sea level. The total pressure exerted by the sample is ideally 760 mm Hg, and oxygen constitutes about 20.9% of the total volume. We may therefore calculate directly that the *partial pressure* of oxygen in the sample is (760)(0.209) = 158.8 mm Hg.

In physiology this relation is important because in many situations two or more gases may be behaving in very different ways, and Dalton's law means that each gas can be considered separately and independently of the other. For example, in the lung, oxygen is moving one way (in) according to certain principles and concentration gradients, and carbon dioxide is moving in the opposite direction (out) in response to different gradients. The analysis of each process can be conducted independently, each gas having no effect on the movement of the other.

Diffusion

The process of diffusion can be grasped intuitively by imagining a room on a zero gravity world with yellow ping-pong balls clustered on one wall and orange ones on the opposite wall. If all are suddenly set in violent random motion, it is not difficult to imagine that in a short time the orange and yellow balls will be more or less uniformly intermixed and evenly distributed throughout the room. So it is that in mixtures of two or more gases gradients will be eliminated over time by the random thermal motion of the molecules. Those gases with smaller molecules and higher kinetic energy will distribute more rapidly and thus have higher diffusion coefficients.

For the purpose of understanding physiology, it is probably sufficient to say that diffusion of gases within other gases is a relatively rapid process. Over short distances diffusion gradients are usually minuscule, but in small-bore tubes of more than a few millimeters length, standing diffusion gradients can become significant.

THE ATMOSPHERE

Our atmosphere is a gas mixture of special significance; within our solar system it is unique in several ways. Earth is at just the right distance from the sun to allow coexistence of liquid water and water vapor, for example. We also appear to have the only oxygen-rich atmosphere in the solar system and have just sufficient planetary mass (and therefore gravity) to prevent the loss of our atmosphere to outer space. Hydrogen atoms, for example, must have a velocity greater than 2100 m sec^{-1} in order to overcome Earth's gravitational attraction. At the prevailing temperatures on Earth, even hydrogen, the easiest gas to lose, is retained.

The composition of the "standard" atmosphere is given in Table 1.2. The most abundant gases are nitrogen and oxygen, but there are many minor components, including carbon dioxide. At sea level the standard atmosphere is somewhat arbitrarily assigned the pressure value of 760 torr, excluding water vapor. Water vapor is, however, an ever-present component of the atmosphere as well as of most physiological gases. The relation between water vapor in a saturated atmosphere and temperature may be calculated using the program TEMPTABL (on the accompanying disk). Of course the proper calculation of ambient atmospheric partial pressures for various gases must take into account the relative humidity and temperature. The correct partial pressure for a given atmospheric gas (P_g) may be calculated as:

$$P_g = [P_B - (RH)(P_W)][F_g] \quad \text{(Eq.1.4)}$$

Table 1.2: Components of atmospheric air, excluding water vapor, at the standard pressure of 760 torr.

Component	% Volume	P, torr	P, kPa
N_2	78.084	593.4	79.10
O_2	20.946	159.2	21.22
CO_2	0.033	0.25	0.033
Ar	0.934	7.10	0.946
	ppm	P, millitorr	P, Pa
Ne	18.18	13.8	1.84
He	5.24	3.98	0.53
CH_4	2	1.5	0.20
Kr	1.14	0.87	0.12
H_2	0.5	0.38	0.051
N_2O	0.5	0.38	0.051
Xe	0.087	0.06	0.008

The minor components are also given as parts per million (ppm). Data from the *Handbook of Chemistry & Physics*.

where P_B is the barometric pressure, RH the relative humidity, P_W the water vapor pressure for a saturated atmosphere at ambient temperature, and F_g the fraction of the atmosphere comprised of that gas.

At altitudes from sea level to about 10,000 m above sea level, the atmosphere is relatively well mixed, so its composition is constant. The total pressure, however, declines according to the equation:

$$P = P_0 e^{-Mgh/RT} \quad \text{(Eq. 1.5)}$$

where P_0 is the sea-level pressure, g is the acceleration of gravity, h is the height, M is the average molecular weight, and R and T have their usual meanings (see Program UNITS). A close approximation is given by $P/P_0 = e^{-0.0001233h}$ when h is in meters. Since the vertical mixing in the atmosphere is more rapid than thermal conduction, there is a decrease in temperature with altitude until a temperature of about –55°C is reached at a height of about 15,000 m and a pressure about 5% of sea level. Above 100,000 m, mixing is slower and gases separate in our gravitational field so that the uppermost layers (above 800,000 m) contain only hydrogen and a little helium. Since the limits of the biosphere extend to only about 10,000 m, the composition of the atmosphere may be considered constant.

PROPERTIES OF LIQUIDS

Compressibility

Most liquids may be considered incompressible within the ordinary range of conditions for life. In the consideration of the gaseous state (above), it was pointed out that the intermolecular distance was about 33 Å, or about ten times the molecular diameters of most gases. For a liquid such as water the molar volume is about 18 cm^3, so each cm^3 will have approximately 3.35×10^{22} molecules. The molecular diameter of water is about 2 Å, and the mean intermolecular distance is only about 3.1 Å. Compared with gases, then, the molecules of a liquid are packed rather tightly together, and increasing pressure has effects that are largely masked by other intermolecular forces. Only at the extreme pressures of the deep sea does the compression of water become important for biology (Swezey & Somero, 1982).

Viscosity

The concept of viscosity in liquids is the same as for gases, except that the absolute magnitude of viscosity in liquids is several orders of magnitude greater than for gases (Table 1.1). One way to think of viscosity is as internal friction, resulting in a force tending to oppose deformation or shear in a liquid.

If flow of a fluid is considered for a planar slice oriented parallel to the direction of flow, the shearing force is tangential to the plane, and shear deformation

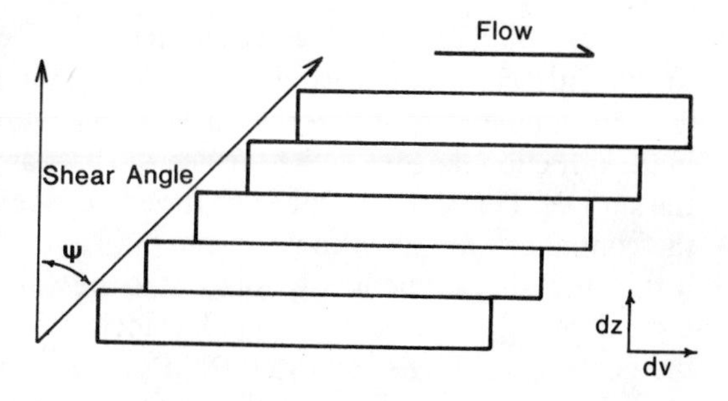

Figure 1.1: Rate of shear is defined as the change in velocity (dv) in a direction normal to the direction of bulk flow (dz). Shear stress is equal to the tangential force (applied in the direction of bulk flow) divided by the area over which the shear (slippage) is occurring.

occurs by slippage of the plane relative to parallel planes above and below it (Fig. 1.1). The fluid is not completely unlike a solid, however, since there is some resistance to flow that can be thought of as a transient elasticity, a frictional resistance between planes, or simply a "lack of slipperiness." This characteristic is the *viscosity*, which varies with the fluid and the temperature.

In Fig. 1.1, the velocity gradient, or shear stress, is represented by dv/dz. Intuitively it is easy to see that greater tangential force per unit area will produce a greater shear stress; the proportionality constant is η, the viscosity. The cgs unit of viscosity is the poise, defined as 1 dyne sec cm^{-2}. The centipoise (1cp = 0.01 poise) is more commonly used, especially as the viscosity of pure water at 20.2°C is conveniently 1 cp. In SI units the derived term Pascal-second is used, but in both cases the fundamental units can be shown to be mass over the product of length and time (see Program UNITS).

In many fluid flow problems a viscosity divided by density term (η/ρ) occurs, so the term *kinematic viscosity* (ν) is often substituted for this quotient. Turbulence, for example, is directly dependent upon kinematic viscosity (see Chapter 4). The kinematic viscosity of most common gases is actually greater than that of water (Table 1.1).

Shear forces can arise in several ways, but in physiological systems we are generally concerned with shear forces generated mechanically, whether by action of cilia, ventilatory apparatus, the heart, or other means. Thermal gradients may also give rise to convective flow due to differences in density and the force of gravity acting upon the fluid, but these thermal convection processes are seldom significant in gas exchange or transport.

Convection: Bulk Flow of Fluids

The physics and mathematics of fluid flow (fluid dynamics) are treated extensively in texts and specialized journals, but an understanding of the fundamentals is necessary for the analysis of many physiological processes. It is particularly useful to understand the basics of flow velocity profiles, the relation between pressure and flow, the difference between laminar and turbulent flow, and the calculation of work.

Flow Velocity Profiles

When a fluid in contact with a surface is subjected to a shearing force parallel to that surface, whether we are considering blood in contact with vessel walls, water in contact with an external surface, or gas flow in a tube, the velocity of flow in the direction tangent to the surface will vary as a function of the distance from the surface. This is due to cohesive forces between the molecules of the fluid and the container wall. If we consider the fluid as divided into planar slices, infinitely thin, that slice immediately adjacent to the wall will have zero velocity, the next layer outward will have a very small velocity with respect to the innermost layer, the next slightly larger, and so forth. If there is a second wall (or for a cylinder), the velocity will be maximal at the center and decline again as the other surface is reached, as shown in Fig. 1.2. For a cross-sectional slice of a blood vessel, the velocity profile across the lumen will have a parabolic shape, with zero velocity at either wall. Furthermore, since the rate of change of velocity with distance in a direction perpendicular to the flow (which is shear, by definition) is least at the center, the shear is least there and greatest at the walls. The layers nearest the walls contribute the greatest *drag*, or resistance to flow. It is not, then, any special property of the wall itself that imparts the viscous drag but an intrinsic property of fluid flow.

Since the proportion of the fluid that is near the wall is greater in a small vessel or channel than in a large one, resistance to flow is greater in small vessels. The familiar Poiseuille-Hagen equation describes this relation for tubular channels:

$$V = (\Delta P \pi R^4)/(8\eta L) \qquad \text{(Eq. 1.6)}$$

where V = volume per second, in cm^3; ΔP = the difference in pressure over the length of the tube, in dynes cm^{-2}; R = the radius of the tube, in cm; L = the length of the tube, in cm; and η = the viscosity of the fluid, in poises (= dynes sec cm^{-2}). This equation states that pressure and flow are linearly related, given constant radius and viscosity. Second, only very small increases (or decreases) in radius are required to cause very large increases (or decreases) in flow, provided pressure and viscosity remain the same. Constant flow can therefore be maintained in the face of falling pressure by very small increases in radius.

The viscosity of most fluids at constant temperature may be considered a con-

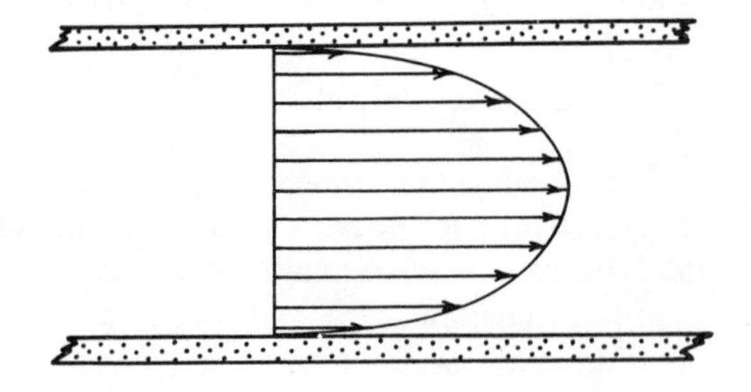

Figure 1.2: When a shearing force (i.e. a pressure) is applied over the area of the tube, a parabolic flow velocity profile results. The rate of change of velocity, dV/dx, increases away from the center, so more viscous drag is contributed by the fluid nearer the walls.

stant, which is to say that the fluids are Newtonian in behavior. Certain fluids of biological importance, however, exhibit significantly nonlinear viscosity behavior, usually caused by dissolved (colloidal) proteins and by particles (erythrocytes) in the fluid. Blood has a highly anomalous relation between pressure and flow, with its apparent viscosity declining at higher flow rates (Burton, 1965)(see Chapter 4).

The Unstirred Layer

In literature on both gas exchange and ion exchange, the term "unstirred layer" is frequently encountered. It is clear from the foregoing discussion that this term is misleading, since it seems to imply that there is some layer of finite thickness in which flow is zero, bordered by some layer further out that is completely stirred or has a high velocity. In fact the flow velocity proceeding outward from a boundary wall increases in a progressive and continuous fashion. What is generally intended by the term "unstirred layer" is that area near a surface in which convective flow is insignificantly low and in which pure diffusion is the rate-limiting process. The concept of an unstirred layer is nonetheless important, since steep gradients may arise in this low–convection zone, and the addition of this thickness to the diffusion pathlength (see Eq. 1.15) has a profound influence on analysis of membrane transfer problems.

Laminar and Turbulent Flow

The flow velocity profile shown in Fig. 1.2 applies to conditions under which flow may be considered in layers, planar slices, or concentric cylinders; it is termed *laminar flow*. Under these conditions, the applied force is translated entirely into shear in the direction tangential to the plane of the slices or cylinders. At higher flow velocities, however, *turbulent flow* may occur, i.e. flow containing eddies and flow currents in directions other than the tangential. The velocity at which the transition from laminar to turbulent flow occurs (V_c, the critical velocity) varies directly with the size of the flow channel and the viscosity of the fluid, and inversely with density, according to:

$$V_c = \frac{R_e \eta}{\rho R} = \frac{R_e \nu}{R} \qquad \text{(Eq. 1.7)}$$

Where R_e is the Reynolds number, a dimensionless number related to size, velocity, and the kinematic viscosity; and R is the channel radius.

Bends in the flow channels and sometimes surface roughness may also lead to turbulence. Because of the excess energy that must be translated into the turbulent component, the transition from laminar to turbulent flow is seen as a break in the linear pressure/flow relation, so that increases in pressure (and work) produce proportionately less increase in flow.

Turbulence is important in many biological situations. Perhaps the most familiar example is the production of so-called Korotkow sounds by turbulence around the heart valves and in the aorta of man. Here some of the flow energy is expended as random low-frequency vibrations detectable as sound. Turbulent flow dynamics become important in the consideration of swimming and flying drag, but these subjects have been reviewed extensively (Webb, 1978) and are outside the scope of this book.

The Work of Pressure and Flow

The definition of work from physics is the energy expended by a force (F) acting over a distance (x), or $W = Fx$. Since pressure is force applied over an area, or $P = F/A$, substituting a pressure term in the work equation gives:

$$W = PAx \qquad \text{(Eq.1.8)}$$

Combining A and x gives units of volume (cm^3), and dividing both sides by a term for time produces the useful equation:

$$W/t = PV/t \qquad \text{(Eq.1.9)}$$

where the units of W are ergs (dyne·cm), t is in seconds, P is in dynes cm^{-2}, and V is in cm^3. This equation is adequate for describing the work done in steady flow at constant pressure, but more often in biological situations the pressure and flow are continuously variable in time, so a slightly more complex equation is needed:

$$W = \int PVdt \qquad \text{(Eq.1.10)}$$

where W is the rate of work per unit time, and the pressure-volume relation is integrated over the time period.

One important aspect of the work calculation is that it provides a numerical estimate of the physical work performed on the fluid, sometimes called the *external work*. It is not the same as the total work done by the organ propelling the fluid, however; calculation of this total metabolic work by the organ must include the efficiency of the organ in translating metabolic work into mechanical work. Even the heart, which is a structurally efficient organ compared to many others, has an efficiency below 10% under normal light-load conditions, so the total work done by an organ to bring about convective flow can easily be an order of magnitude greater than the external work that results.

Dissolution of Gases in Liquids

In an idealized two-phase system of a gas and a liquid in contact, if the gas is soluble in the liquid the system will come to an equilibrium condition in which the gas pressure in the two phases is equal (Fig. 1.3). At equilibrium, by definition, the number of gas molecules leaving the gas to enter the liquid phase will just equal the number of gas molecules leaving the liquid and entering the gas. If the gas phase is a mixture of several gases, each with its own partial pressure, at

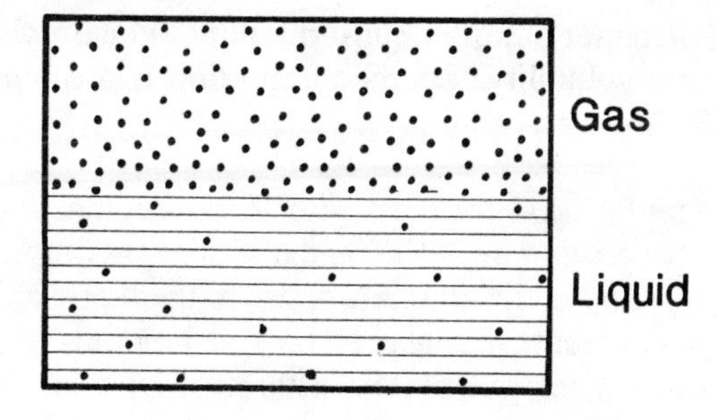

Figure 1.3: When a liquid and gas phase are in equilibrium with respect to a gas, the gas molecules enter and leave the fluid at exactly opposing rates. The partial pressure in both phases is equal, but the quantity or concentration per unit volume (represented by the relative dot density) is not.

equilibrium the system will have equal total gas pressures in the liquid and gaseous phases; the partial pressure law will also hold, and the partial pressure of each of the components in the gaseous and liquid phases will be equal (Fig. 1.3). This law is physiologically important, since one is frequently dealing with situations that require a complete understanding of gas/liquid equilibria. To extend the previous example, if a liquid is in equilibrium with air at sea level, the partial pressure of oxygen can be calculated by knowing that if the partial pressure in the gas phase was 158.8 mm Hg it will be the same in the liquid.

Knowing that a gas is soluble in a liquid and what the partial pressure in the liquid will be is not sufficient information from which to calculate the mass of the gaseous substance that will be present in a given volume of the liquid. For this calculation an additional piece of information relating the partial pressure in the liquid and the mass present is needed. The relation is generally stated mathematically as:

$$Q = \alpha P \qquad \text{(Eq.1.11)}$$

where Q is the mass or amount, P is the partial pressure, and α is the *solubility coefficient*. A great deal of confusion arises at this point, however, since there are many ways in which the solubility coefficient can be expressed and lamentably little standardization in the physiological literature. Most of the time, the solubility is expressed as a quantity of gas contained in a volume of liquid under some convenient set of conditions, and the units of quantity are also chosen according to various conventions or expedience. There is unfortunately no symbol that can be used without some ambiguity; the Greek alpha (α) is most frequently used. Bunsen originally defined this entity as the absorption coefficient, the volume of dry gas reduced to standard temperature and pressure (0°C, 760 mm Hg, dry, or STPD) present in one volume of liquid. The symbol α is encountered frequently with different definitions, which are not always clearly stated. Dejours (1975) used the Greek beta (β) for solubility but called it "capacitance"; various other authors have employed a potpourri of symbols, even occasionally employing α or β as nonlinear functions for blood! In this book, the symbol α will be used to mean the volume of dry gas at STP contained in a liter of liquid, subscripted to indicate which gas is meant, e.g. αO_2 denoting the solubility of oxygen. Equation 1.11 can now be re-stated as:

$$V = \alpha P \quad \text{(Eq. 1.12)}$$

replacing Q for mass with V for volume, keeping in mind the relationship between volume at STP and moles of gas (approximately 22 L $mole^{-1}$)(Table 1.1).

Just as the amounts of gas in the liquid and gas phases will be quite different at the same partial pressure, the amounts of gas present at equilibrium between two liquids may also be different if the solubility coefficients are different. In non-equilibrium cases, we may have circumstances in which gas will diffuse **up** a concentration gradient, while diffusing **down** a partial pressure gradient. This idea is illustrated in Fig. 1.4. By setting P_1 equal to P_2, then:

$$V_1/\alpha_1 = V_2/\alpha_2 \quad \text{(Eq. 1.13)}$$

In other words, although the system is at equilibrium and the partial pressures are the same, the amounts of oxygen in the two liquids are different. This point cannot be overemphasized: Equilibrium occurs at equal partial pressures, not necessarily equal concentrations. It is furthermore a common problem, e.g. when assessing exchange between water and blood, or blood and cerebrospinal fluid.

The solubility of gases is influenced by many factors, only one of which, the nature of the solvent, has been mentioned. Temperature and the solubility coefficient are inversely related but not linearly (see Program TEMPTABL). Increasing ionic strength reduces the solubility of most but not all gases in a nonlinear fashion. The solubility of gases is for practical purposes insensitive to the presence of other gases and insensitive to the nature of the electrolytes present. The

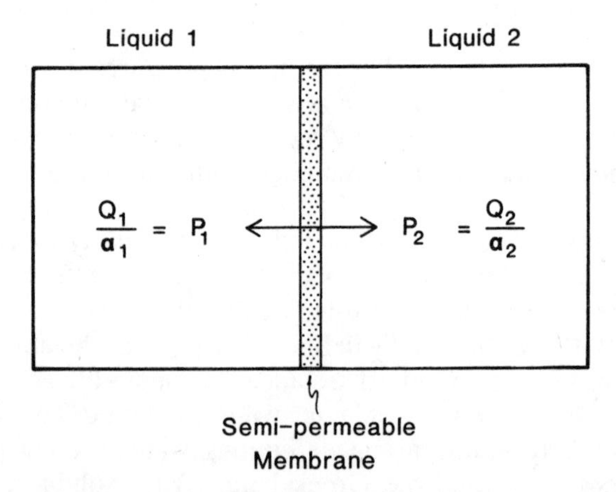

Figure 1.4: When two liquids of unequal properties are separated by a semipermeable membrane, a dissolved gas at equilibrium will reach the same partial pressure in each liquid. That is, $P_1 = P_2$, but the concentrations (Q_1 and Q_2) will differ in proportion to the solubility coefficients in each liquid.

solubilities of some common gases in water are given in Table 1.3, and some more detailed tables of gas solubility have been provided in the Program TEMPTABL.

Diffusion of Gases in Liquids

The equations describing diffusion in three dimensions are quite complex (see Rashevsky, 1960) but for physiological purposes a one-dimensional treatment is often adequate and serves to illustrate the principles. The term "Fick's law" is nowadays applied to a variety of diffusion equations of the general form:

$$dQ/dt = -AD(dc/dx) \qquad \text{(Eq. 1.14)}$$

This equation states that the rate of change (or net movement, d/dt) of a substance, Q, with time (t) is equal to the product of the area (A) through which the substance is diffusing, the diffusion coefficient (D), and the concentration gradient (dc/dx). This equation has little utility, since an exact description of the concentration gradient can seldom be mathematically stated for biological systems. We therefore usually make the further simplifying assumption that the differential dc/dx may be replaced by $\Delta c/x$, where Δc is simply $(c_i - c_o)$ and x a linear distance measurement. In other words, we assume that the concentration gradient is linear (Fig. 1.5) which reduces the equation to:

$$dQ/dt = -AD(\Delta c/x) \qquad \text{(Eq. 1.15)}$$

The sign of the right-hand term signifies that with a greater concentration at the point of reference the net movement will be outward.

The diffusion of gases in liquids is not very different physically from "self-diffusion" (diffusion within a homogeneous gas) or from diffusion in gaseous mixtures. In liquids, however, the diffusion coefficients are even more dependent

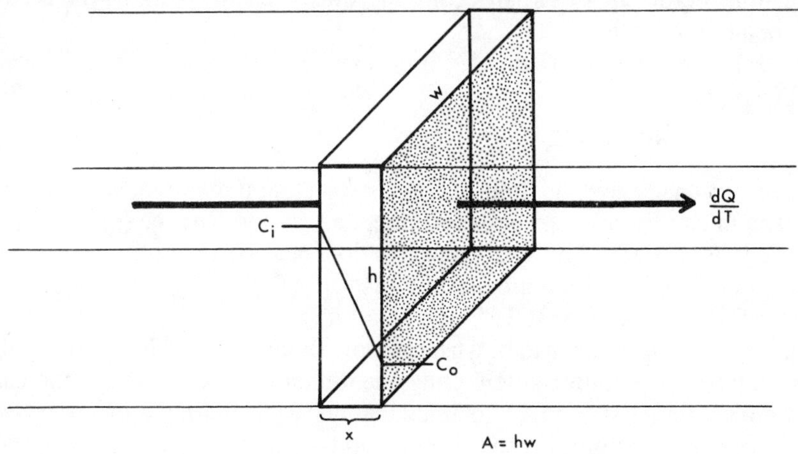

Figure 1.5: Simplified representation of diffusion in one dimension, with a linear concentration gradient through a barrier of thickness x. The rate of diffusion is a function of x, the area, the concentration difference, and the diffusion coefficient (D)(see text).

Table 1.3: Relative solubility of various common gases in water at 20°C.

Gas	α	Relative Solubility	Diffusion Coefficient
He	0.0088	0.28	
Ne	0.0147	0.47	
A	0.0337	1.07	
H_2	0.0182	0.58	
N_2	0.0155	0.49	
O_2	0.0314	1	1.98×10^{-5}
CO	0.0232	0.74	
CO_2	0.872	27.8	1.77×10^{-5}
NH_3	715.4	22783.	1.77×10^{-5}

The Bunsen coefficient (α) is defined as the volume of gas (STPD) dissolved per volume of liquid at a gas pressure of 1 atm, and the relative solubility as divided by the solubility of O_2. The units for diffusion coefficient are cm^{-1}. Data from various sources.

upon the relative strength of Van der Waal's forces, not only among molecules of gas but between gas and solvent molecules. In aqueous solution the relative diffusion rates of O_2 and CO_2 depart from the inverse square root rule and differ from their ratio in gaseous media, the actual O_2 to CO_2 being about 1.39. Another important difference is that the absolute rates are slower in water than in gases by about a factor of 10^5, as shown in Table 1.3.

For the example of O_2 and CO_2, although the diffusion coefficients of both are much lower in water than in gaseous media, their ratio is not very different in the two media (1.17 versus 1.39). The solubility of the two gases, however, is very different, CO_2 being about 28-fold more soluble in water (Table 1.3). Because concentration is equal to the partial pressure times the solubility, or $C = \alpha P$, the D in the diffusion equation (above) is often replaced by a new term, D', which is equal to αD, where α is the Bunsen solubility coefficient. D' is often called the permeation coefficient, or Krogh's constant, since the early work he did on diffusion dealt in these terms.

The diffusion equation (Eq. 1.13) may now be restated in a physiologically useful form:

$$dQ/dt = -AD'(\Delta P/x) \quad \text{(Eq.1.16)}$$

D' is conventionally given in units of $cm^2\ sec^{-1}$, so that if the other variables are expressed in cm^2 for area, seconds for time, and cm for distance (x), the equations are independent of units of concentration, provided the units of Q and c are the same. The units for α are usually cm^3 gas STPD cm^{-3} liquid $torr^{-1}$, so the resultant D' has units of cm^3 STPD $sec^{-1}\ cm^{-1}\ torr^{-1}$.

Equation 1.15 has proved to be a useful approximation, provided the simplifying assumptions are not forgotten: Only one dimension is considered; the concentration gradient has been approximated as linear; the permeation coefficient is different from the diffusion coefficient; and the units must be internally consistent.

Two additional points regarding diffusion and permeation coefficients are worth emphasizing: The temperature sensitivity of D is about 2% per °C, but because solubility is inversely proportional to temperature and D′ combines these two parameters, the result is a temperature sensitivity for D′ of only about 1% per °C. The second point is that diffusional gradients that may be effective over distances of 10 cm in air will only be equally effective over distances of 1μm or so in aqueous media, due to the 10,000-fold difference in the absolute diffusion coefficients. Aquatic animals obviously must provide much more intimate contact of the respiratory medium with the exchange surface. Diffusion in aqueous solutions is a short-range phenomenon.

PROPERTIES OF WATER AND AQUEOUS SOLUTIONS

Life itself depends on the anomalous characteristics of water as a solvent. The water molecule is highly polar, which gives it a high dielectric constant, a considerable tendency to bind not only other water molecules but solute molecules, a high specific heat, and a density minimum at 4°C. The latter property dictates that ice will not sink to the bottom but will form on the surface of bodies of water in winter, thereby protecting deeper-lying waters from freezing completely. The various properties of water also lend it a very high surface tension, a high wetting tendency, and high capillarity; each of these properties is important in maintaining cellular integrity. Some of the salient physical and chemical characteristics of water are given in Table 1.4 and by Program TEMPTABL.

Colligative Properties

Several properties of water are influenced by solvent particles dissolved in it, and these solute-dependent properties are termed the *colligative properties*. The addition of solutes to water lowers the freezing point, raises the boiling point, and decreases the vapor pressure. In the 1880s Raoult observed that the change in

Table 1.4: Characteristics of Water.

Property	Value
Density @ 0°C	0.999968 gm ml^{-1}
@ 3.98°C	1.000000 gm ml^{-1}
@ 20°C	0.998234 gm ml^{-1}
Thermal conductivity	5.14 cal hr^{-1} cm^{-1} $°K^{-1}$ @ 20°C
Viscosity @ 0°C	1.79 centipoise
@ 20°C	1.002
@ 40°C	0.65
@ 100°C	0.28
Specific heat	0.9988 cal gm^{-1} $°C^{-1}$
Dielectric constant	80.1
Surface tension	72.8 dynes cm^{-1}

Values given are for 20°C unless otherwise stated.

these properties was proportional to the number of particles in the solution, so that a completely dissociable solute such as NaCl should add 2 moles (i.e. two times Avogadro's number of particles) to the solution when completely dissociated. The addition of 1 mole of active particles to water increases the boiling point by 0.54°C and depresses the freezing point by 1.86°C, and a solution that changes these values by those amounts is said to be 1 *osmolar* (Osm).

Activity versus Concentration

Early investigations of such solution properties led to some apparent anomalies. For example, the addition of 1 mole of NaCl to 1 kg pure water should produce 2 Osm of ions when completely dissociated into Na^+ and Cl^-, and should therefore lower the freezing point by 1.86°C. The actual reduction is only 1.58°C, which was originally interpreted to mean that the NaCl was only 85% dissociated. In fact very dilute solutions of nonelectrolytes (less than 0.2 M) behave in a reasonably ideal way, but dilute solutions of electrolytes, which include most physiological fluids, deviate significantly from ideal behavior. Ions in solutions are subject to the influence of attractive forces between solute and solvent and to the effects of electrical fields on movement of charged particles. In the early 1900s Gilbert Lewis described a thermodynamic theory that accounted for these and other solution anomalies. He derived two parameters, the *fugacity* and *activity* of a species in solution. The fugacity, literally an escaping tendency, is defined as a change in free energy compared to a reference state and is a general property of any chemical species in any phase system. The activity is defined as the ratio of the fugacity in a particular state to the fugacity of the reference state. Activity expresses a true "chemical potential" which takes into account the nonideal influences of interionic attraction.

The strength of the various intrasolution forces is dependent upon the concentration of each solute and of the total solutes in solution, and so the relation

Table 1.5: Activity coefficients for some common ionic species in aqueous solutions.

Compound	Activity Coefficient
HCl	0.796
HNO_3	0.791
KCl	0.770
KNO_3	0.739
KOH	0.798
LiCl	0.790
$MgSO_4$	0.150
NaCl	0.778
NaH_2SO_4	0.744
NaOH	0.766

All values are for 25°C and a concentration of 0.1 molal. Data from the *Handbook of Chemistry & Physics*; and Hills, 1973.

between activity and concentration is complex. For any chemical equilibrium of the type:

$$W \longleftrightarrow Y^- + Z^+ \quad \text{(Eq.1.17)}$$

the equilibrium constant is calculated as:

$$K = \frac{[Y^-][Z^+]}{[W]} \quad \text{(Eq. 1.18)}$$

the true equilibrium constant (K′), should be calculated using activities of each reaction species, rather than concentrations. In most physiological systems we work with the apparent dissociation constant, K_a', which incorporates the activity coefficients. K_a' can be derived from the value for K and the *ionic strength* (μ) by the Debye-Hückel equation:

$$K_a' = K \pm 0.52(\mu)^{0.5} \quad \text{(Eq.1.20)}$$

Ionic strength is defined as:

$$\mu = (C_1 z_1^2 + C_2 z_2^2 + \ldots) \quad \text{(Eq.1.21)}$$

where C_i is the concentration of the *i*th species, and z_i is its valence (Hills, 1973).

The case of H^+ ion is a particularly important one and is discussed in Chapter 6. Activity coefficients for a variety of common solutes may be found in Table 1.5.

Chapter 2

Metabolism and Energetics

Measurements of the rate at which animals consume oxygen are as old as the discovery of oxygen itself, but it has only been in this century that we have come to understand the biochemical basis for it. All animals must obtain chemical energy for everything from cellular growth and repair to locomotion. For the majority of animals chemical energy is obtained by combustion of organic substrates using atmospheric oxygen (the "Fire of Life," Kleiber, 1961). Animals in this category are called chemoorganotrophs, distinguishing them from other classes of organisms that obtain energy either from light or from inorganic compounds such as H_2S. Among the higher animals the vast majority are aerobic, i.e. require atmospheric oxygen, but there are some interesting classes of animals that can survive for long or short periods without oxygen. In the following sections we examine the metabolic basis for respiration and the quantitative relations between respiration and factors such as temperature, body size, and activity.

THE METABOLIC BASIS FOR GAS EXCHANGE

The broad term "metabolism" includes both anabolism (synthesis) and catabolism, the latter term meaning those processes involved with breaking down organic molecules to produce, ultimately, carbon dioxide and stored energy in the form of high energy phosphate bonds. A flow chart of the principal catabolic pathways for three main classes of organic substrates— proteins, carbohydrates and fats— is shown in Fig. 2.1. Complete catabolism can be roughly divided into several stages. An initial series of preliminary reactions break large molecules into smaller subunits; proteins are first hydrolyzed into constituent amino acids, polysaccharides to hexoses and pentoses, and fats to fatty acids and triglycerides. A second series of reactions feeds each group into the acetyl group of acetylcoenzyme A (CoA), which in turn feeds the further steps in the chain. Successive steps in the tricarboxylic acid (TCA) cycle feed into oxidative phosphorylation and the electron transport chain, ultimately producing CO_2, water, and ATP.

In each series of catabolic reactions there is a decrease in the free energy of the reaction products compared to the reactants, and this change in free energy is what fuels the animal. Part of the free energy change is captured and stored by coupling the catabolic reactions to the synthesis of high energy phosphates, primarily ATP, from inorganic phosphate and an organic precursor, usually ADP. Part of the energy is also dissipated as heat, which can be measured directly by calorimetry. To take the metabolism of glucose as an example, the initial breakdown of glucose into two pyruvate molecules can be described by:

$$\text{Glucose} + 2P_i + 2\text{ADP} + 2\text{NAD}^+ \rightarrow \text{Pyruvate} + 2\text{ATP} + 2\text{NADH} \quad \text{(Eq. 2.1)}$$

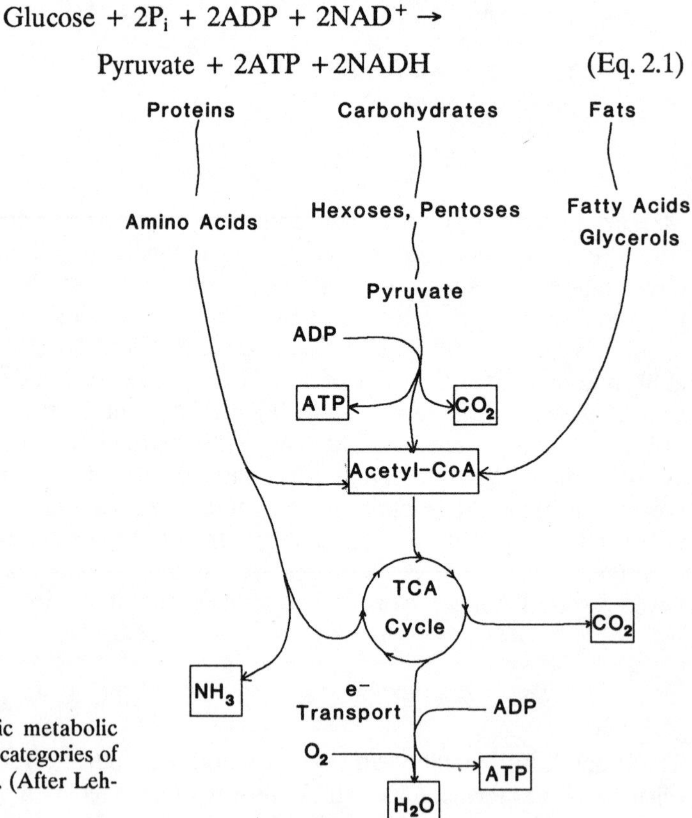

Figure 2.1: Aerobic metabolic pathways for major categories of metabolic substrate. (After Lehninger, 1975.)

If metabolism stops with this glycolytic portion, the pyruvate reacts further to lactate, consuming the 2NADH. The change in free energy ($\Delta G°$) for the breakdown of glucose into lactate is –47 kcal mol^{-1}, and for the synthesis of ATP it is + 14.6 kcal mol^{-1}. The net energy decrease is therefore –32.4 kcal mol^{-1}, and the 14.6 kcal stored in the high energy phosphate bond of ATP represents 31% efficiency of energy storage from the reaction. Up to this point no oxygen is consumed; and if the pyruvate is shunted to lactate, no CO_2 is produced. This portion of the catabolic process, then, is anaerobic and is the most common pathway for temporary energy production in animals.

If pyruvate is instead directed further down the chain of catabolic reactions into the TCA cycle and the electron transport chain, the overall energy balance becomes:

$$\text{Glucose} + 6O_2 + 36P_i + 36ADP \rightarrow 6CO_2 + 36ATP + 42H_2O \qquad \text{(Eq. 2.2)}$$

with a net free energy yield of –686 kcal mol^{-1}. The energy storage as ATP is 7.3 $\times$ 36 = +263 kcal mol^{-1} for an overall efficiency of 38%.

Oxygen acts as the terminal electron acceptor in the catabolic chain and must arrive at the inner mitochondrial membranes to serve this role. The beautifully elaborate cellular machinery extracts energy from food substrates in a stepwise fashion, involving cyclic formation and degradation of acetyl CoA, NADH, and ATP. Oxygen enters into the process at the very end, accepting hydrogen from the electron transport chain, the spatially linked cytochromes of the mitochondrial cristae, to form water. The balance shown in Eq. 2.2, then, is misleading, since the oxygen in the CO_2 produced arises from the substrate and not directly from atmospheric oxygen.

Respiratory Quotient

The processing of noncarbohydrate metabolic substrates such as fats and proteins follows similar pathways (Fig. 2.1), with some differences in the stoichiometry of CO_2 and O_2 and the different energy yields. The respiratory quotient (RQ) is defined as the ratio of CO_2 produced to O_2 consumed. From Eq. 2.2 it is obvious that catabolism of glucose (and other carbohydrates) will produce an RQ of 1.00. Catabolism of proteins, however, results in an RQ of 0.80, and of fats 0.71. For a human on a normal diet, the average RQ is 0.82. The energy yield from different substrates also varies from 5.05 kcal per liter of O_2 consumed for carbohydrates to 4.69 for fats and 4.46 for protein (Lehninger, 1975). The average "oxycalorific equivalent" is usually taken as 4.8 kcal per liter of oxygen (see Program UNITS for conversion to other energy units).

A distinction should probably be made between the true metabolic RQ and the apparent respiratory RQ. The two are the same if the animal is in a steady state, e.g., with oxygen entering the animal at the same rate as it is being consumed by metabolism. There are many non-steady-state situations, however, in which the metabolic and respiratory RQs may differ. In the initial stages of activity, for example, or during a temporary bout of anoxia internal oxygen stores may sustain aerobic metabolism at rates higher than the rate of oxygen exchange with the environment. Similarly, during the initial stages of acidosis, bicarbonate stores may be converted to CO_2, elevating the apparent respiratory RQ to values well above the actual metabolic RQ (see Chapter 6).

Anaerobic Metabolism

Under conditions of oxygen limitation, energy can still be produced from organic substrates but in limited amounts and by different pathways than those shown in Fig. 2.1. The most common causes of anaerobic metabolism are muscle activity and environmental hypoxia or anoxia. During brief bursts of muscle activity, particularly in striated skeletal muscle, the exergonic yield of glycolysis is used to produce ATP and lactate from glycogen granules stored in the muscle tissue. This is a relatively fast and efficient source of energy but has the dual disadvantages of producing excess H^+ ions and of leaving much of the potential energy unutilized. This pathway is therefore most commonly encountered as a temporary expedient. After activity, for example, lactate can either be transported via the blood to the liver for reprocessing into glycogen or, in some cases, reprocessed in situ (Hochachka, 1973). Some animals have enormous tolerance of circulating lactic acid (Jackson & Heisler, 1983; Jackson & Ultsch, 1982)(see Chapter 12), and can therefore utilize this pathway for relatively long periods of time. Few animals excrete significant amounts of lactate, however.

Several groups of animals have evolved to exploit habitats that are either permanently or cyclically anaerobic. Some examples are the parasitic intestinal worms, intertidal molluscs, and some benthic marine annelids. In these groups, several metabolic pathways have evolved that utilize mainly glucose and aspartate to produce succinate, propionate, acetate, alanine, octopine, alanopine, and strombine in addition to lactate (Fig. 2.2)(Zandee et al., 1980; Zwaan et al., 1982; Hochachka & Mommsen, 1983; Pörtner et al., 1984a, c, 1987). The energy yields from these alternate pathways range from 2.5 to 5.0 moles ATP per mole metabolized, versus 2.0 moles for metabolism of lactate. In addition to the energy advantage, these pathways result in little intracellular acidification, according to the most recent studies (Pörtner et al., 1984b, c; 1987).

A final anaerobic system worth noting is the sulfur-based metabolism of animals discovered near rift vents on the ocean floors. At the rift vents, water that has percolated downward through the sea bed emerges from channels after being heated considerably by thin underlying areas of the earth's crust. During its passage through the rocks it becomes anaerobic and dissolves substantial quantities of H_2S and metal sulfides (Corliss et al., 1979). Quite rich communities of animals are found near the vents, contrasting with the otherwise barren abyssal plain. It is a little difficult to say whether some of the animals are truly anaerobic, since the jets of heated anaerobic water are mixing with the surrounding water, which

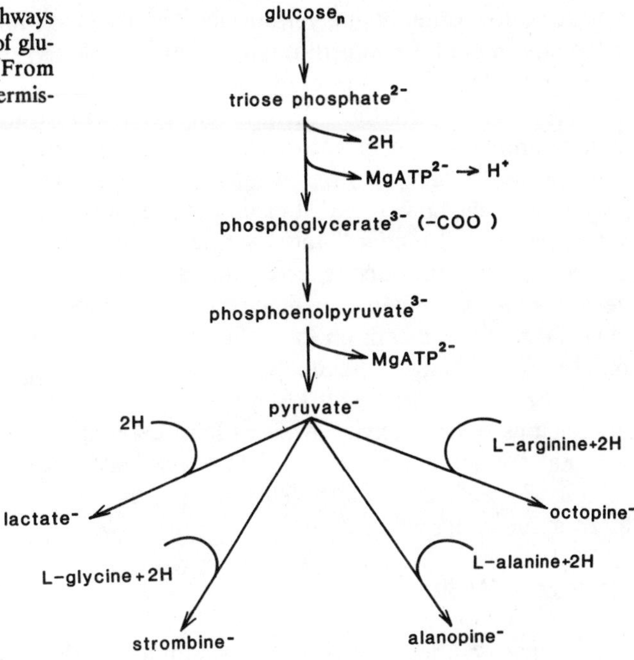

Figure 2.2: Alternate pathways for anaerobic metabolic of glucose in invertebrates. (From Pörtner et al., 1984b, by permission, ©Springer-Verlag.)

is cooler and aerated. Some animals do appear to live directly in the rift plume, but all of the animals of the rift vent community directly or indirectly take advantage of the energy source presented by sulfides. Some of them contain symbiotic sulfur-reducing bacteria that supply organic substrates to the host as a supplement to normal heterotrophy (Hand & Somero, 1983a; Arp et al., 1984).

RATE OF METABOLISM AND BODY SIZE

Definitions of Metabolic Rate

In the literature on metabolic rate, a number of terms are encountered whose meanings require some definition and comment. Based primarily upon early studies with homeotherms, a "standard metabolic rate" (SMR) or "basal metabolic rate" (BMR) was defined as the rate obtained with no visible activity in a recently starved subject. For man, BMR usually means a measurement performed with the subject supine, early in the morning before breakfast. The BMR or SMR is thought to represent the minimum metabolic energy costs of cell turnover, body temperature maintenance, etc., with no provision for the energy expenditures required for growth or activity.

With man or other higher vertebrates it is also possible to elicit the maximum sustainable metabolic rate through strenuous voluntary activity. Rates so obtained are called "active" rates, or sometimes "$V_{O_2\text{-max}}$".

In the poikilotherms these definitions are much more difficult to defend. Many authors employ the term "standard" for fish (Brett & Groves, 1979), whereas others prefer the term "resting" when fasted animals are tested at zero visible activity. One finds, however, that given long periods in the dark without stimulation or food, the "standard" rate continues to decline, and there does not seem to be any easily reproducible level of metabolism the way there is with a homeotherm. The measurements are much more dependent upon experimental conditions and to some extent on the behavior of the species. The term "resting" is something of a hedge, intended to acknowledge the operational difficulties of standardizing measurements. Many species of animals exhibit a certain level of spontaneous motor activity, even without external stimuli, and here the term "routine" metabolic rate (Fry, 1957) has applied. Each of these terms has utility when used carefully, but terms such as "standard metabolic rate" tend to have different meanings when applied to different groups of animals.

Effects of Body Size

For any given species, it has been known for a long time that the rate of metabolism (rate of oxygen consumption) rises with body size, although not in a linear fashion. For a very large number of animals ranging in weight from about 1 μg to over 1000 kg, the rate of metabolism (MR) has been found to obey the relation:

$$MR = aW^b \qquad \text{(Eq. 2.3)}$$

where the value for *b* averages 0.74 (Zeuthen, 1953; Prosser, 1973). This relation appears to hold for individual species and for a range of species (Fig. 2.3). The absolute level of metabolism, however, as represented by *a* in Eq. 2.3 varies by about a factor of 20 to 50 when comparing homeotherms and poikilotherms of the same body mass. A salmon of 1 kg weight, for example, has a resting metabolic rate at 20°C of approximately 80 ml O_2 kg^{-1} hr^{-1}, whereas a typical 1 kg mammal has a metabolic rate of about 700 ml O_2 kg^{-1} hr^{-1}, almost 10-fold higher.

Although this relation between body weight and metabolic rate has been observed time and again, there is still no general agreement upon why it is true. If the metabolic rate rose in a manner proportional to surface area, for example, one might expect a value for the exponent *b* in Eq. 2.3 to be closer to 0.66. Exact proportionality would produce a *b* value of 1.00. The observed average of 0.7 to 0.8 is true whether the metabolic rate is expressed on the basis of wet weight, dry weight, total body nitrogen, or even total number of cells in the organism (Zeuthen, 1953). There are variations in the exponent for a given species, however. The weight exponents for actively swimming sockeye salmon, for example, approach 0.97 to 1.00; *b* values for feeding fish are lower than for starved fish (Brett, 1979); and *b*-values vary from 1.0 to 0.6 during mammalian develop-

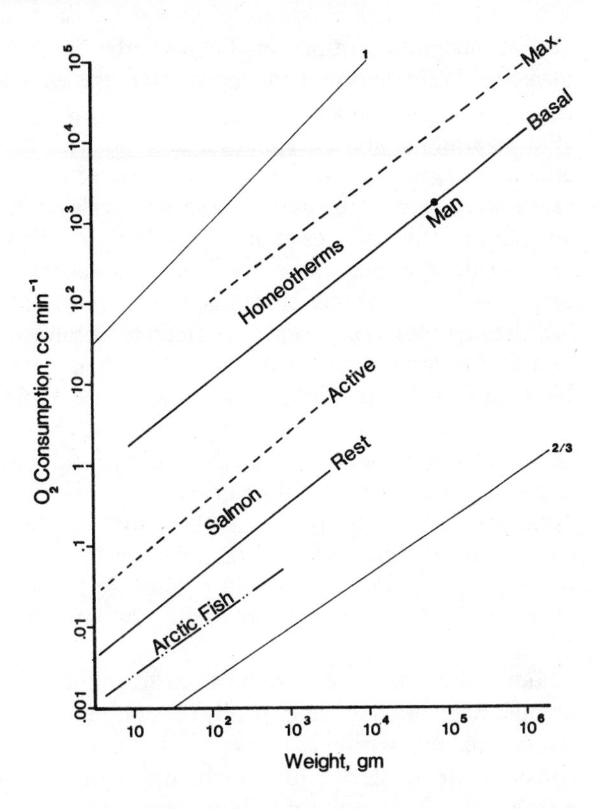

Figure 2.3: Metabolic rate vs. size. Average regression lines are shown for homeotherms at rest and at maximum metabolic rate (from various sources); for salmon at rest and during maximum sustained exercise (Brett, 1965), and for an Arctic fish (Holeton, 1974). Slopes of 1 and ⅔ are shown for reference. The slopes for most metabolic data fall between 0.75 and 0.90.

ment (Prosser, 1973). The fact that gill surface area in fishes varies with a similar 0.8 exponent is probably evidence only for the matching of gas exchange surfaces to metabolic requirements and is not a reason per se for the relation (Price, 1931).

TEMPERATURE EFFECTS ON METABOLISM

Poikilotherms

The first measurements of the effects of temperature upon metabolic rate in a poikilotherm appear to have been those of Ege and Krogh (1914). They found that the oxygen consumption rate of a goldfish declined precipitously at low temperature, and that even on a log scale, the curve skewed downward at low temperatures. Subsequent investigations of many species of several different phyla have confirmed this general relationship (Scholander et al., 1953; Fry, 1957). Ege and Krogh's goldfish data were taken as a sort of standard for a long time, but it now appears that more eurythermal animals (i.e., animals that tolerate a wide temperature range) do not have such a steep metabolism-temperature relation.

One intriguing question has been whether animals living in cold environments have metabolic rates that are depressed to the same extent as would be predicted from studies of temperate and tropical forms. That is, if the curves for warm-climate animals are extrapolated to 0° to 5°C, how do rates of cold-adapted animals compare to the extrapolated values? An upward shift in the metabolic rate with respect to the warm-climate animals has been termed "metabolic cold adaptation" (Scholander et al., 1953). In their original study, which included insects, spiders, crustaceans, and fishes, Scholander et al. (1953) found no clear evidence for metabolic cold adaptation in any of the groups. In later work, Wohlschlag (1964) and others claimed to demonstrate substantial cold adaptation in Arctic and Antarctic fishes, but their conclusions were challenged by Holeton (1974) on methodological grounds. Holeton pointed out that comparisons were made between laboratory studies of temperate fish and field studies of the polar fishes, and that the methods employed in field studies tend to produce higher metabolic rates. He also criticized the use of Krogh's (1916) "standard" curve for goldfish as the primary criterion for judging cold adaptation, as subsequent studies of the same species have shown much lower rates and not as steep a temperature relation (Fig. 2.4). In Holeton's study of various Arctic fishes, no evidence for metabolic cold adaptation was found (Fig. 2.5) (Holeton, 1974).

Interest in the general subject seems to have declined, and other groups of animals have not been as critically reexamined. At this point it is not clear if metabolic cold adaptation is a real phenomenon at the whole animal level. There is considerable evidence for biochemical adaptation of enzyme systems, but whether this is a response to temperature per se or a response to changing intracellular pH with temperature is not clear (Somero, 1986)(see Chapter 6).

Homeotherms

Presumably one of the great evolutionary advantages of homeothermy was to free the metabolism of animals from the constraints imposed by low temperature. By maintaining a constant, high internal temperature an animal is able to optimize its energetics and to perform at high work output levels regardless of ambient temperature. Nonetheless, there is a metabolic cost to homeothermy which is dependent upon ambient temperature.

The relation between ambient temperature and metabolic rate in a mammal is interesting and complex. Some data for various terrestrial and marine mammals are shown in Fig. 2.6, and although the marine sea otter has a somewhat higher basal metabolic rate and better insulation than many mammals, the patterns observed are similar (Morrison et al., 1977). Over the range of ambient temperature from –20 to + 19°C, metabolic rate is constant. This is the so-called thermoneutral range. The actual temperature range varies considerably for different homeotherms but represents the ambient temperature range over which internal heat production is more than sufficient to balance the rate of heat loss

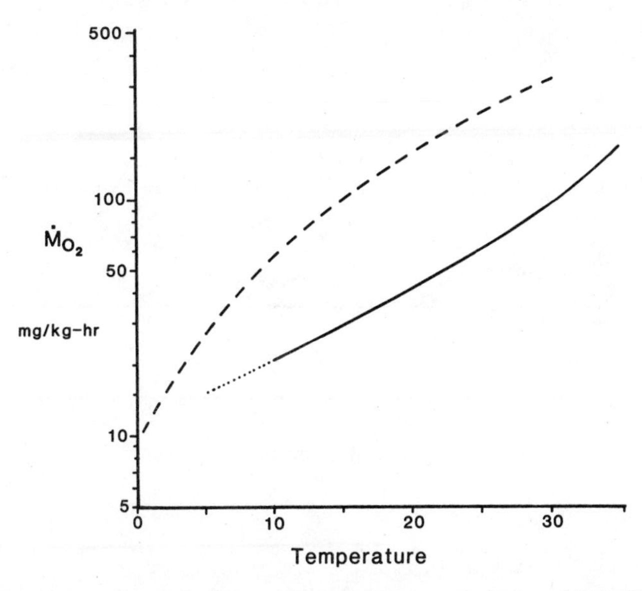

Figure 2.4: Comparison of "standard" metabolism *vs*. temperature in the goldfish. The upper line is from the original study of Ege and Krogh (1914). The lower line, with a much flatter slope, was taken from a more recent study using proper control of handling and acclimation (Beamish & Mookherji, 1964). (Figure redrawn after Holeton, 1974.)

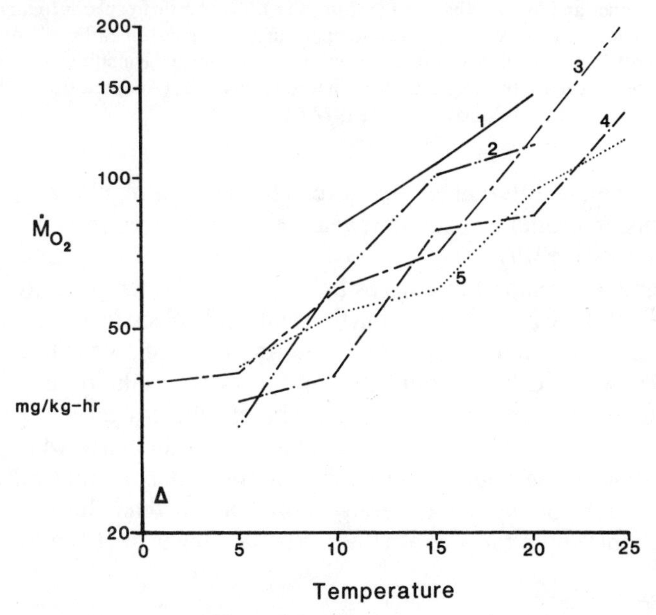

Figure 2.5: A comparison of "standard" metabolic rates for various temperate salmonids with data from fully cold-acclimated Arctic char (triangle, lower left). Holeton's (1973) analysis did not reveal any metabolic cold adaptation in the Arctic forms (1,2,3: brook trout; 4,5: rainbow trout). (Original data from various sources; redrawn from Holeton, 1973, by permission.)

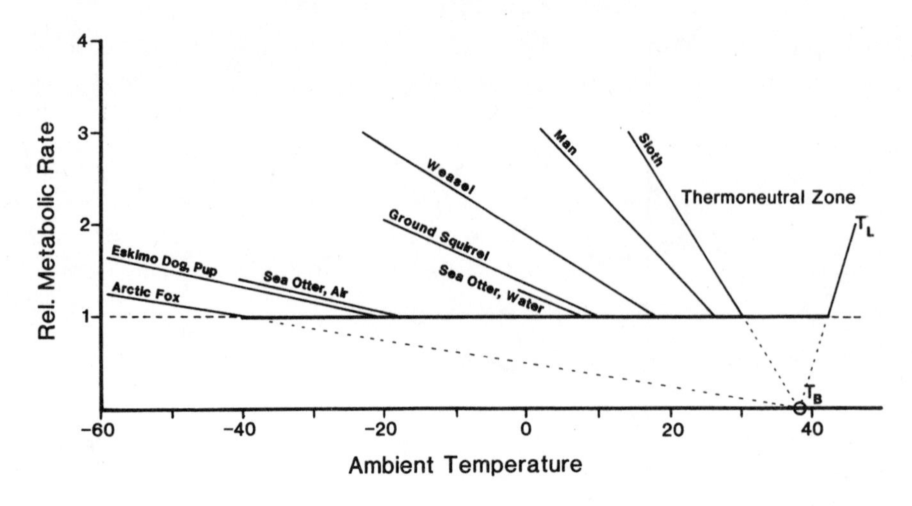

Figure 2.6: Metabolic rate in mammals stays constant over the "thermoneutral zone" of ambient temperature (indicated for the sloth) but rises below some critical lower and upper temperatures, usually with slope that intersects zero at the body temperature (T_B). The critical temperature and slope of the metabolic rise are functions of the basal metabolic rate and the body insulation, as are values of upper and lower lethal temperatures (T_L). The Arctic fox needs to increase metabolism only slightly to cope with the lowest temperatures occurring on earth, whereas man and the tropical sloth have a small thermoneutral zone. Conductivity of the medium is also important, as shown by the difference between sea otters in water and in air. (Adapted from Scholander et al., 1950; sea otter data from Morrison et al., 1977.)

to the environment. Internal temperature in this range is regulated by a number of mechanisms including panting, sweating, and changes in skin circulation (Cooper & Veale, 1979).

When ambient temperature drops below the critical temperature, heat loss begins to exceed the basal heat production and metabolism increases in order to regulate body temperature (Fig. 2.6). Some Arctic mammals can tolerate temperatures below –70°C by a combination of increased insulation and metabolic heat production (Morrison et al., 1977), whereas the critical temperature for naked man is about +27°C (Scholander et al., 1950). Similarly, when temperature rises above some upper critical temperature, metabolic rate will also rise. Mechanisms for cooling are generally more limited than those for heating, however, so this upper range is usually narrow, ending in heat death.

Hibernation

Perhaps the most interesting phenomenon associated with metabolism and temperature is hibernation, which in its strict definition pertains only to mammals, and excludes torpor states occasionally encountered in other groups. Hibernation is primarily an adaptation to low winter food supply, and it involves a long period of lowered body temperature and low metabolism (Hoffman, 1964). The Arctic ground squirrel, for example, curls itself into a tight ball in a well-insulated burrow and allows its body temperature to drop to around 6°C, reducing the thermal gradient from perhaps as high as 100°C to only 10° or 20°C. The reduction in energy expenditure so achieved allows this animal to spend almost three-fourths of its life in hibernation (Morrison & Galster, 1975). Although the animals are unresponsive when hibernating, physiological regulation is still effective, as shown by regulation of body temperature and respiratory variables (Malan et al., 1973)(see Chapter 14).

ACTIVITY METABOLISM

Virtually all animals must have the capability of increasing metabolic rate in order to supply energy for activity of various sorts. The difference between the resting or standard metabolic rate and the maximum active rate is called the "scope for activity" or "aerobic capacity" and represents the extent to which additional metabolic energy is available for activity. Not much information is available for the simpler invertebrates, but the studies of Pörtner et al. (1984c) indicate that aerobic capacity in the marine worm *Sipunculus nudus* is low, perhaps only allowing a doubling of oxygen consumption during digging activity. In the blue crab (*Callinectes sapidus*) oxygen consumption rises a maximum of 6-fold during maximum sustained swimming activity (Houlihan et al., 1985). In teleost fishes the scope for activity is about 5 to 10 times the resting rate, although there is a great deal of variability depending upon habits and body form (Brett & Groves, 1979). Data are somewhat scarce for birds, as measuring metabolism in flight presents obvious technical difficulties, but Tucker (1968) found that in-flight metabolism in the budgerigar rose to about 5-fold the resting rate, and Jones and Johansen (1972) found up to a 12-fold increase in various studies of other bird species. For man the maximum sustainable aerobic metabolism is about 7-fold the BMR, but rates up to 12 times BMR can be sustained for very short periods, as in sprints (Ruch & Patton, 1965; Wasserman et al., 1979).

Some of the highest rates of metabolism for animal tissue and the highest values for aerobic scope have been found in insects' flight muscles. In honeybees, locusts, and other strong flyers, rates of oxygen consumption during flight may exceed 100 cc g^{-1} hr^{-1}, nearly 20 times their resting metabolism (Krogh, 1941; Weis-Fogh, 1964; Withers, 1981).

Cost of Transport

An interesting comparison among different groups of animals is the amount of work energy expended (in the form of metabolism) in order to transport a

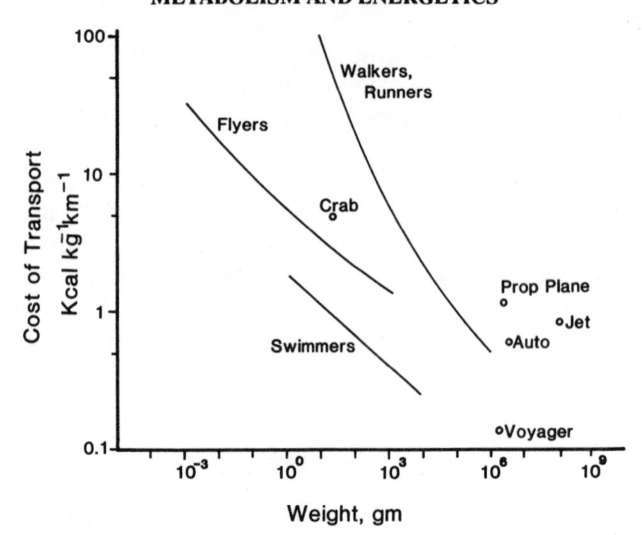

Figure 2.7: Metabolic "cost of transport." Adapted from Tucker (1972) with additional data on swimmers (fish) from Beamish (1978), the swimming blue crab from Booth et al. (1982), and calculations from the round-the-world Voyager flight using various news reports of distance and fuel consumption.

kilogram of body mass a given distance. Tucker (1972) originally assembled data in this form, both from his own studies and others, and coined the term "cost of transport" as a comparative measure of locomotory efficiency and metabolic capability. Data from Tucker's study, which included data for various man-made vehicles, as well as a few more recent studies, are shown in Fig. 2.7. Some interesting generalities emerge. Mode of transport is clearly the most important determinant of energy costs: Running and walking are least efficient, flying somewhat better, and swimming best of all. For any given mode, there is surprising little species variation, which may mean that evolution has favored convergence to the most efficient mechanical arrangements among many species. Finally, there is a strong inverse correlation between the cost of transport and body size, with energy expenditure reaching very high levels in the smallest species employing each mode.

The man-made vehicles do not fare well in the comparison, at least using 1970 figures for energy consumption. My calculations for the round-the-world voyage of the experimental aircraft Voyager, however, show an order of magnitude improvement over other propeller-driven aircraft, and a per-kilogram efficiency about as good as that of birds.

The study of activity metabolism embraces many subjects and subdisciplines, ranging from ecological energetics to sports medicine and biomechanics. The intent of the foregoing brief discussion has been to define the limits over which the respiratory systems of animals must work. It is a very broad range indeed, spanning perhaps six orders of magnitude on a weight-specific basis, from the near cessation of metabolism in hibernation and "cryptobiosis" to the intense combustion in insect flight muscle.

CONCLUSIONS

Respiration is concerned with the exchange of the gases produced by metabolism between the animal and its environment. Animals usually operate aerobically, consuming oxygen and producing carbon dioxide. Anaerobic metabolism is able to sustain most animals temporarily, however, and a few animals permanently. The overall rate of metabolism and respiratory gas exchange is affected by many variables, of both internal and environmental origin. They include temperature, activity, anoxia, hibernation, and feeding. Since metabolism is an integrative measurement of all energy-requiring activity in an animal, it follows that almost anything that affects behavior or function in an animal will be reflected in metabolic rate. The respiratory system must be designed for each animal to meet the entire spectrum of its metabolic requirements.

Chapter 3

Gas Exchangers

Although in the simplest animals gas exchange takes place over the general body surface, most animals have evolved specialized gas exchange tissues or organs. These are extremely diverse, but can be categorized on the basis of the way they function and the structures from which they are derived. In spite of the morphological diversity of gas exchangers, their function can usually be described and analyzed by one of a few relatively simple models. In the following discussion, an attempt is made to describe a variety of gas exchangers from various phyla, and to describe the general approach to characterizing the way they work. There is also a conventional system of symbols and notation that is required for the understanding of more detailed descriptions in the later chapters.

DIFFUSIVE GAS EXCHANGERS

From a respiratory standpoint, the simplest animal is a single spherical cell, producing CO_2 and consuming O_2 at a uniform rate throughout the stationary cytoplasm. In such an animal, pure diffusion alone must suffice to maintain an adequate O_2 supply at the center of the cell and to prevent CO_2 from building up. For such a single cell the concentrations of the diffusing substances follow the equation (Rashevsky, 1960):

$$c_i = c_o + (qr_o/3h) + (q/6D_i)(r_o^2 - r^2) + (qr_o^2/3D_e) \qquad \text{(Eq. 3.1)}$$

where:

c_i = concentration at a point *i* along a line drawn from the cell center outward,
c_o = concentration of the outside medium at infinite distance
q = rate of production (consumption) of the diffusing substance
h = permeability of the cell boundary (membrane)
r_o = cell radius
r = distance from the cell center
D_i = diffusion coefficient of the inside medium
D_e = diffusion coefficient of the outside medium

Values for q are positive if the substance is being produced, and negative if being consumed. The shape of the gradient inside the cell will be a parabola with max-

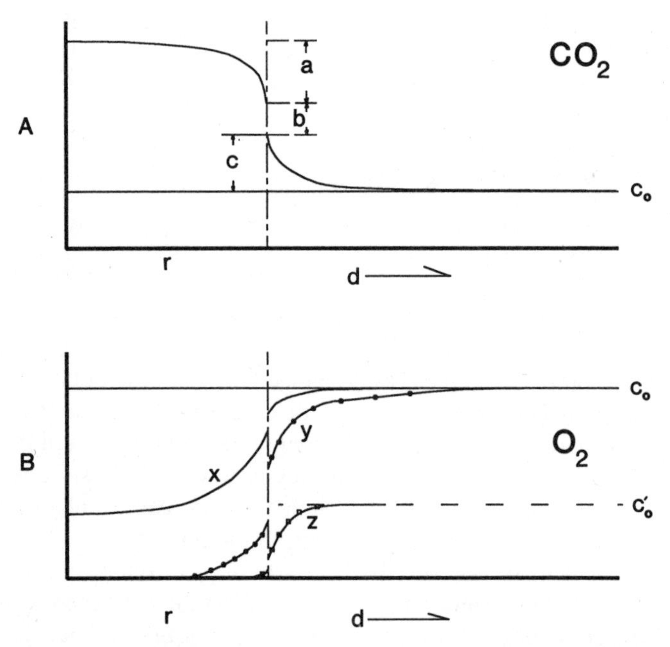

Figure 3.1: Diffusion gradients for oxygen and carbon dioxide. The x-axis represents distance from the center of a single cell (at left) of radius r, and the vertical line represents the cell membrane. In the upper panel, the diffusion gradient from the cell center to the external bulk concentration (c_o) is divided into portions a, b, and c according to the membrane thickness and the ratio of radius to production rate. The value for b is given in the text; $a = qr_o^2/6D_i$, and $c = qr_o^2/3D_e$. In the lower panel, the solid line (x) represents normal oxygen profiles for a cell consuming oxygen. If the consumption rate is increased (y, line plus circles), the gradient across the membrane increases and part of the cell interior drops to zero oxygen concentration. Decreasing the external concentration to c_o (z, line plus squares) has a similar effect. See text for further explanation. (Adapted from Rashevsky, 1960.)

imum CO_2 concentration at the cell center, as shown in Fig. 3.1A. The discontinuity at the cell membrane is equal to $qr_o/3h$, and the concentration then drops proceeding outward until the infinite bulk solution concentration is reached. If c_o is increased or reduced, the whole curve of Fig. 3.1A is shifted upward or downward without any change in its shape.

For oxygen the situation is exactly reversed (Fig. 3.1b) except that if q (the rate of consumption) is too high or r_o (the cell radius) too large the curve described will reach negative values at some point away from the cell center (lower curves, Fig. 3.1B). This statement is obviously absurd, as negative concentrations are impossible, but it means that there will be a zone of the interior of the cell in which oxygen concentration is zero.

If we assume that zero oxygen concentration in the cell interior cannot be tolerated, we can then calculate, for any given rate of consumption and outside oxygen concentration, what the maximum radius of a cell can be. Because we have seen that on a weight-specific basis small organisms have much higher rates of metabolism than larger ones (see Chapter 2), a more relevant exercise for the purposes of the later chapters is to use a rate of oxygen consumption of a larger animal and calculate the maximum allowable cell radius. The formula of Eq. 3.1 can be reduced in this case to:

$$r = [(c_o \times 6D)/q] \qquad \text{(Eq. 3.2)}$$

Using a value of 1 cc O_2 per gm-hr (2.778×10^{-4} cc gm^{-1} sec^{-1}), an outside oxygen concentration of 5.712 cc O_2 cm^{-3}, and a value for D of 1.98×10^{-5}, the maximum cell radius possible without an anoxic zone at the center is 0.5 mm. This value corresponds fairly well with observations in nature, since it is rare to encounter single spherical cells larger than 1 mm diameter. Single cells do get larger than that, but by assuming irregular shapes the maximum diffusion path length is kept small. In highly active cells, the mitochondria are sometimes arrayed around the internal margins of the cell, effectively reducing the mean diffusion path. With only slight modifications the same analysis can be applied to cylindrical cells or tissue sections with similar results, i.e. a prediction that 1 mm represents a practical upper limit imposed by diffusion limitations (see Chapter 4).

As with the CO_2 curve (Fig. 3.1A) changes in the external concentration of oxygen displace the entire curve of Fig. 3.1B up or down. As external oxygen concentration drops a "critical tension" will be reached when the center of the cell falls to zero concentration. Further reductions lead to a progressive but nonlinear increase in the cell volume that is anoxic, allowing a mathematical prediction of the relation between external oxygen concentration and the rate of oxygen consumption. Since volume is proportional to $4r^3/3$, we might expect oxygen consumption to decrease as a function of $4(r_o^3 - r_1^3)/3$ and the gradient (Eq. 3.1), where r_o is the cell radius and r_1 the anoxic zone radius. In fact such a relation provides an approximate fit to some experimental data, but a much better fit is obtained if an additional term is included to account for the dependency of oxygen consumption upon oxygen concentration. The mathematical treatment becomes rather complex, however, and differs for various substances according to the kind of concentration limitation encountered (Rashevsky, 1960).

In the overall scheme of respiratory gas exchange, diffusion in aqueous media is a very short-range phenomenon, since even steep gradients can develop over distances of less than 0.1 mm. Clearly an animal of any appreciable size must rely on mechanisms other than diffusion to transport respiratory gases to and from the environment. In air, on the other hand, the rates of diffusion are about 10,000-fold higher than those in water (see Chapter 1), which means that for the same concentration difference the mass transfer achieved through 1 μm in water can be achieved through 10 cm in air.

CONVECTIVE EXCHANGERS

There are several classes of propulsive mechanisms employed in the respiratory systems of animals. Convective flow of the external medium is called "ventilation" and of the internal medium "perfusion." Depending on the species and the medium, ventilatory mechanisms may range from simple cilia to the highly complex apparatus of the bird lung.

In aquatic animals ventilatory machinery can be roughly categorized as either pumps or paddles. Cilia are the most common type of paddle in the lower invertebrates; and although they do not seem very efficient, they can be employed to considerable effect. Although sponges can reach fairly large mass, their entire bodies are made up of fine water channels arranged in such a way that each cell

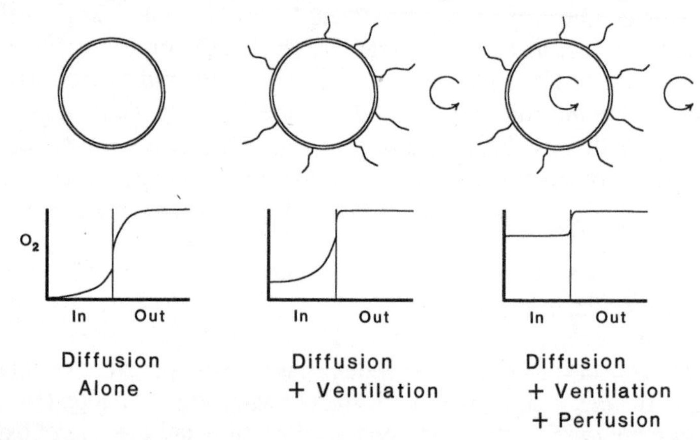

Figure 3.2: Effects of external and internal convection on the diffusion gradients for oxygen. In the left panel, with diffusion alone, a considerable gradient develops both within and without the cell. In the center, with external convection provided by cilia, the external gradient is reduced, and oxygen concentration inside the cell is increased. At the right, the further addition of internal convection e.g., cytoplasmic streaming, has the effect of reducing the internal gradient as well, supporting a higher rate of oxygen consumption.

borders a water channel. No part of the tissue is more than two or three cells thick, and a high rate of water flow is provided to all the channels by cilia contained in specialized cells called choanocytes (Barnes, 1980). This arrangement is an extreme case of development of the ventilatory system in the absence of any internal system for perfusion, but cilia are employed in many other groups of animals. Molluscs have well-developed gills that often serve for both respiration and feeding; the substantial water flows through the siphons of clams are propelled entirely by beating of the cilia on the gills (Borradaille et al., 1963).

Water flow can be provided by other means in invertebrates which lack obvious ventilatory structures. Stonefly larvae, for example, position themselves in water currents in streams in order to have a continuous flow of oxygenated water past their outer surfaces. Many marine tube-dwelling worms maintain water flow through their tubes by peristaltic movements of the body wall (Wells, 1949; Mangum, 1985).

Beginning with the cephalopod molluscs a variety of pump mechanisms have evolved to provide water flow over gills contained in an enclosed chamber. The nautilus and the octopus, for example, have enclosed mantle cavities lined with smooth muscle layers that rhythmically contract, producing a pulsatile flow of water through the cavity (Johansen & Lenfant, 1966; Houlihan et al., 1982). The water current again serves more than one function, in this case respiration and locomotion. Most of the aquatic arthropods have rigid enclosed gill chambers that are ventilated by the pumping action of a paddle-like appendage (see Chapter 9). The ventilatory pump apparatus is more complicated in the fish, involving an enclosed gill chamber and the coordinated action of the mouth and the coverings of the gill chambers (operculi).

Whatever means are employed for ventilation, the importance of both external and internal convection is highlighted by considering the oxygen gradients in the presence or absence of convection (Fig. 3.2). Keeping in mind that gas transfer increases as the gradient across the external surface increases (Eq. 1.14), it is easy to see why there has been such extensive evolution of more complex and efficient systems for gas exchange.

CUTANEOUS GAS EXCHANGE

Many animals possess no specialized structures for gas exchange, but rely on the surface of the skin or cuticle. When animals are very small, diffusion directly from the external environment to internal tissues or organs may supply a significant portion of the total oxygen demand, and only a slight supplement to cutaneous exchange is needed. The circulatory system provides a convective "boost" for transport of gases between the surface and internal regions. The numerous phyla of small worms, for example, provide for most of their gas exchange needs by this combination of direct diffusion and circulatory transport. In most cases there is neither a need for nor a means of providing external ventilation.

In larger animals external ventilation may be accomplished in a variety of ways, even when there is no particular structure devoted to that function. Stonefly (Plecoptera) larvae, for example, position themselves in running water so that a constant supply of oxygen is ensured. Simple swimming or crawling movements may suffice to stir the water for many other species, and in the case of tube-dwelling worms peristaltic movements of the body wall maintain water flow through the tubes (Barnes, 1980).

Cutaneous respiration may also be quite important in much larger and more complex animals, including vertebrates (Feder & Burggren, 1985). In these cases, however, the cutaneous respiratory mode may be a secondary development; i.e., these animals mostly have evolved from forms that were much more dependent upon gills or lungs (Romer, 1972). Modern amphibians present the best examples: Some have well-developed lungs, others have well-developed gills, and still others have neither and depend entirely upon skin breathing (Johansen, 1972; Guimond & Hutchison, 1976; Piiper et al., 1976). Cutaneous respiration is also quite important in various fishes (Johansen, 1972; Toulmond et al., 1982) and reptiles (Standaert & Johansen, 1974).

As in the invertebrates that rely on cutaneous respiration, a variety of means are employed for stirring the external medium. They may include behavioral positioning in the moving medium or active body movements. There is also some degree of control of the cutaneous exchange by alterations in blood flow to the skin; increasing the stirring of the water, for example, increases the blood flow to the skin of the bullfrog (Burggren, 1988a), and cutaneous conductance may change in response to variations in CO_2 (Burggren & Moalli, 1984; Malvin & Hlastala, 1986).

There are obvious limitations to cutaneous respiration, primarily the risk of desiccation and the diffusion limitation imposed by the relatively small surface. Animals relying heavily on cutaneous exchange are either aquatic or inhabit moist environments.

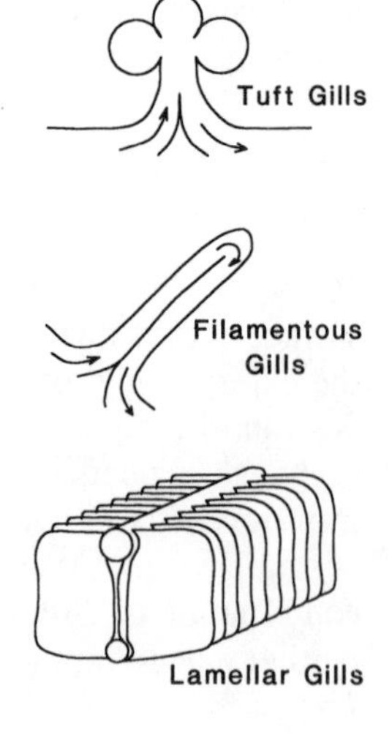

Figure 3.3: Basic types of gill. Tuft gills are found in many invertebrate groups, including various worms and echinoderms, and usually do not have very well organized vascular systems. Filamentous gills, such as those in many Crustacea, may sometimes be present by the thousands, and may or may not be confined to a gill chamber. The lamellar gills shown are from a crab; lamellar gills of fishes have further levels of subdivision and can provide very large exchange surfaces.

GILLS

A gill may be loosely defined as any elaboration of the external surface of the animal for the purpose of gas exchange. The definition is actually somewhat arbitrary, since we would not define the infoldings of the cell membrane of an ameba as a gill; but if the outward extension is a little more pronounced, forming a bud or tuft like those found in many marine worms and echinoderms, the term gill may apply. Gills can be categorized as tuft-like, filamentous, or lamellar (Fig. 3.3). Although the various types appear to have arisen independently in various groups of animals, there is a decided trend toward lamellar gills in the more active and advanced animals.

Tuft-Like Gills

Tuft-like gills, formed by outpouching of the body wall, undoubtedly supplement gas exchange in a wide variety of invertebrates. In fact their diversity is so great that only some very general statements can be made. In the simplest ones there may be only a thinning of the body wall and a slight protrusion or tubercle, while in the more complex forms beautiful arrays of pinnate or arboreal gills may be found (Fig. 3.4). The gills may be paired or not; located on all segments, the head, or the tail; placed singly or multiply on segments; and may or may not have feeding functions in addition to respiration.

Figure 3.4: Invertebrate gills that have additional non-respiratory functions may be quite elaborate and beautiful. Two Caribbean polychaete fanworms are shown here, a species of the family *Sabellariidae* on the left and of the *Sabellidae* on the right. (Photographs courtesy of C. L. Kitting).

The circulation in most of these gills has been poorly described, but the simplest forms seem to be tidally perfused. That is, they fill when muscular contractions of the body wall force blood into them and empty when the body wall relaxes (Mangum, 1985). Some are perfused with blood, but many invertebrates lack a true vascular system and perfuse their gills with coelomic fluid. Other, better-developed gills do have networks of vessels providing directed flow of oxygenated and de-oxygenated blood (Barnes, 1980), but there are many vascular arrangements for connection of the vessels (Mangum, 1985).

The importance of tuft-like gills in gas exchange has not been determined in most groups of animals. Most of the animals are very small, which makes experiments difficult, and aside from anatomical and taxonomic investigations there has not been a great deal of interest shown in them. In animals such as echinoderms, which have very thick and presumably impermeable cuticles, the tube feet and other gill-like projections are surely important (Johansen & Peterson, 1971), but for many of the smaller worms the whole of the external surface is no doubt important in gas exchange, making the gills less important.

Filamentous Gills

In one sense a filamentous gill is no more than an elongated tuft-like gill, but in certain groups, notably the arthropods, there are advances in the arrangement of both internal and external flow that warrant a separate category. The common freshwater crayfishes are the best-studied animals with gills of the filamentous type and probably show the limits of performance that can be achieved.

The structure of the crayfish gill filament is shown in Fig. 3.5 (Burggren et al., 1975); hundreds of filaments extend from the basal and medial portions of gill arches called arthrobranchs, pleurobranchs, or podobranchs, depending upon the segment from which they originate (McLaughlin, 1982). The gill arches and filaments are contained in a partially enclosed gill chamber formed by lateral ex-

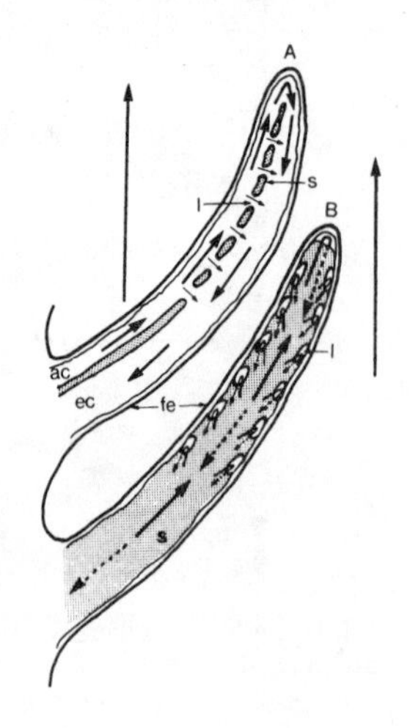

Figure 3.5: Cross-sections of two gill filaments of a crayfish, one (B) cut in a plane parallel to the internal septum (s), the other (A) perpendicular to it. The flow arrangement is more or less cross-current, since blood flows out one side (path ac) of the internal septum, crosses to the other side through internal openings (I), and returns along the other side of the filament (ec). The filamental epithelium (fe) is covered with a thin chitin layer. The large arrows indicate the direction of water flow. (From Burggren et al., 1975, by permission.)

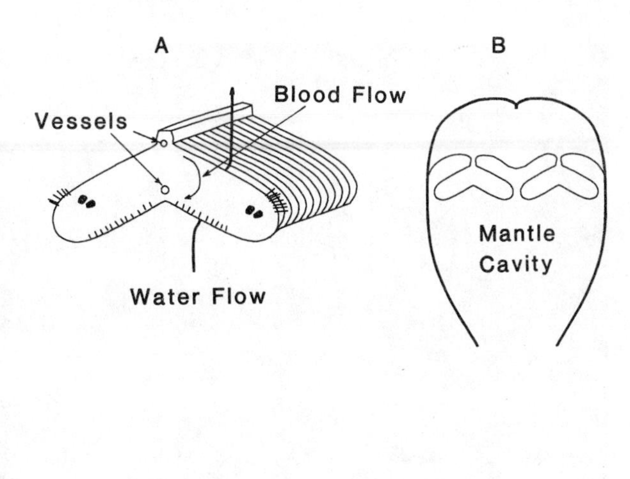

Figure 3.6: Lamellar gills of a primitive bivalve mollusc. Water flow is driven opposite to the direction of blood flow by cilia on the surfaces of each lamella, but particularly by large marginal cilia. The orientation and spacing of the lamellae are maintained partly by patches of cilia on adjacent surfaces that interlock with facing lamellae, and with various points of attachment in the mantle cavity (at right). Other mollusc groups show a wide variety of patterns that are modifications of this basic gill plan.

tensions of the carapace and are ventilated by the beating motion of a specially modified mouthpart appendage, the scaphognathite (*scaphos* = shovel + *gnathos* = tooth). Water flows roughly parallel to the orientation of the filaments, from the base toward the tip. Inside each filament there is a transverse septum running from the base almost to the tip, dividing the afferent and efferent blood. In addition to a communicating channel around the septum at the distal end of the filament, there are also small openings on either side along the length of the septum, allowing some crossover from afferent (= flowing toward) to efferent (= flowing away) flow channels.

The analysis of gas exchange in these filamentous gills becomes quite complex, since the water flows parallel (co-current) to the afferent stream and counter to the efferent stream. The afferent channel is only about half as large as the efferent channel in cross-section, however, which would minimize the time for co-current exchange and maximize the contact time during the counter-current portion (see below). There is a further possibility that there is some diffusion of oxygen from the efferent stream back into the afferent stream across the central septum, raising the possibility of some loop multiplication (see Chapter 7). In the absence of detailed local flow measurements and gas tensions from different regions of the filament under different conditions, one can only get a general idea of the processes occurring in the crayfish gill.

Lamellar Gills

A more efficient arrangement of gill exchange surfaces is found in the more advanced invertebrates and in both bony and cartilaginous fish, i.e., lamellar gills (Fig. 3.3). In the simpler gills, flat platelets (lamellae) extend to either side from a supporting gill arch, forming box-shaped channels through which water flows. In most cases these gill arches are enclosed in a branchial chamber, with a pump-

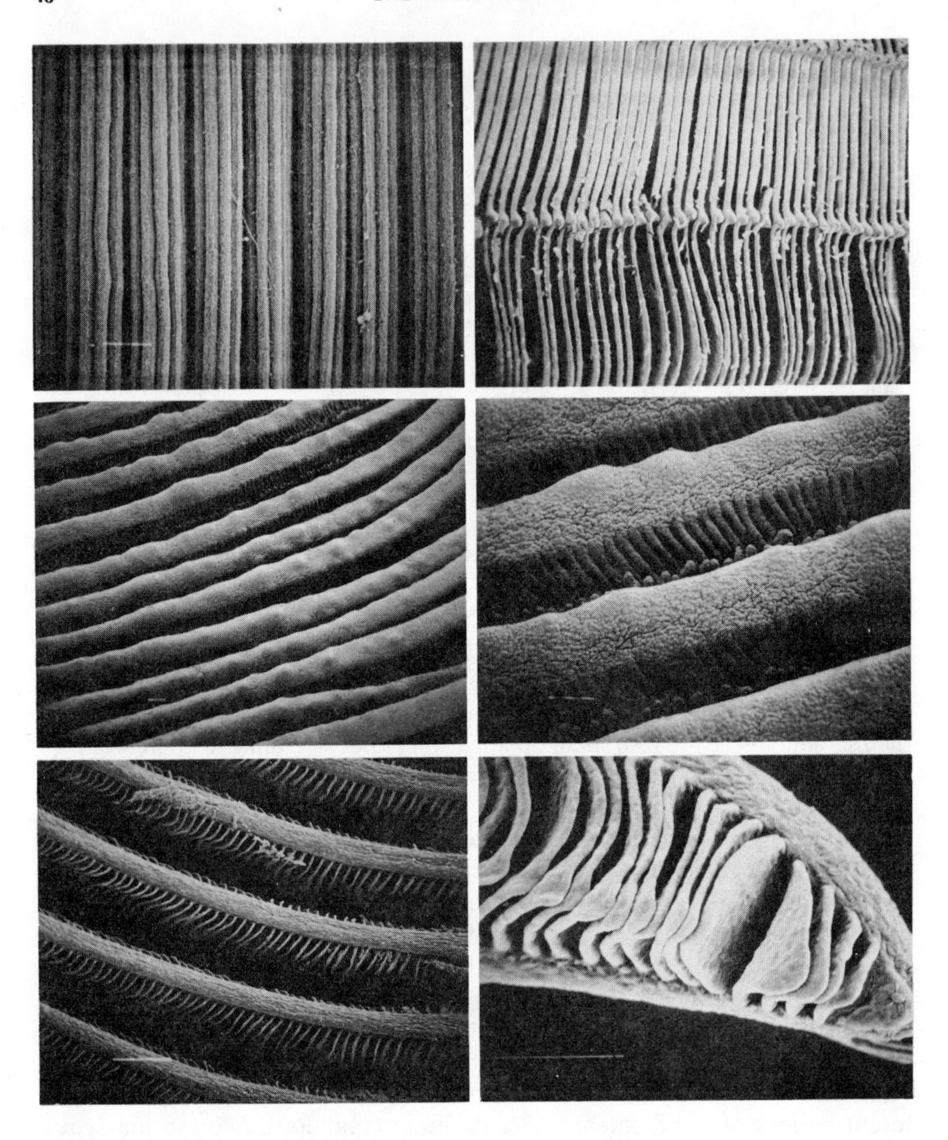

Figure 3.7: Montage of scanning electron micrographs of lamellar gills of a variety of aquatic animals. (Upper left) A portion of the feeding gills of an advanced bivalve mollusc (oyster, *Crassostrea virginica*)(54). (Upper right) Lamellar gills of a decapod crustacean (blue crab, *Callinectes sapidus*)(16). (Middle left) Lamellar gills of an elasmobranch. The large filaments running horizontally correspond to the lamellae of the crab gill; rows of perpendicularly stacked secondary lamellae may be seen between the filaments (cow-nosed ray, *Rhinoptera bonasus*)(16). (Middle right) Close-up of the elasmobranch gill, showing the secondary lamellae. (54). (Lower left) Gills of a perciform teleost (pinfish, *Lagodon rhomboides*)(54). (Lower right) Close-up of the tip of a teleost gill filament (215).

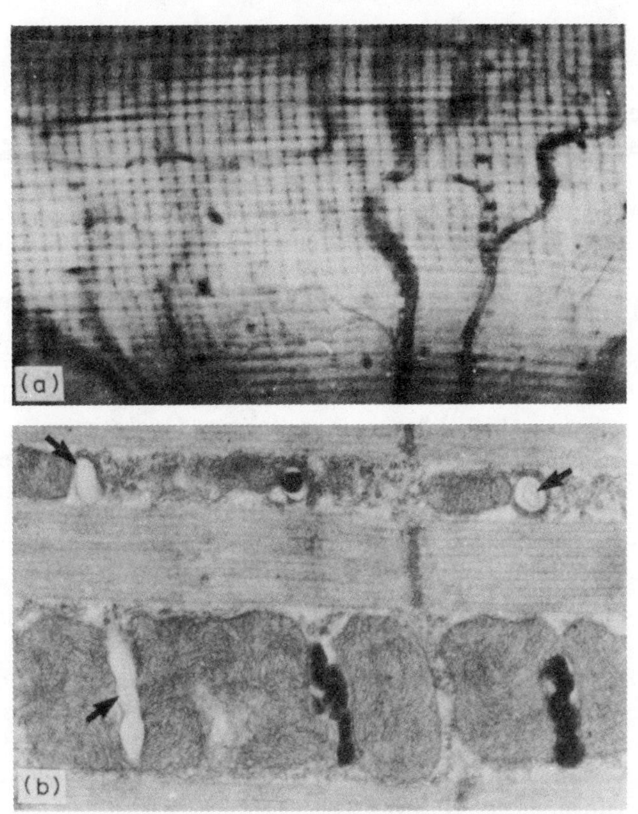

Figure 3.8: The tracheal supply to the flight muscles of the insect *Pieris* after injection of the tracheoles with black contrast medium. Top: The tracheoles are visible in optical sections as black dots, with larger branched tracheoles out of the plane of focus. × 1000. Bottom: A similar section seen in the electron microscope, showing injected tracheoles (black) between the mitochondria; uninjected tracheoles indicated by arrows. × 27,000. (From Wigglesworth, 1984, by permission, ©Chapman & Hall, Ltd.)

ing mechanism to direct water through the lamellar sieve. The gill sieve of a mollusc is shown in Figs. 3.6 and 3.7. Molluscs are somewhat unusual in that the sieve is ventilated by the combined action of many cilia (see above), rather than a paddle or pressure pump arrangement.

Blood channelling is more effective in lamellar gills than in the filamentous or tuft-like types, since afferent and efferent vessels lead to either end of the lamellar blood space and provide blood flow directed opposite to the water flow.

In the fish the lamellar organization is extended further: Each structure corresponding to the primary lamella in the arthropod and molluscan gills is extended laterally and has secondary lamellae extending to either side, forming yet another subdivision of the water channels (Fig. 3.7).

TRACHEAE

A unique respiratory system has evolved in the insects and spiders, consisting of a network of branching tubes that lead from the outside air to the respiring tissues. The tracheae are derived from the cuticle and share its structure, although there is no waxy layer and the chitin layer is thin in most of the tracheae and absent in the finest branches. The largest tracheae open to the exterior and in many species are guarded by spiracles that control the opening size. The tracheae branch into smaller and smaller tubes, finally ending with diameters of 0.2 to 0.5 μm. In active tissues such as flight muscle the fine endings actually penetrate the muscle cells, running adjacent to the mitochondria (Fig. 3.8)(Wigglesworth, 1984). The finer branches of the tracheal system (tracheoles) are 200 to 400 μm long, a distance over which diffusion is adequate to supply the gas exchange requirements of even very active tissues. The intimate contact of these air supply tubes and the oxidative machinery of the mitochondria supports oxygen consumption values of muscle during flight that exceed 500 cc gm^{-1} hr^{-1} (Weis-Fogh, 1964).

At various locations in the abdomen and thorax the tracheae are enlarged to form air sacs, and the muscular movements of the body serve to ventilate these air sacs during activity (Weis-Fogh, 1967). Opening and closing of the spiracles are controlled by respiratory demand, and are sensitive to CO_2. Under normal circumstances the spiracles are kept open just enough to supply gas exchange, helping to reduce water loss. At the internal end of the tracheoles there is also some evidence for control, since the finest endings appear to be filled with fluid at rest, and with air during activity (Wigglesworth, 1984).

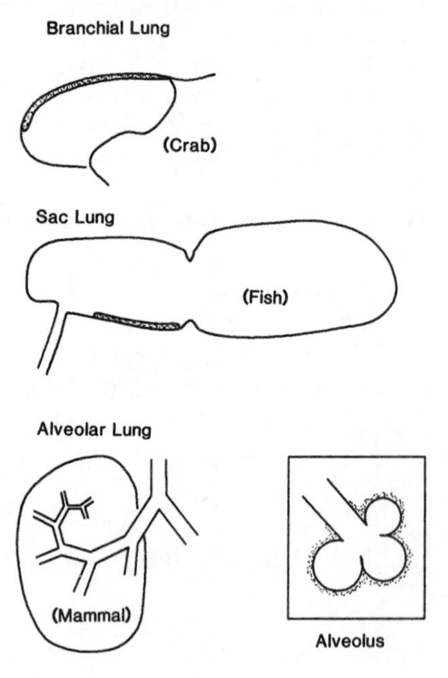

Figure 3.9: Basic types of lung. In many crabs the lung consists only of a vascularized area of the walls of the branchial chambers (stippled area). This area may or may not be elaborated to increase its surface area. These branchial lungs are ventilated by a one-way flow. In fish, some amphibians, and some reptiles the lung consists of a blind sac that may have no internal partitions and only a limited area of vascularization (stippled area). These lungs are ventilated periodically through a trachea or air duct. The alveolar lungs are characterized by a repeated airway branching pattern, ending in small, blind sacs called alveoli. Gas exchange takes place in the alveoli and in a limited area of thin-walled terminal airways.

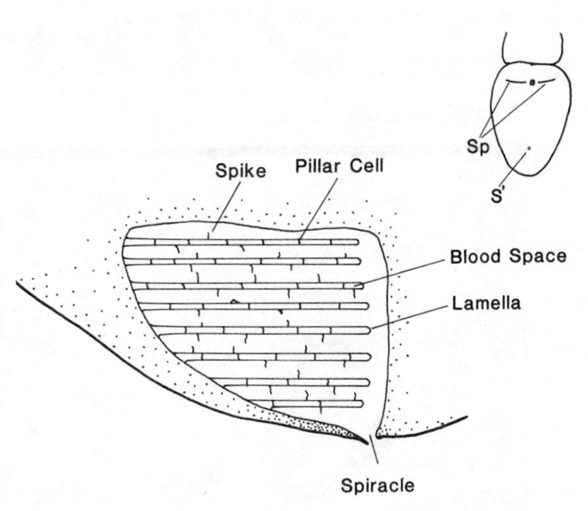

Figure 3.10: (Upper right) The slit-like spiracles (Sp) opening into the book lungs are usually located as shown on the ventral surface of the abdomen. There are two pairs of book lungs in some spiders, as well as variable numbers of normal spiracles (S') guarding the trachea. (Lower) Book lungs cut in dorsoventral section in a plane parallel to the midline with anterior to the left. The book lung consists of a stack of thin lamellae whose blood spaces are maintained by pillar cells. Chitinous spikes extend into the interlamellar air spaces to maintain spacing.

LUNGS

Perhaps the best distinction between a lung and a gill is that a gill arises from evaginations of the body surface, whereas a lung arises from invaginations. Both are epithelial structures designed to optimize the surface for gas exchange. While gills are sometimes enclosed, lungs always are. Almost all of the gills described above are ventilated unidirectionally, whereas lungs are usually ventilated tidally, i.e. by alternate filling and emptying along the same channels. Lungs can be categorized somewhat arbitrarily on the basis of the geometry and complexity of their air and blood channels (Fig. 3.9). The simplest might be termed sac lungs, the more complicated alveolar lungs, although there is quite a spectrum phylogenetically (Tenney & Tenney, 1970).

Although most lungs are air-breathing structures, the outstanding exception is the "respiratory tree" of the holothurians. This arboreally branching internal structure appears to serve the function of gas exchange (Winterstein, 1909; Brown & Schick, 1979), but owing to the delicate nature of the animals (they tend to eviscerate when handled) it has not been studied very much.

An in-between structure, the "book lung," is found in spiders (Figs. 3.10, 3.11). In structure it resembles the crab gill, consisting of a stack of lamellae or platelets with blood channels on the inside and air channels between. The book lungs are enclosed in an abdominal chamber whose opening is controlled by a spiracle that can be opened and closed to regulate gas exchange and water loss (Paul et al., 1987). The book lung is ventilated, but the tidal volume is so small that the lung clearly functions by diffusion with no significant convective component. Similar structures are found in aquatic horseshoe crabs and in isopods, but in these animals the platelets are not enclosed and are ventilated by being waved back and forth in the water.

The earliest aerial lungs in the chordate evolutionary line appear to have evolved in the primitive fishes. All of the extant Holosteans, for example, as well

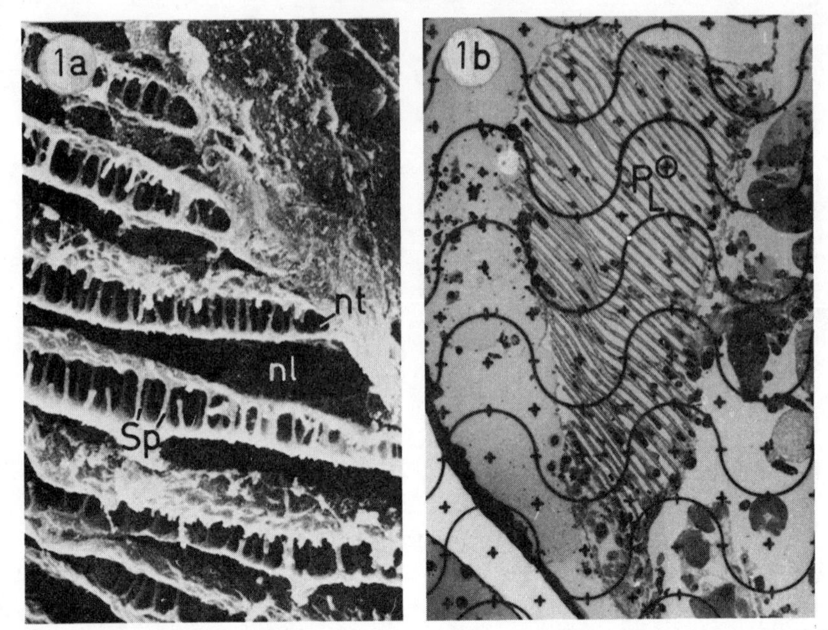

Figure 3.11: Scanning electron micrograph (1a) and semi-thin section of Spurr-embedded samples (1b) of lungs of house spiders (*Tegenaria* spp.). The medial surface is to the right and the dorsal surface at the top in all three photographs. Note the chitinous spikes (Sp) that extend into the air spaces (nl) and the internal bridges of each leaflet that maintain the width of the blood space. In 1b the grid and other markings were for morphometric measurements. (From Strazny & Perry, 1984, by permission, ©Alan R. Liss, Inc.)

as the preponderance of Chondrosteans and all Dipnoans, are air-breathers, but many groups of teleosts possess air-breathing organs as well (see Chapter 11). The typical (modern) fish lung consists of a single elongated sac connected to the upper esophagus by a pneumatic duct. In most fish there is only rudimentary division of the lung into smaller air chambers, and the total surface area is small. The walls of the lung are well provided with capillaries from an afferent circulation deriving from the dorsal aorta and draining back into the venous circulation (see Chapters 7 and 10). In the vast majority of air-breathing fishes the lung is ventilated periodically by a combination of smooth muscle in the wall and the action of the buccal/opercular pump. As the fish approaches the surface the smooth muscle of the lung wall contracts, forcing air out through the pneumatic duct. When the surface is broken, a gulp of air is taken, the mouth and operculi are closed, and the floor of the buccal cavity is raised to produce a positive pressure, forcing air back into the lung. For only one fish, the Amazonian osteoglossid *Arapaima gigas*, is there evidence of aspiration breathing, *i.e.*, a contribution of negative pressure produced by contraction of a smooth muscle diaphragm lying below the lung (Farrell & Randall, 1978).

Although there is a strong morphological resemblance between primitive fish lungs and the swimbladders of extant fishes, current evolutionary thinking is that the lung arose earlier and that the development of the swimbladder as a buoyancy organ came somewhat later.

Amphibian lungs are difficult to characterize in a general way, since a complete spectrum exists within the group from animals with no lungs at all, e.g. the Plethodont salamanders, to those with well-developed and finely subdivided lungs, e.g. *Xenopus*. In those amphibian lungs with "alveoli" (small air sacs at the termini of branching airways), the smallest mean alveolar diameter is about 1 mm. The amphibians are positive-pressure breathers, taking air into the buccal chamber, then pumping it in and out of the lung. Most amphibians and especially the aquatic ones still carry on an important part of their gas exchange through the skin (Krogh, 1904; Guimond & Hutchison, 1976), so the skin surface must be counted as part of the total respiratory surface (Tenney & Tenney, 1970).

In the reptiles there is also great diversity in the development of the lung, ranging from poorly divided sac-like lungs to well-developed and finely partitioned ones. The total lung surface area varies with body weight in precisely the same fashion as does metabolic rate (see Chapter 2); i.e., it follows a relationship of the form $S = aW^b$ where $b = 0.75$ (Tenney & Tenney, 1970). Aspiration breathing does appear in various reptiles, but positive-pressure inflation via a buccal pumping mechanism predominates (Gans, 1970).

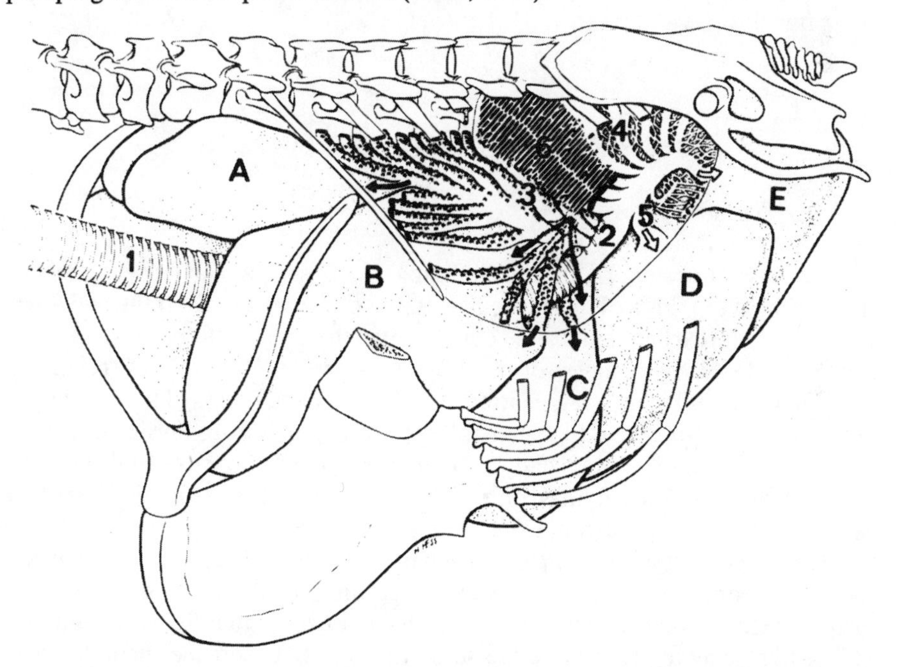

Figure 3.12: Major features of the lung and air sac system of a large bird. Air enters through the trachea (1) and is distributed in a complex way among the lungs and air sacs. See text and Chapter 13 for further discussion. (A) Cervical air sac, (B) Interclavicular air sac, (C) anterior thoracic air sac, (D) posterior thoracic air sac, (E) abdominal air sac, (2) primary bronchus, (3–6) secondary airways. (From Duncker, 1972, by permission, ©Elsevier-North Holland.)

The lungs of birds are highly developed for meeting the extreme metabolic demands of flight and have structural modifications that significantly alter the way they function compared to other types of lungs. Air enters the trachea, branches into the two major bronchi, and then branches into tertiary airways leading to five large air sacs and to the exchange sites, the parabronchi (Fig. 3.12)(Duncker, 1972). The pressure, flow, and volume changes during the ventilatory cycle have been the subject of considerable controversy, but it is now generally agreed that the net effect of the concerted action of the air sacs and ventilatory musculature is to produce a mostly one-way flow through the parabronchi (Bretz & Schmidt-Nielsen, 1971; Scheid, 1979)(see Chapter 13).

Finally, in the mammals we see a similar extreme development of the lung as a high-surface-area, high-capacity gas exchanger. A sphere of 1 cm^3 volume has a surface area of 4.8 cm^2, but 1 cm^3 of the rat lung contains an alveolar surface area of 661 cm^2 (Weibel, 1972), achieved by very fine division of the terminal air sacs by septa containing capillary nets (see Chapter 14). Interestingly, in mammals the total alveolar surface area has a weight exponent of about 1.0, rather than the 0.75 usually found in such allometric relations (see Chapter 2). Mammals ventilate their lungs by aspiration, or negative-pressure inflation, and only the major airways are ventilated; diffusion provides the mechanism for exchange in the fine airways and alveoli.

CHARACTERIZATION OF EXCHANGERS

Major Exchanger Types

A useful basis for categorizing and analyzing convective gas exchangers is by the orientation of the flows of internal and external exchange media. If the two flows are parallel, the exchanger is called co-current; if in opposite directions, counter-current; and if at right angles (or nearly), cross-current (Fig. 3.13). At relatively equal flows of both media, large differences in the effectiveness of the exchanger types are readily apparent. In the co-current exchanger, the gas tension of the internal medium leaving the exchanger cannot exceed that of the external medium; under ideal conditions it may only approach equilibrium (Fig. 3.13, second panel). In the counter-current exchanger, since the effluent internal medium is in exchange contact with the *incoming* external medium (Fig. 3.13, third panel), under ideal conditions its tension may approach that value. The cross-current exchanger has characteristics in between the former two, so that the effluent internal medium may approach a value somewhere in between the incoming and effluent external medium (Fig. 3.13, lower panel). Finally, some exchangers are categorized as "uniform pool" types, but they are really a special case of the others with ventilation effectively infinite, holding the external medium value constant. This latter model applies to the human lung, in which the alveolar air reaches a steady state by diffusion from the major airways, and to skin breathing, in which the external medium is the atmosphere, an infinite source or sink (Piiper, 1985).

Primary Parameters of Exchangers

Gas exchange lends itself nicely to mathematical analysis and, borrowing heavily from the engineering analysis of flow and heat exchange, a considerable literature has developed. As shorthand for such analysis, a large number of terms and symbols have been adopted to denote the important parameters of exchange. Although many of these terms and symbols are standardized (Pappenheimer, 1950; Bartels et al., 1973), there is less than perfect adherence to the conventional notation. Accordingly, Table 3.1 and Fig. 3.14 illustrate most of the symbols employed in gas exchange analysis in this book. Others are defined in the relevant text sections. In the following section, each of major parameters affecting exchange is described, with relevant equations and terms presented.

Flow

The flow of the external medium, ventilation, is usually represented by $\dot{V}$, but there is considerable variation. Tidal ventilation is often shown as $\dot{V}_T$, and if the external medium is water, the term $\dot{V}_G$, or even $\dot{V}_g$ is often encountered. For the internal medium, flow through the exchanger is denoted by either $\dot{V}_b$ (Dejours, 1975) or $\dot{Q}$ (Bartels *et al.*, 1973). In cases where the flow through the exchanger

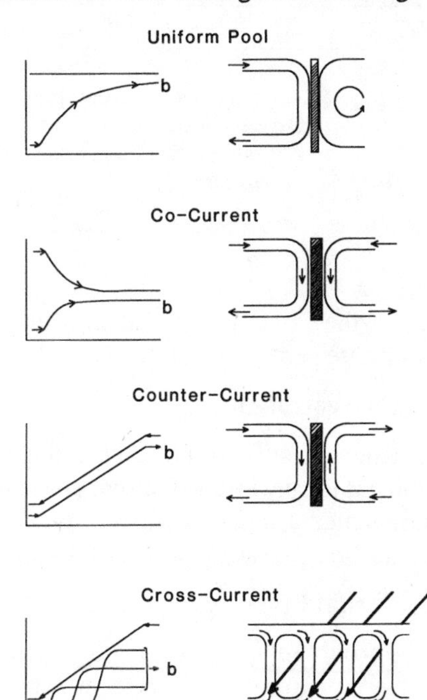

Figure 3.13: Classification of gas exchangers on the basis of the orientation of flow in blood and respiratory medium. The graphs at the left show the profiles of oxygen partial pressure in blood (b) and respiratory medium as a function of distance along the exchanger surface. Note that the partial pressures in the efferent blood may theoretically exceed those in the efferent medium only in the counter- and cross-current exchangers.

Table 3.1: Conventional symbols used in respiratory physiology.

Primary Symbols			
A	Area	M	Quantity, mass
BW	Body weight	P	Pressure
C	Concentration	RQ	Respiratory quotient
D	Diffusion coefficient	S	Fractional saturation
F	Fractional concentration	V	Volume
H	Heart	α	Solubility coefficient
Hb	Hemoglobin	β	Buffer value
Hc	Hemocyanin	$\dot{}$	Super dot; time integral
Hct	Hematocrit		

Secondary Symbols			
a	Arterial	G	Gills
A	Alveolar	I	Inspired
b	Blood	L	Lung
E	Expired	W	Water
f	Frequency	τ	Harmonic mean thickness
g	Gas	T	Tidal

Examples	
Pa_{O_2}	Arterial oxygen (partial) pressure
$\dot{M}_{O_2}$	Oxygen consumption per unit time
$\dot{V}_A$	Alveolar ventilation per unit time
fH	Heart rate

is not the same as total cardiac output, the latter term is imprecise, so an appropriate subscript should be used.

Concentration

Concentration (C) of exchanging substances can be expressed in a variety of units; it is only important that consistency is observed between internal and external media, and with other parameters such as solubility coefficients. The concentration of oxygen in inspired water or air would be denoted as $C_{I_{O_2}}$.

Partial Pressure

Partial pressures at various locations are qualified with a location designator (arterial, expired, *etc.*) and a subscript to indicate the gas. Thus $P_{I_{O_2}}$ would indicate the partial pressure of oxygen in the inspired medium.

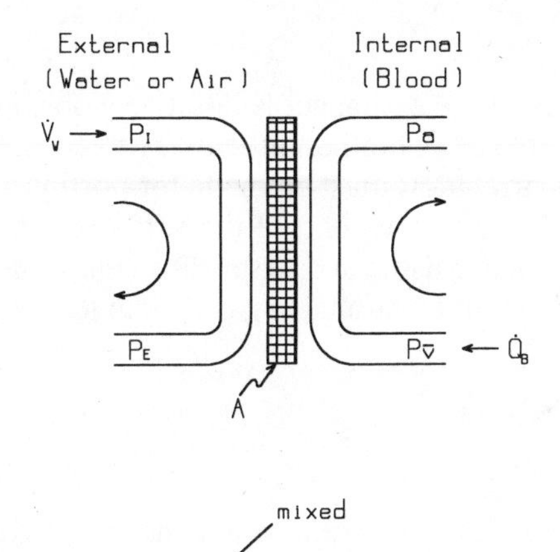

Figure 3.14: Some symbol conventions in respiratory physiology. The various symbols listed in Table 3.1 refer to locations as indicated in the upper panel, and complex symbols are constructed as shown in the lower portion.

Solubility

Many authors currently employ β to denote the "capacitance coefficient" as defined by Piiper et al. (1971; *cf.* Dejours, 1975). This is confusing on two counts: The symbol β is used to denote buffer value in acid-base physiology (see Chapter 6); and β as they define it is employed simultaneously as a simple (Bunsen) solubility coefficient for water and as a complex function designator for blood. In this book the symbol α is used as the Bunsen solubility coefficient, i.e., to designate the solubility of gas per unit volume of liquid per unit partial pressure of the gas (see Table 1.3; Program TEMPTABL). The same term is useful when applied to blood, since its meaning is clearly the solubility of the gas exclusive of the chemical binding due to carbonic acid reactions (see Chapter 6) or binding to respiratory pigments (see Chapter 5). To the extent that the binding curves can be expressed mathematically, the notation $\theta(O_2)$ will be used to denote the transfer function of partial pressure to concentration according to the binding relation.

Capacity

The capacity of respiratory media means the quantity contained per unit volume at standard atmospheric pressure. Thus for oxygen in air the capacity is simply the fraction of oxygen in the atmosphere, 0.2094. For water, the capacity is the solubility coefficient multiplied by the partial pressure of oxygen in the atmosphere, or αP (see TEMPTABL for solubility values). A more complex relation holds for blood, however, since the capacity is determined by the sum of dissolved oxygen plus that bound chemically to hemoglobin.

Area

Although the concept of surface area of the exchanger is intuitively obvious, actual determination of the functional area can be rather difficult. The geometry of the surface is usually complex, and corrections must be made for portions of the total area that are not functional. For example, part of the surface area of most gills and lungs is underlain by connective tissue and supporting cells rather, than by blood capillaries, and is therefore not functional in gas exchange.

Derived Parameters

The Exchange Equation

From simple mass flow considerations, the oxygen removed (consumed) from the respiratory medium equals the flow times the concentration difference between inflow and outflow, or:

$$\dot{M}O_2 = \dot{V}_A \times (C_I - C_E) \qquad \text{(Eq. 3.3)}$$

and for the blood:

$$\dot{M}O_2 = \dot{Q} \times (Ca - Cv) \qquad \text{Eq. 3.4)}$$

Since concentration equals solubility times partial pressure ($C = \alpha P$), the equations for water can be written as:

$$\dot{M}O_2 = \dot{V}_A \times \alpha \times (P_I - P_E) \qquad \text{(Eq. 3.5)}$$

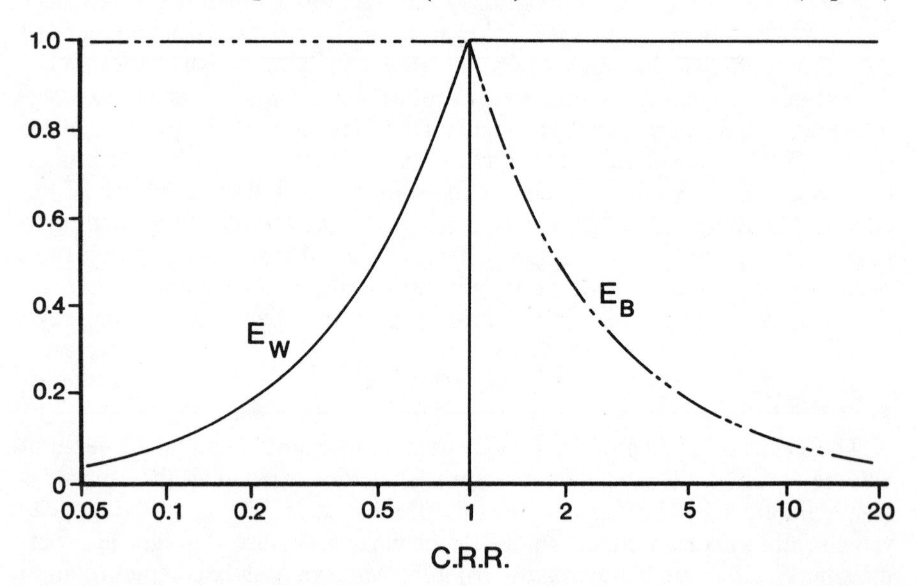

Figure 3.15: Theoretical relation between effectiveness and the capacity-rate ratio (CRR), as defined in the text. It is the classical relation derived from heat exchanger analysis and does not necessarily apply to nonlinear systems.

and for blood:

$$\dot{M}O_2 = \dot{Q} \times [\theta(Pa) - Pv] \qquad \text{(Eq. 3.6)}$$

Equation 3.4 is often written as:

$$\dot{M}O_2 = \dot{Q} \times \beta(Pa - Pv) \qquad \text{(Eq. 3.7)}$$

but as discussed above, this equation is incorrect since β is a nonlinear function and not a constant as Eq. 3.7 would imply. In particular, $\beta(Pa - Pv)$ is not equal to $\beta Pa - \beta Pv$, as normal mathematics would lead us to think.

These equations are often referred to as "Fick" equations, although their relationship to Eqs. 1.14 – 1.16 is indirect, and the original work by Fick seems to have been lost.

Ventilation-Perfusion Ratio

The ventilation-perfusion ratio, written as $\dot{V}_A/\dot{Q}$ for a lung or $\dot{V}_G/\dot{Q}$ for a gill, is a dimensionless index of the degree of flow matching in a gas exchanger. To take an extreme example, if the blood flow were nearly zero and the ventilation volume extremely high, it is clear that the efficiency of the gas exchanger will be low; the ventilation is much higher than needed for exchange with the small blood flow. For any exchanger there is an optimal ratio of flows of opposing media, and changes in the ventilation-perfusion ratio are indicators of changing efficiency.

Capacity-Rate Ratio

The ventilation-perfusion ratio does not take into account situations in which the capacities of the two exchanging media are different, so another useful index of gas exchanger performance is the so-called capacity-rate ratio, first adopted from heat exchanger analysis (Kays & London, 1958; Holman, 1981) by Hughes and Shelton (1962; Hughes, 1964):

$$CRR = \frac{\dot{V}A \times \alpha}{\dot{Q} \times \theta(sat.)} \qquad \text{(Eq. 3.8)}$$

where θ(sat.) is the oxygen content when blood is fully saturated. Actually, Hughes & Shelton's equation is inverted, with the blood capacity-rate as the numerator, but it is more consistent with the ventilation-perfusion ratio and more common to present it as in Eq. 3.8. Theoretical and experimental analyses of both heat and gas exchangers have shown that the maximum exchanger efficiency is obtained when the CRR is 1 (Fig. 3.15). These analyses, however, assume linear solubility functions and when one medium (blood) has a strongly nonlinear dissociation curve, the analyses produce significantly different results (Malte & Weber, 1985). These effects are explored in Chapter 7.

Effectiveness

The effectiveness of transfer is simply the ratio of gas (or heat) actually transferred to that which could be transferred under ideal conditions. Thus for water flowing in a gill the effectiveness (E_w) would be the ratio of the oxygen lost from the water to the amount of oxygen that would be lost if the expired water were completely equilibrated with the incoming venous blood, i.e., counter-current flow (Fig. 3.13). It is equal to:

$$E_W = (P_I - P_E) / (P_I - P_V) \quad \text{(Eq. 3.9)}$$

(Hughes, 1964). (To denote quantities, α would appear in both numerator and denominator but may be omitted.) This term is also defined by Piiper & Scheid (1972) as a fractional resistance to gas transfer, which for the ventilation side is denoted as r_{vent}, and in Piiper and Scheid (1975) as Δp_{vent}.

A similar index of effectiveness for the transfer to blood is defined as the ratio of oxygen actually transferred to the quantity that would be transferred if arterial blood were completely equilibrated with the inspired medium. The effectiveness for blood (E_b) as defined by Hughes is:

$$E_b = (P_a - P_V) / (P_I - P_V) \quad \text{(Eq. 3.10)}$$

This same definition is also used by Piiper and Scheid (1972) as the fractional resistance to gas transfer for blood, r_{perf}. Unfortunately, the derivation of this index is flawed. In the case of water, the quantity transferred can be written as:

$$\dot{M}O_2 = \dot{V}G \times (C_I - C_E) \quad \text{(Eq. 3.11)}$$

and since content $= \alpha P$,

$$\dot{M}O_2 = \dot{V}G \times \alpha \times (P_I - P_E) \quad \text{(Eq. 3.12)}$$

but for the blood the corresponding equation should be:

$$\dot{M}O_2 = \dot{Q} \times (C_a - C_V) \quad \text{(Eq. 3.13)}$$

and in this case no solubility constant can be factored out, since tension and content are related by the non-linear function θ. The proper expression for effectiveness of blood oxygenation must be kept in terms of content and thus becomes:

$$E_{b\text{-mod.}} = (C_a - C_V) / [\theta(P_I) - C_V] \quad \text{(Eq. 3.14)}$$

where $\theta(P_I)$ is the blood content when equilibrated with the incoming oxygen tension (P_I). I have termed this $E_{b\text{-mod.}}$ to differentiate it from E_b as defined by Hughes (1964) and r_{perf} as defined by Piiper and Scheid (1972).

Diffusing Capacity

The diffusing capacity is a measure of the ability of an exchanger to transfer gas per unit gradient across it; it is a function of the total surface area, diffusion coefficient of the gas in the barrier and media, and the thickness of the barrier. From Fick's equation:

$$\dot{M}O_2 = -(AD\alpha/x)(\Delta P) \quad \text{(Eq. 3.15)}$$

we see that the diffusing capacity is equal to $-AD\alpha/x$, where x is the thickness of the barrier in cm, A the area of the exchanger in cm^2, D the diffusion coefficient in $cm^2\ sec^{-1}$, and α the solubility coefficient in ml gas ml $fluid^{-1}\ torr^{-1}$. This index is sometimes called the "transfer factor" (T_{O_2})(Randall, 1970; Piiper, 1982), but Randall defined transfer factor as:

$$T_{O_2} = \dot{M}_{O_2}/[\tfrac{1}{2}(P_I + P_E) - \tfrac{1}{2}(P_a + P_{\bar{v}})] \qquad \text{(Eq. 3.16)}$$

Although this equation sometimes provides reasonable results, the nonlinearity of the blood oxygen dissociation curve renders this formula meaningless under many conditions in gas exchangers. The problem is that the arithmetic term for the mean gradient (the denominator of Eq. 3.16) does not provide an accurate estimate (see Chapter 7).

Conductance

Using Ohm's law as an analogy, where current equals the voltage (or potential) divided by the resistance ($I = E/R$), we can think of $\dot{M}_{O_2}$ as the current, the partial pressure or concentration difference as the potential and, from Eq. 3.11, V as 1/R. Since the reciprocal of resistance is conductance, the flow times the capacity is the conductance. If the conductances for blood and ventilatory medium are similarly defined as G_{perf} and G_{vent}, the ratio G_{vent}/G_{perf} is simply the CRR (above).

Many other derived parameters have been defined by various authors in the course of analysis of gas exchange systems but after differing symbols are reconciled, what appear to be very different expressions can almost always be related to the fundamental parameters described above.

DEAD SPACE AND SHUNTS

Early on in the study of gas exchangers it was recognized that some portions of the ventilatory flow do not, for various reasons, participate effectively in gas exchange, so the nonexchanging portion is called "dead space." A shunt, on the other hand, refers to a route for either ventilatory or circulatory flow that bypasses the exchange surface. The results of dead space and shunting are similar in that they reduce the efficiency of exchange.

A circulatory shunt usually involves specific vessels that bypass the exchange organ or at least bypass the active exchange surface. Shunting can be as far from the exchanger as the heart. In animals with three-chambered hearts, e.g. amphibians and reptiles, a crossover of systemic blood returning to the right atrium shunts the blood to the systemic circulation without going to the lungs, effectively a shunt past the lungs (see Chapters 7 and 12). In the fish gill there is a complex intralamellar circulation (Vogel, 1978; Boland & Olson, 1979; Laurent, 1984) that diverts some blood entering the gills into a non-exchanging path that is returned to the heart.

Ventilatory shunting in water-breathing animals is rather common, the most obvious form of which is water that spills between gill arches rather than entering the lamellar sieve (Randall, 1970b). It seems likely that at normal resting ventilation rates this spillage is negligible, but at high swimming speeds and when ventilation is greatly elevated it may be substantial. The gill arches and filaments have muscles and a sympathetic innervation that control their position relative to the water flow, affording some control over this ventilatory shunt (Saunders, 1961; Pasztor & Kleerkoper, 1962). In the lung-breathing vertebrates the concept of a ventilatory shunt does not apply.

There is, however, some application of the term shunt to situations in which units of the gas exchanger are ventilated but not perfused. In this case the ventilated medium is not exchanging, and the effect is like that of a shunt. It is really more appropriate to call this a functional or physiological dead space, rather than a shunt. In a water gill there is evidence that not all of the lamellae are perfused all of the time (Davis, 1972), so any water flowing past these non-perfused lamellae represents functional dead space. Similarly, in a lung there may be regions or individual alveoli that are not perfused, and the portion of ventilation going to these areas also represents functional dead space. The converse is also true, i.e., that blood flow to regions of the exchanger that are not ventilated represents functional dead space on the circulatory side.

A more interesting kind of functional dead space results not from total lack of ventilation or perfusion, but from local mismatching of the flows. It is well known, for example, that neither the ventilation nor the perfusion of the human lung are uniform, and that the $\dot{V}_A/\dot{Q}$ ratio varies considerably within the lung (West, 1962). If, for example, in some portion of the lung the $\dot{V}_A/\dot{Q}$ ratio is much higher than the optimum value, a portion of the ventilation does not participate fully in gas exchange and can be regarded as functional dead space.

Finally, in all exchangers there is anatomical dead space, i.e. areas of the exchanger that are ventilated or perfused, but not both. This anatomical dead space includes the major airways in lungs, the surface that is underlaid by non-exchanging tissues such as supportive and connective tissue, and any other structures whose thickness is so great as to preclude any significant exchange.

DIFFUSION BARRIERS

One of the most important parameters of any exchanger is the nature of the physical barrier between the opposing exchange media. For simple animals that respire through their body walls, there must be a delicate compromise between making the wall thick and strong enough for body support and making it thin enough to support sufficient gas exchange. For an animal of any size the support strength requirement alone would seem to strongly favor the development of specialized areas for gas exchange, i.e., gills or lungs. Within the gas exchange organs themselves, however, there is also a requirement for some structural support. In crab and fish gills the opposite walls of the exchanging lamellae are supported by "pillar cells," which are connective tissue containing some actinomyosin-like elements (Bettex-Galland & Hughes, 1973)(Chapter 10). Between these elements the epithelium is very thin in order to minimize the resistance to diffusion.

Within the animal kingdom there is considerable variation in the thickness and composition of the gas exchange barrier. In crabs and other arthropods the barrier consists not only of the epithelial layer but an outer chitin layer as well. The combined thickness of this barrier can be as little as 2 to 3 μm in relatively active crabs to as much as 12 to 15 μm in cold-water forms with lower metabolic activity (author, unpublished data). The homeothermic vertebrates, birds and mammals, have high gas exchange requirements, and in these groups the mean thickness of the diffusion barrier ranges from 0.4 down to 0.1 μm (Weibel, 1972). Assuming a uniform diffusion coefficient, this range of barrier thickness provides for a roughly 150-fold difference in the rate of gas transfer per unit surface area among animal groups. The actual difference is even larger, since the diffusion coefficient in chitin is about only about 10% of the values for muscle or connective tissue (Krogh, 1919a).

In lungs the surface of the exchanging epithelium may not be dry, but may be covered with a thin layer of water that contributes to the diffusion resistance. Mammalian lungs are thought to be comparatively dry, but in groups such as turtles, which have higher blood pressures in their lungs, the water layer may be a significant portion of the external diffusion barrier (Burggren, 1982a).

At the other end of the blood transport pathway, the capillary wall, interstitial fluids, and cellular material form an internal diffusion pathway for movement of oxygen from erythrocyte to mitochondria. The rates of diffusion of oxygen and carbon dioxide are not very different in tissue and water (Krogh, 1919a; Randall, 1970b), so the problem becomes simply one of minimizing the path length. Tissue diffusion gradients are discussed more fully in Chapter 4.

Chapter 4

Circulatory Systems

In the broadest sense a circulatory system is any system designed to provide convective flow of internal fluids. The purposes of circulation are several, but gas exchange is one of the most important. As size and complexity of animals increase, the circulatory system takes on an increasingly important role in transporting gases and other materials from internal tissue sites to the sites of exchange, ingestion, and excretion. The physiology of circulation is based upon basic physics of pressure and flow, but within these constraints evolution has resulted in an interesting array of circulatory systems.

PRINCIPLES OF PRESSURE AND FLOW APPLIED TO CIRCULATION

Static Pressure

In any standing column of fluid (i.e., a liquid or a gas) there is a static pressure (in dynes cm^2) equal to ρgh, or the product of density (ρ, gm ml^{-1}) times the acceleration of gravity (g, 980 cm sec^{-2}) times the height (h, in cm). A second general principle is that within the fluid all points on a horizontal plane have equal static pressure. For animals living in water these considerations are unimportant, since

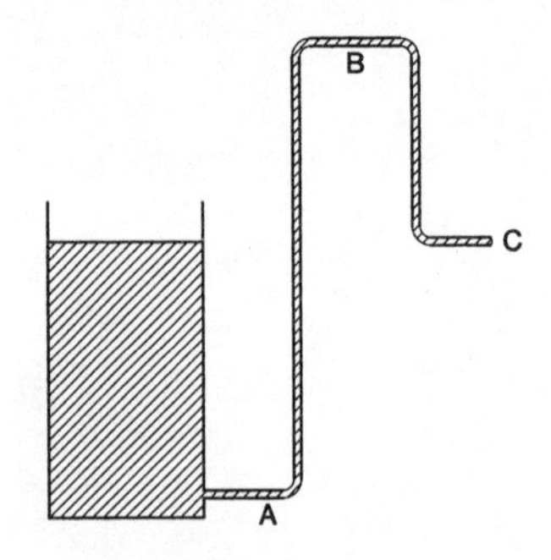

Figure 4.1: Comparison of the standing hydrostatic pressure differences and driving pressure differences that cause flow, using a U-tube as an example. There is a standing hydrostatic pressure difference between A and B; but if there is no pressure difference between the inlet and B, there will be no flow.

as one moves from top to bottom of an animal the increase in hydrostatic pressure inside and out is equal. For animals in air, however, there will be a considerable difference between static pressures at the top and bottom. For a typical human there might be a difference of 50 cm between brain and heart, and + 130 cm between feet and heart. An average pressure of 100 mm Hg at heart level, then, will produce a pressure in the brain arteries of about 50 mm Hg and in the feet arteries of 230 mm Hg. The situation is further complicated by movements of the body, which can produce large, rapid changes in pressure. Obviously, terrestrial animals must have effective means of dealing with the effects of variable pressure in the vessels.

At first glance it might appear that all the blood would flow to the head in this hypothetical human, since the pressure difference (100 – 50) would seem to favor that. When considering the flow from the heart and back again, however, it is the difference in pressure at the same level that is important, due to the siphon principle. That is, since blood is contained within a closed tube system, the static pressure differences at various heights are not the significant pressure differences (Fig. 4.1). Similarly, blood pressure measurements in air must be referred to a common level in order to determine if a driving pressure for flow exists; these measurements are conventionally referred to the level of the heart.

Dynamic Pressure (Flow and Side Versus End Pressure)

Poiseuille's equation for flow in tubes was given in Chapter 1 along with some general discussion of the relation between pressure, viscosity, tube radius, and flow. The term for pressure in this relation refers to the pressure at either end of the tube, or the gradient along the axis of the tube. Another pressure that is important, however, is the side pressure exerted against the wall of the tube. According to Bernoulli's principle, as flow increases and the kinetic energy of the fluid increases, the sum of the kinetic plus the static (pressure) energy remains the same. As the flow velocity increases, then, the side pressure exerted against the wall will decrease by a factor equal to ρv^2.

At the low velocities encountered in most poikilotherms this effect is unimportant, but in vertebrates' systems with high flow rates some peculiar things occur. Narrowing of a vessel will produce a region of higher flow and therefore lower pressure exerted against the wall. Due to the elasticity of vessel walls (see below), the vessel may constrict further, causing a further increase in velocity and reduction of pressure. At some point this cycle results in a reduction of flow, an increase in upstream pressure, and an increase in the diameter of the constriction. An unstable oscillation is set up, causing flutter in the vessel. The same sequence of events causes a reed to vibrate in a wind instrument, or soft tubing to flutter with a high rate of water flow through it.

A final consequence of the kinetic pressure effect is that pressures measured with a T-cannula in a vessel will be lower than those measured in line with the flow by a factor related to the flow velocity.

Turbulence and Noise

The differences between laminar and turbulent flow were discussed in Chapter 1 but are also relevant to the discussion of circulation. In the invertebrates and most of the poikilothermic vertebrates, flow rates in the circulation are such that flow is probably always laminar. In more metabolically active animals such as birds and mammals, however, Reynolds numbers in the heart and major vessels exceed the range for laminar flow, and turbulence results. One physiological consequence of turbulence is a sharp increase in the resistance to flow and hence in the work required for a given flow. A second consequence of some clinical usefulness in man is that turbulence induces vibrations that are transmitted through tissue as noise. The so-called Korotkow sounds serve as a diagnostic tool in the measurement of blood pressure, and the sounds of heart valves provide characteristic clues in the diagnosis of various heart diseases. Turbulence is especially likely to develop in the region of irregularities in the flow channels, e.g., in normal or diseased valves, vessel constrictions, or major arterial branches (Burton, 1965).

Viscosity

In Newtonian fluids the usual relations between viscosity, pressure, and flow apply, and there is a regular relationship between the flow and tube radius proportional to the fourth power of the radius. Early studies of the rates of flow of vertebrate blood through tubes of varying radii, however, produced anomalous results. Specifically, at tube radii approaching the diameter of the small vessels, the calculated viscosity of blood decreases in an anomalous manner, called the "Fahraeus–Lindqvist effect" (Snyder, 1973). This effect is known to result from the suspended erythrocytes, partly because they make a discontinuous flow profile across the vessel diameter (the "sigma effect"), and partly because they tend to concentrate more in the center of the vessel. The tendency of the erythrocytes to stream in the middle of the vessel means that a region nearer the wall is relatively depleted of cells and that small side-vessels may receive a flow of blood with much lower hematocrit than the average. The effects of this "plas-

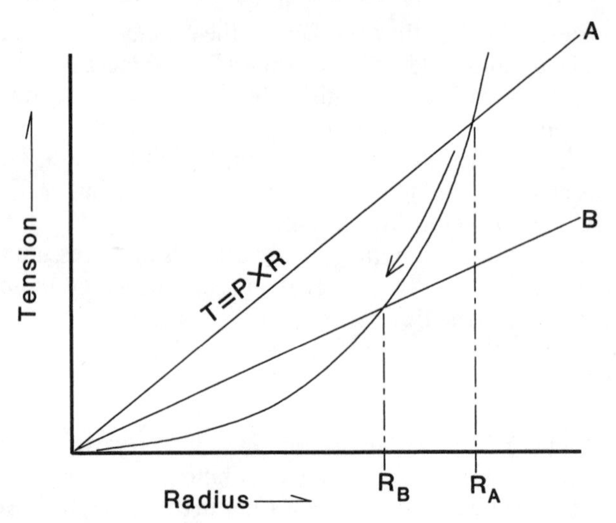

Figure 4.2: Tension-radius diagram for a blood vessel. Stretching of the vessel wall under normal pressure causes the radius to increase until a balance is reached at line A. If the pressure decreases, the radius will decrease to R_B, at which point a new tension-radius steady state is reached along line B. (From Burton, 1951, by permission, ©Amer. Physiol. Soc.)

ma skimming" are well known in capillary beds and may also be important in the function of other organs such as the kidney. The physical cause of plasma skimming is rotation of the cells as they move, resulting in forces that push them away from the walls and toward the center.

Properties of Blood Vessels

The nature of the materials making up walls of blood vessels impart important properties of elasticity to the vessel, and these properties in turn affect the relation between pressure and flow. For a perfectly elastic cylinder the increase in cross sectional area is directly proportional to the change in pressure:

$$k = \Delta P/\Delta a \qquad \text{(Eq. 4.1)}$$

where a is the area and k is the *bulk modulus*, or elasticity constant. The vessel walls of vertebrates are actually made up of mixtures of collagen and elastin; apparently as the elastin approaches its elastic limits the stiffer collagen fibers begin to stretch, resulting in nonlinear pressure/area curves.

At increasing strain in the vessel wall, the wall tension, or force in a circumferential direction, increases according to Laplace's law:

$$P = T/R \qquad \text{(Eq. 4.2)}$$

where T is the total tension and R is the radius. By combining these two relations we can arrive at a relation between pressure, radius and tension (Fig. 4.2). The nature of the relation dictates a stable equilibrium for any particular vessel under the influence of elastic tension alone. For example, the line A intersects a tension-length diagram for the vessel at a radius R_1. If pressure (and thus tension) is reduced in the vessel to a value represented along line B, the radius will decrease to R_2, balancing the opposing forces of pressure and tension within the vessel.

HEARTS

Invertebrate Hearts

The heart as a principal pumping force for circulation is hardly developed in the lower invertebrate phyla. In nemertines, for example, there are only two principal longitudinal blood vessels; in those species with hemoglobin, the blood can be observed flowing sometimes in one direction and sometimes in the other (Hyman, 1951). In the echinoderms the true vascular system is very small with no functional heart, the main functions of circulation having been assumed by the "water vascular system" (Hyman, 1955). In some annelids the first closed vascular systems appear along with a well-developed system of blood vessels and contractile pumps. There are usually one or more thickened, muscular regions of the principal dorsal vessel that show contractile activity and function as hearts,

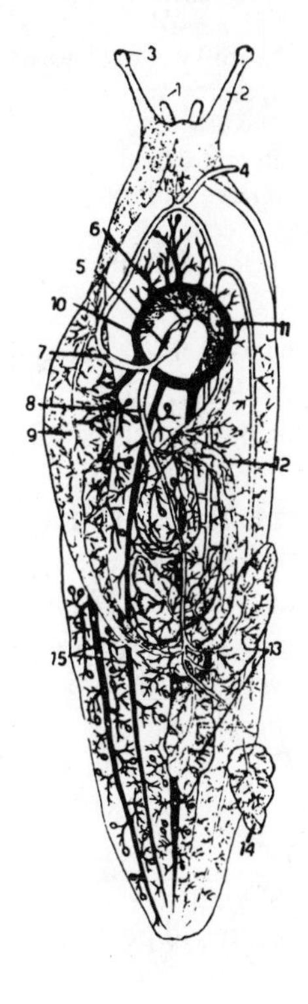

Figure 4.3: Circulatory system of a pulmonate snail. 1, anterior tentacle; 2, posterior tentacle; 3, eye; 4, genital artery; 5, ventricle; 6, atrium; 7, anterior aorta; 8, posterior aorta; 9, stomach; 10, venous circle; 11, pneumostome; 12, intestine; 13, midgut gland; 14, ovotestis; 15, venous channels. (From Hyman, 1967, by permission.)

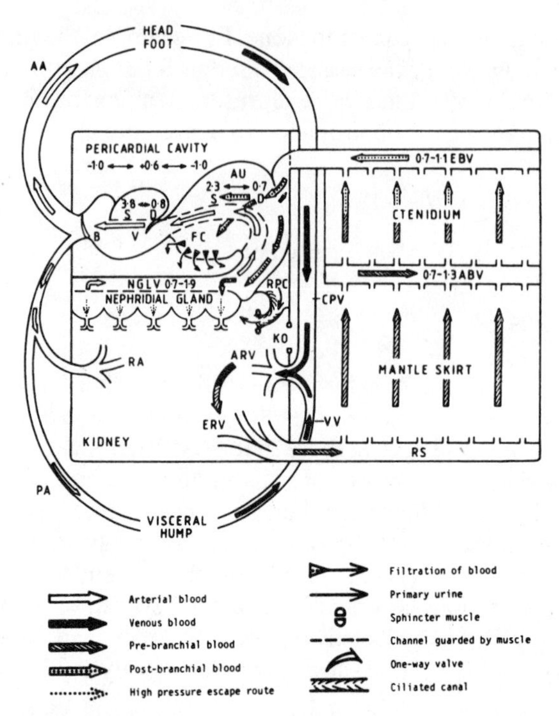

Figure 4.4: Circulatory system of the periwinkle, *Littorina littorea*, a gastropod mollusc. The fluid pressures are expressed in cm H_2O. AA, anterior aorta; ABV, afferent branchial vein; AU, atrium; ARV, afferent renal vein; B, bulbus aortae; CPV, cephalopedal vein; D, diastole; EBV, efferent ctenidial vein; ERV, efferent renal vein; FC, filtration chamber; KO, kidney opening; NGLV, nephridial gland vein; PA, posterior aorta; RA, renal artery; RPC, renopericardial canal; RS, rectal sinus; S, systole, V, ventricle; VV, visceral vein. (From Andrews & Taylor, 1988, by permission, ©Springer-Verlag.)

as well as paired lateral hearts in many segments. Blood pressures as high as 25 cm H_2O can be developed in some giant earthworms, and the blood flow velocity in the vessels is fairly high (Johansen & Martin, 1965). Little physiological information is available for most annelid circulatory systems, however, due to the small size and delicate nature of the animals.

The molluscs are the first invertebrate group as a whole in which the circulatory system is well developed. The pulmonate snails have an elaborate circulatory system consisting of a heart, an arterial vessel system, a venous vessel system, and various accessory sinuses (Fig. 4.3)(Hyman, 1967). Examination of the circulatory system of a gastropod mollusc reveals some interesting features (Fig. 4.4). The circulation consists of a single circuit, with a single atrium and ventricle, but the output of the heart is divided into parallel paths that serve various regions of the body. Return flow passes through a renal portal system and finally traverses the gills before returning to the heart. The heart is enclosed in a semirigid pericardium, so the sequential contraction of the atrium and ventricle

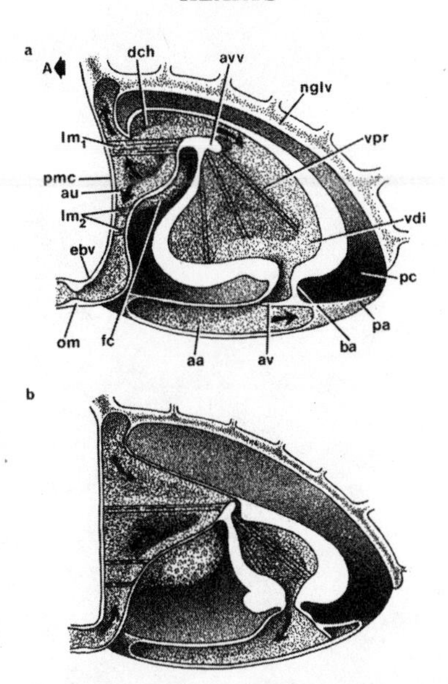

Figure 4.5: The heart of *Littorina* in sagittal section drawn from the left side. Cut surfaces are white. Arrows indicate direction of blood flow. Only a few trabecular muscles are shown. (a) Auricular systole. (b) Ventricular systole. A, anterior; aa, anterior aorta; au, atrium; av, aortic valve; avv, auriculo-ventricular valve; ba, bulbus aortae; dch, dorsal channel of atrium; ebv, efferent ctenidial vein; fc, filtration chamber; lm1, longitudinal muscles forming floor of dorsal channel; lm2, longitudinal muscles controlling reflux to efferent ctenidial vein; nglv, nephridial gland vein; om, occlusor muscle; pa, posterior aorta; pc, pericardial cavity; pmc, posterior wall of mantle cavity; vdi, distal chamber of ventricle; vpr, proximal chamber of ventricle. (From Andrews & Taylor, 1988, by permission, ©Springer-Verlag.)

provides a filling boost by suction, augmenting the very low venous return pressure (Fig. 4.5)(Andrews & Taylor, 1988). The heart develops as a thickened contractile region of the aorta, has nervous synchronization, and is equipped with valves, qualifying it as a functionally complete heart (Jones, 1983). Although the heart and vessels of the molluscs lack a true endothelial lining (Nold, 1924, in Hyman, 1967), they are lined with a layer of connective tissue and surrounded by first a layer of circular muscles, a layer of longitudinal muscles, and finally an outer layer of "chondroid" epithelial cells. The heart has the same structure, differing only in shape and relative thickness of the various layers.

An unusual degree of circulatory system development is evident in the cephalopod molluscs: the squid, octopi, and nautili. Although the heart has the same anatomical structure as that of the other molluscs, it is generally larger and the muscle layers thicker. Its pumping action, however, is supplemented not only by a general contractility of the arteries but also by contractile "branchial hearts," thickened contractile regions of the veins that provide a boost of venous return flow to the gill vasculature (Fig. 4.6)(Wells, 1983). The muscle structure of the walls of the systemic ventricle is similar to that of arteries, but the branchial

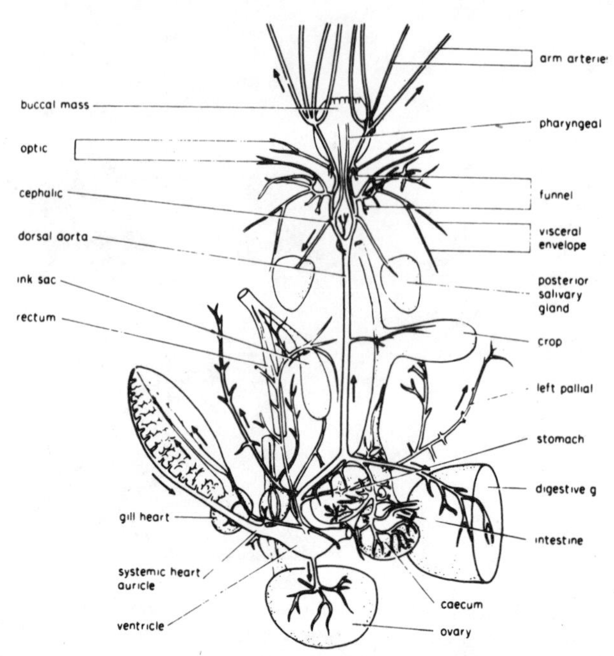

Figure 4.6: Vascular system of the octopus *Eledone*. (From Wells, 1983, by permission, ©Academic Press.)

(venous) hearts have a spongier structure with some muscle fibers oriented across the lumen.

Yet another kind of structure is evident in the hearts of crustaceans, as illustrated in Fig. 4.7. The heart in these animals is a roughly box-shaped organ, often suspended in a semirigid pericardium by a series of suspensory ligaments. The heart wall consists of an inner epithelial lining, muscle layers of varying orientation, and an outer lining tissue layer. The heart fills by several pairs of ostia, valve slits in the heart walls. In order to fill, the heart relaxes, the ostia open, and the elastic suspensory ligaments cause it to expand and fill. The ostia then close as the heart contracts, expelling the blood into the arterial system, and the negative pressure produced in the rigid pericardium draws blood from the connect-

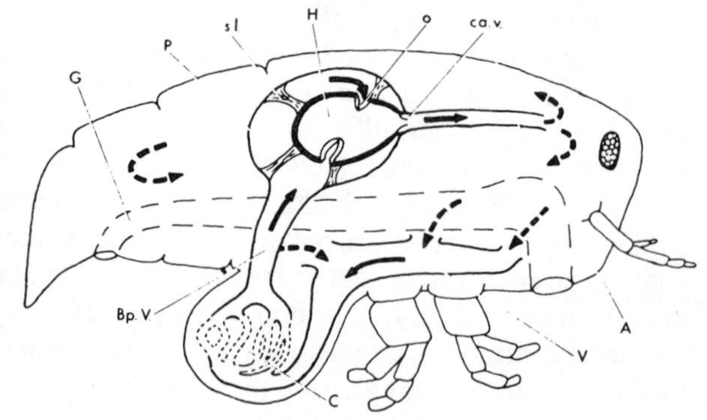

Figure 4.7: Circulatory system of a crustacean. G, gut; P, pericardium; s.L., suspensory ligament; o, ostium; ca.v., cardioarterial valve; A, anterior median aorta; V, ventral venous sinus; C, capillary bed of respiratory surface, Bp.V., branchiopericardial vessel. (From Maynard, 1960, by permission, ©Academic Press.)

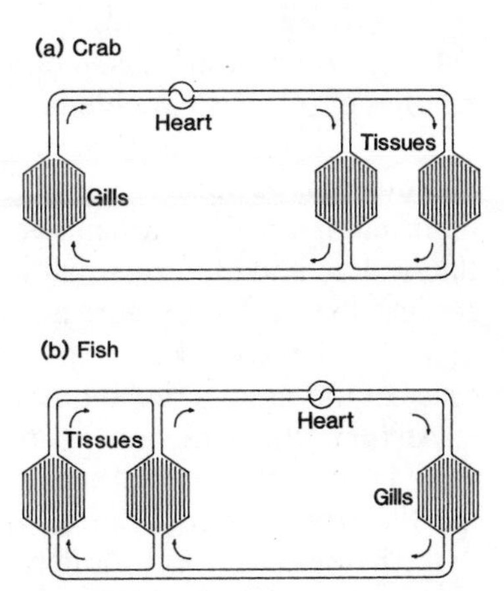

Figure 4.8: Invertebrate (crab) and fish circulatory systems, showing that both are single-circuit but that the gills are upstream in the fish and downstream in the crab.

ing venous sinus system. The degree of vascular development in crustacea is highly variable, with some animals having only a few short vessels and others an extensive vascular system.

In all of these invertebrate hearts and vessel systems there is an absence of smooth muscle; rather, the muscle layers are unmodified striated muscle. Hearts can be classified on the basis of where the excitation for each beat originates. In some hearts the muscle cells themselves originate the beat impulse, usually in a specialized region of cells. These hearts are termed *myogenic*, whereas those hearts whose beat orginates in specialized neurons or ganglia are termed *neurogenic*. All vertebrate hearts are myogenic, but in the invertebrates there is no general pattern, with some taxonomic groups showing both kinds of organization (Hyman, 1951, 1955; Jones, 1983; McMahon & Wilkens, 1983; Wells, 1983)(see Chapters 8 and 9).

The Vertebrate Heart

Fish: The Undivided Heart

The basic plan of the circulation in fish is quite similar to both the arthropods and molluscs. The heart provides the primary propulsive force for blood flow and the path for blood consists of a single circuit. One important difference is that the blood flows first through the gas exchange organs, the gills, and thence through major vessels to the systemic circulation, finally returning to the heart. In the invertebrate groups the order is reversed, with blood flowing first from the heart to the systemic circulation, and finally through gills just prior to returning to the heart (Fig. 4.8). Another important difference is that the entire circulation is contained within closed vessels lined with endothelium, hence the distinction between "open" and "closed" circulatory systems. The heart itself consists of four principal structures, the first three of which are the sinus venosus, the atrium, and the ventricle. The fourth structure consists of either a bulbus arteriosus in teleost fish or a conus arteriosus in elasmobranchs (Fig. 4.9), although a few primitive families of teleosts have elements of both a conus and a bulbus (Satchell, 1971).

Sinus Venosus. The great veins, particularly the ductus Cuvieri from each side and the hepatic portal veins in the midline, converge at the rear of the pericardium, emptying into the sinus venosus through valves. The sinus venosus itself is thin-walled and of fairly small volume, but it does possess a contractile layer of cardiac muscle. Its structure and location are similar in elasmobranchs, although in these fish the weak pulse pressure is superimposed on a negative baseline because of the rigid pericardium.

Atrium. When the sinus venosus contracts, blood flows through the sino-atrial valve into the thin-walled atrium, which lies mainly above the ventricle. In many fish the atrium extends dorsally and anteriorly to the ventricle and lies adjacent to the bulbus arteriosus. A muscular sphincter assists the atrioventricular valves, providing unidirectional flow for the contractile sequence. Two radiating layers of cardiac muscle invest the wall of the atrium, pulling its roof downward during contraction. The principal function of the atrium is to fill the ventricle during diastole, so the pressure generated in this chamber is also rather small, on the order of a few millimeters of mercury.

Ventricle. In both teleosts and elasmobranchs the ventricle is thick-walled and highly muscular, with a range of shapes including tubular, conical, and pyramidal. Because of the superior position of the atrium, the incurrent and excurrent openings are sometimes rather close together, but both are protected by competent valves. The muscular structure of the fish heart varies considerably among species: In less active forms nearly all of the muscle is arranged radially to form a trabecular mass with only a small open lumen at the center (Cameron, 1975a; Santer & Greer-Walker, 1980; Tota, 1983). The outer capsule consists entirely of connective tissue, and the coronary artery is missing altogether. In active fish, such as tuna, there may be a considerable outer layer of muscle called the corti-

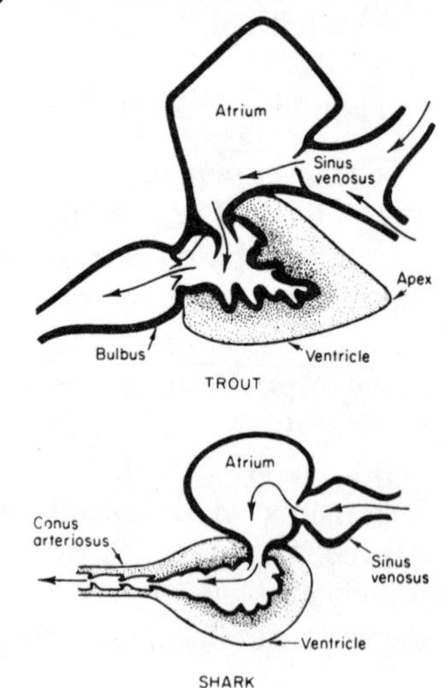

Figure 4.9: Chambers of the fish heart: teleost (upper) and elasmobranch (lower). (From Randall, 1968, by permission, ©Amer. Soc. of Zoologists.)

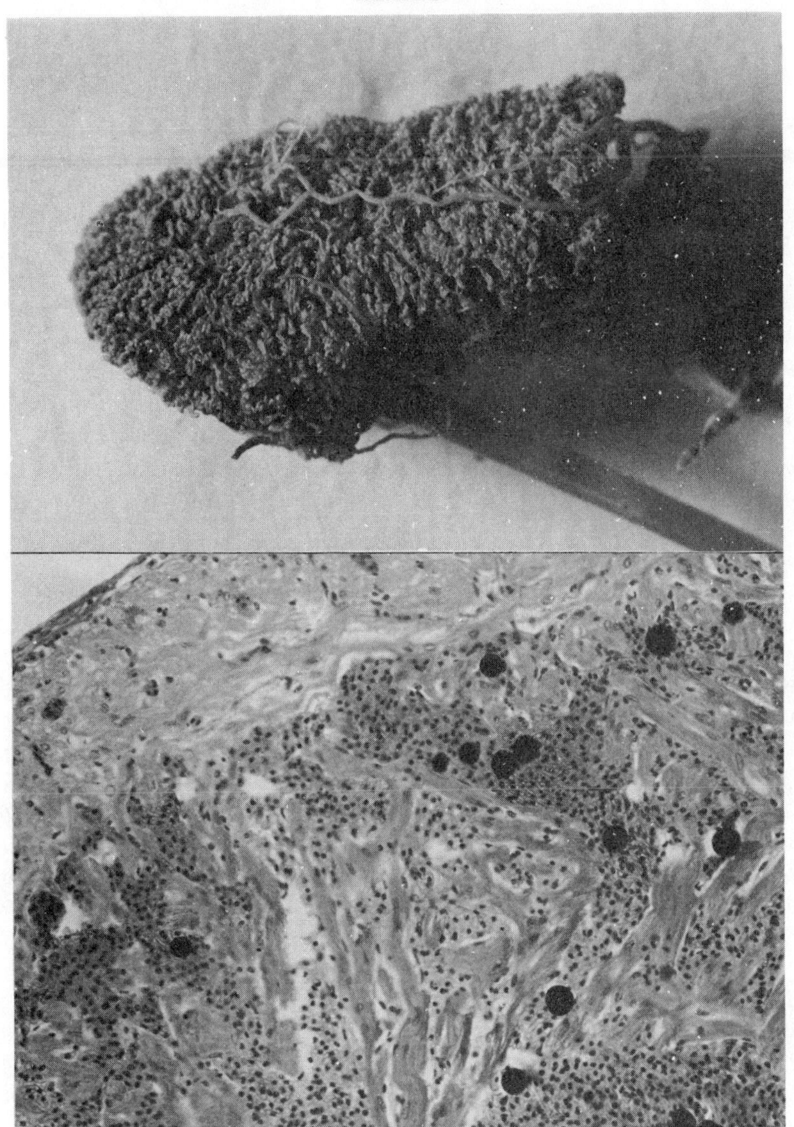

Figure 4.10: Plastic cast of the vascular spaces of the teleost heart (upper) and a section through the outer portion of the ventricle (lower), showing that most of the heart is spongy, trabecular tissue receiving its blood supply from the lumen, i.e., venous blood. The large dark spheres in the lower photograph are microspheres injected luminally. (From Cameron, 1975a, by permission.)

cal layer. This muscle layer is oriented circumferentially, does not connect with the inner trabecular layer, and receives its blood supply from a well-developed coronary arterial system. Other fish show intermediate structure, with about 25% of the muscle mass cortical in fish such as trout, the balance belonging to the inner spongy layer (Fig. 4.10)(Cameron, 1975a).

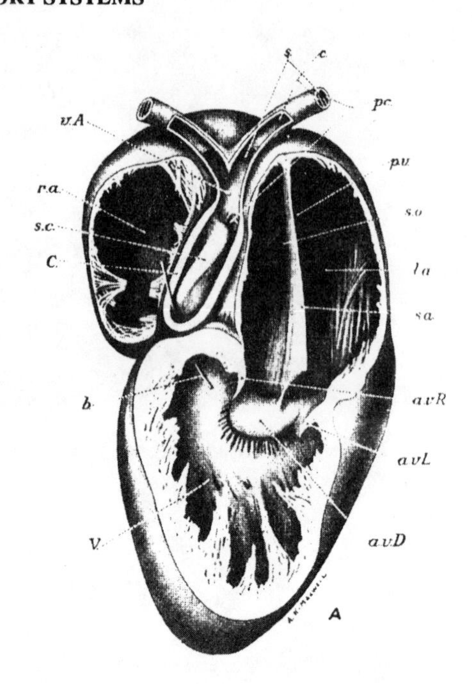

Figure 4.11: Amphibian heart (frog). s., systemic cavity; c., carotid; p.v., opening of pulmonary vein; s.o., opening from sinus venosus; l.a., left atrium; s.a., atrial septum; a.v.R., right atrioventricular valve; a.v.L., left atrio-ventricular valve; a.v.D., dorsal atrio-ventricular valve; V, ventricle; b., bristle passed from ventricle into conus; C., conus; s.c., septum of conus; r.a., right atrium; v.A., ventral aorta. (From Goodrich, 1930, ©Dover Press.).

Conus Arteriosus and Bulbus Arteriosus. The elasmobranch conus arteriosus appears to have two functions corresponding to its two main tissue types: The considerable thickness of elastic tissue in the walls provides a pressure smoothing ("Windkessel") effect, and the cardiac muscle present provides additional propulsive power. Variable numbers of valves are present in different species, no doubt aiding in both these functions. Teleosts, on the other hand, have only a thickened elastic structure called the bulbus arteriosus. It functions solely as a Windkessel, providing pressure and flow smoothing before blood enters the gills.

Amphibians: The Partially Divided Heart

Evolutionary developments in the terrestrial vertebrates are to a certain extent foreshadowed by the development of the heart in the primitive lungfish, the Dipnoi. Here the return vessel from the air-breathing organ enters the sinus venosus separately, with its own valve, and remains unmixed in much of the atrium due to an intraauricular septum arising from the atrial wall. The ventricle is also partially divided by a plug-like intraventricular septum, although this particular structure seems unique to the lungfish, and is quite different from the septa that arise in other groups. The conus arteriosus is highly modified in lungfish, having a spiral twist and a complex arrangement of valves only superficially similar to those found in amphibians. (The term *bulbus cordis* is sometimes applied to this structure.) The sum effect of these various modifications is at least the beginnings of separation of the pulmonary and systemic circuits. Since the partitioning in the heart is incomplete, there is obviously some crossover of

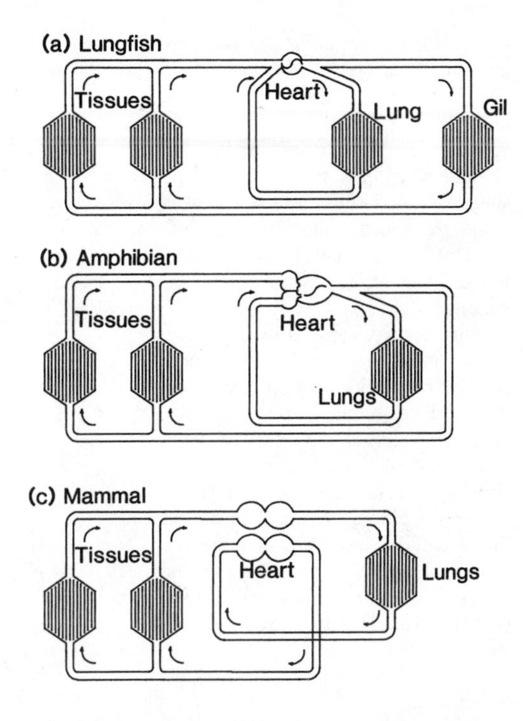

Figure 4.12: Circulation in a lungfish, an amphibian, and a bird or mammal. See text for discussion.

oxygenated blood coming from the air-breathing organ back into the pulmonary circuit (left-right shunt) and some admixture of deoxygenated blood returning from the systemic circuit back into that circuit (right-left shunt). Whether there is, in fact, any functional separation is still a matter of much debate (Johansen, 1972; Ishimatsu & Itazawa, 1983), even though structural studies suggested it long ago (Goodrich, 1930).

The structure of the amphibian heart may be seen as a progression of the trends from teleost to dipnoan, i.e., more complete separation of the venous return circuits (pulmonary and systemic), septation of the atrium into a complete right and left, partial division of the ventricle, and more complete division of the bulbus cordis into the various systemic and pulmonary trunks (Fig. 4.11). There remains the problem of incomplete separation of oxygenated and deoxygenated streams of blood, and there is again a persistent controversy regarding the effectiveness of that separation (cf. Johansen, 1963, and Boutilier et al. 1986). Functional separation has been proposed based upon both timing of flow from the various returns vis-à-vis the ventricular contraction and valve actions and upon mechanical flow channeling due to the shapes and orientations of ridges and other structures within the heart.

From the standpoint of ventricular blood supply and energetics, the amphibia appear to represent a step backward. That is, all of the ventricular muscle is of the spongy trabecular type, and coronary arteries appear to be entirely lacking in modern groups (Grant & Regnier, 1926). The implication is that ventricular

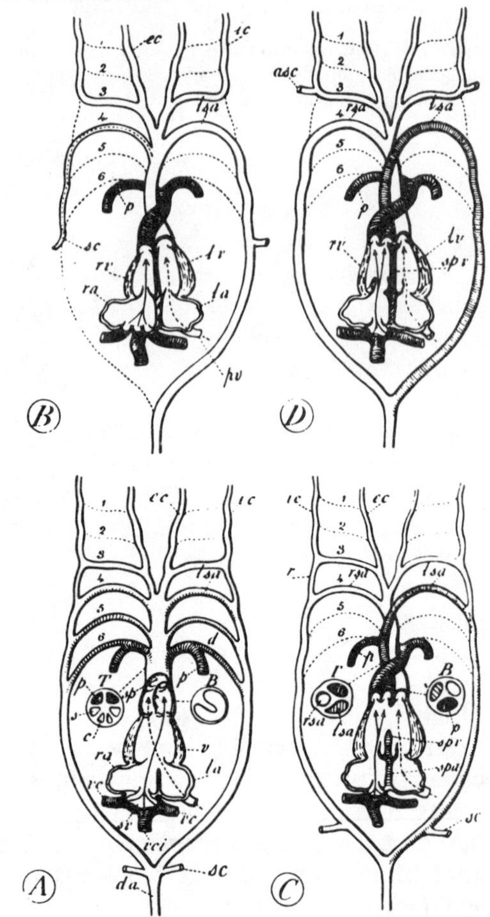

Figure 4.13: Heart and aortic arches in an amphibian (A), a mammal (B), a reptile (chelonian) (C), and a crocodile (D). The hearts are shown in ventral view, untwisted so that the chambers lie in a single plane, with sinus venosus behind and ventricle in front. *asc*, anterior subclavian; *d*, ductus Boalli; *ec*, external carotid; *ic*, internal carotid; *la*, left atrium; *lsa*, left systemic arch; *lv*, left ventricle; *p*, pulmonary artery; *r*, portion of lateral aorta remaining open; *ra*, right atrium; *spa*, interauricular septum; *spv*, interventricular septum; *sv*, sinus venosus; *v*, ventricle; *vc*, vena cava superior; *vci*, vena cava inferior. Arrows from the sinus venosus indicate the main stream of venous blood; arrows with dotted line from the left atria indicate the stream of arterial blood. 1-6, original series of six embryonic aortic arches. (From Goodrich, 1930, ©Dover Press.)

power output must be somewhat limited by the entirely venous blood supply to the working heart muscle.

Reptiles and Birds: Complete Division of the Heart

Examination of a series consisting of a chelonian reptile, a crocodilian, and a bird or mammal reveals a continuation of the evolutionary trends seen in the lungfish and amphibians (Figs. 4.12, 4.13). That is, there is increasing separation of both the venous return circuits and the arterial supply circuits, accompanied by increasing septation and separate development of right and left atria and right and left ventricles. In the turtles, lizards, and snakes the ventricular septation is incomplete (Johansen, 1959; Burggren & Johansen, 1982), but in the adult crocodilians the interventricular communication closes completely by a developmental sequence identical to that found in the birds. The sinus venosus, continuing a trend begun in the amphibians, has almost completely disappeared by being incorporated into the right auricular wall, and the bulbus cordis is divided right back to separately valved openings on the respective ventricles. The heart has reached complete development as two side-by-side pumps, a right atrium-right ventricle system receiving supply from the systemic circuit and pumping de-oxygenated blood to the lungs, and a left atrium-left ventricle system receiving supply from the pulmonary circuit and pumping oxygenated blood to the systemic circuit (Fig. 4.12). The persistence of a left aorta in the reptiles, however, still provides a connection between deoxygenated blood from the right atrium and the systemic circulation.

The extant crocodilians are actually somewhat anomalous. Crocodiles have retained both the right and left systemic aortae and have an opening, the foramen Panizzae, that joins the two just distal of the ventricular valves (Fig. 4.13D). Although on an anatomical basis it appears that this arrangement would lead to deoxygenated blood entering the left aorta, and a considerable right-left shunt, physiological studies indicate that both aortae are filled primarily from the left ventricle and that shunting is usually small (Grigg & Johansen, 1987). The output of the right ventricle is directed mostly to the pulmonary artery.

In these groups the coronary arteries reappear to supply the increasingly cortical cardiac muscle. In fish, however, the coronary artery arises as a single retrograde branch of the first gill arch artery. Now that gill arches have been lost there appears to have been a heartward movement of the origins of the coronary artery; a right and a left coronary artery arise from near the base of the aorta, and both spread across the various structures of the heart, branching in a somewhat variable pattern (Grant & Regnier, 1926; Berne & Rubio, 1979). In birds (and also mammals) very little of the cardiac muscle relies on the (venous) luminal blood, the arterial supply evidently becoming much more important for the increased power output of the heart in these groups. Even temporary blockage of the coronary supply has drastic effects, emphasizing the differences in metabolic organization of the hearts of birds and mammals compared with those of fishes and frogs.

The Mammalian Heart

Little remains to be said about the structure of the mammalian heart, except to point out that certain structural and functional similarities to the birds (Fig. 4.13B) may obscure differences in development. Although both groups have complete separation into left and right ventricles, the septation arises differently, arguing for a common ancestor at the amniote level, rather than a direct evolutionary lineage (Goodrich, 1930). During the embryonic development of the human heart it is possible to identify stages similar in turn to an annelid, a fish, an amphibian, and (superficially) a reptile. That is, the heart progresses from a simple contractile tube (annelid-like), to a slightly twisted, two-chambered heart (fish-like), gradually acquiring septation and separation of arterial and venous connections (amphibian-like), and finally reaching its complete dual-pump configuration (van Mierop, 1979). Naturally, the human heart is of particular interest; structural and functional aspects of the human heart have been reviewed elsewhere (Berne, 1979)(see Chapter 14).

VASCULAR SYSTEMS

Invertebrate or "Open" Systems

The principal difference between open and closed circulatory systems is whether there are capillaries interposed between arteries and veins. The crustacean circulatory system certainly appears to exemplify what is thought of as a "typical" open circulatory system. The blood volume is high, around 30% of body weight (Prosser, 1973), the pressures in the system are usually low, at least compared with vertebrates, and circulation times tend to be fairly long (McMahon & Wilkens, 1983). We would like to have a complete picture of what the relative volumes are in various parts of the circulation, the pressures and flow velocities at all points, and consequently the distribution of flow resistance in different parts of the circulatory system. Unfortunately such information is sketchy, mostly due to technical difficulties of working in small animals whose blood clotting cannot be inhibited. During systole the peak ventricular pressure ranges from 10 to 30 mm Hg in most crustaceans (McMahon & Wilkens, 1983) but can be as high as 50 mm Hg in some terrestrial crabs (Cameron & Mecklenburg, 1973). There are only one or two published measurements of arterial pressure in crustaceans, and they appear to be similar to peak systolic pressure. After traversing the tissue beds, blood pressure falls to about 40% of the systolic pressure and, after passing through the gills, reaches nearly zero. Evidently the total peripheral resistance is about equally apportioned between systemic tissues and gills. Little or no pulse pressure is evident in the postbranchial circulation, but movements of appendages can cause fairly large pressure excursions (Cameron & Mecklenburg, 1973; J. N. Cameron, unpublished data). Pericardial pressure is sometimes negative, particularly during systolic contraction of the heart. Virtually nothing

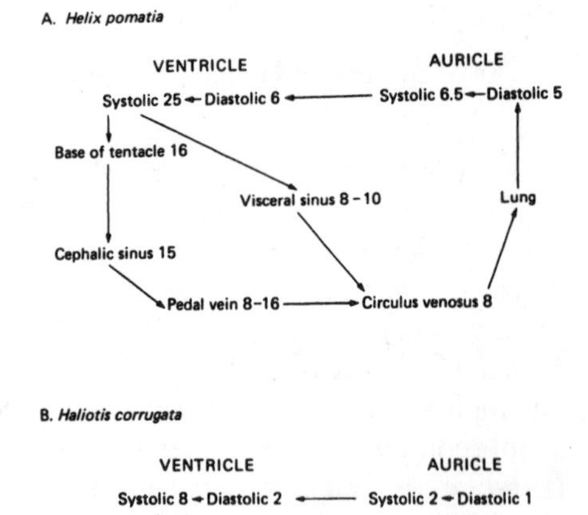

Figure 4.14: Pressure at various points in the circulation of two mollusc species. (From Jones, 1983, by permission, ©Academic Press.)

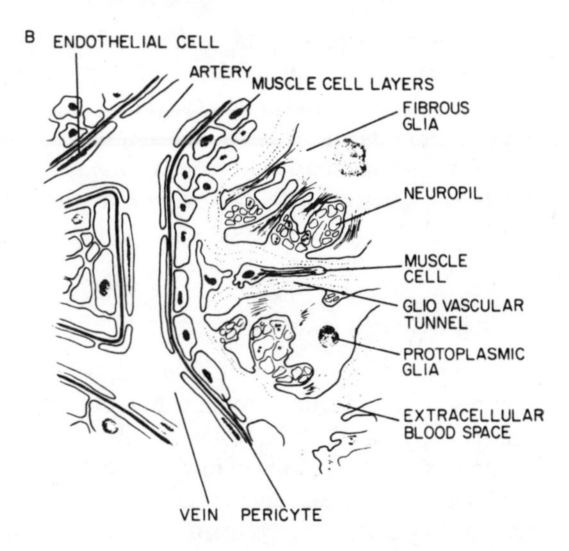

Figure 4.15: Structure of capillaries in the octopus. Note that the blood channels are lined with endothelial cells, a basement membrane, and various pericytes, forming a complete capillary. (From Wells, 1983, by permission, ©Academic Press.)

is known of blood pressure regulation; since there is no smooth muscle in vessel walls there must be a mechanism different from the vertebrate system. A few valves have been described, including apparent valve action in the gills of crabs (Taylor & Taylor, 1986).

The molluscs are the only other large group of animals with "open" circulatory systems that have been studied to any extent, but within the group there is considerable diversity in the size and functional characteristics of the circulatory systems (Jones, 1983). Leaving aside the cephalopods for the moment, blood volumes range from about 60% in bivalves to 25 or 30% in some gastropod groups. Following a general inverse relation between pressure and volume, the peak systolic pressures developed in most of the molluscs appear to be only about 4 mm Hg, although some of the pulmonate gastropods have peak systolic pressures as high as 20 mm Hg (Jones, 1983). Calculations by Bourne and Redmond (1977) indicated that peripheral resistance is apportioned about 60% to various systemic components and 40% to the gills (Fig. 4.14), similar to the figures for crustaceans. The actual resistance values they calculated, however, were quite high, suggesting that the tissue channels do not allow as unstructured a flow as might be supposed.

The distinction between open and closed systems becomes even more blurred in the cephalopods. The "capillaries" differ somewhat from those in vertebrates, but they possess endothelial lining cells (albeit in an incomplete layer), a basement membrane, and an outer layer of pericytes that have only very small gaps between (Fig. 4.15). The system is sufficiently tight to exclude hemocyanin aggregates, at least in some species (Wells, 1983). There is also a functionally and anatomically identifiable extracellular space not connected to the blood space. The systolic pressures in various species at rest range from 30 to 45 mm Hg but can increase at least 3-fold during exercise or disturbance (Wells, 1979), reaching values comparable to those in vertebrates. Venous pressures are difficult to characterize, since most of the venous system in the cephalopods is contractile and under neural control (Wells, 1983). In general, the pressures are fairly low, a few millimeters of mercury, but they can vary locally and in time. Estimates of peripheral resistance do not seem to have been made, but judging from pressures and flows they are probably comparable to those of vertebrates.

Vertebrate Systems

Single Circuit systems

Using a fish as an example (Fig. 4.8), we see that there are really two arterial systems, the first leading from the heart to the gills and the second serving to distribute oxygenated blood from the gills to the rest of the body. The first arterial system is short, consisting of only the ventral aorta and its branches, the afferent branchial arteries and filamental arteries. These vessels are thick-walled, and elastic; they carry a high pressure and flow and have little capacitance. The gills, on the other hand, are thin-walled and still carry the high pressure of the arterial system, damped somewhat by the Windkessel effect of the bulbus or conus and the ventral aorta. The compliance of the gills appears to be limited by the pillar cells (Bettex-Galland & Hughes, 1973), without which the gill lamellae would balloon out. The blood pressure drops about 30% across the gills, from peak values in the 60 to 90 mm Hg range to 40 to 60 mm Hg in the dorsal aorta.

Pressures in veins in fish are difficult to measure, so little information is available. In the few measurements that have been made, however, it is apparent that

Table 4.1: Distribution of blood volume in the circulation of a 12 kg dog.

Vessel	Diameter	Volume,ml	Percentage
Systemic arteries	>0.3 mm	109	17
	0.1 - 0.3 mm	10	2
Systemic vessels	<0.1 mm	112	18
Systemic veins	0.1 - 0.3 mm	64	10
	>0.3 mm	345	54
Systemic, total:		640	
Pulmonary arteries	>0.3 mm	40	18
	0.1 - 0.3 mm	28	13
Pulmonary vessels	<0.1 mm	11	5
Pulmonary veins	0.1 - 0.3 mm	27	12
	>0.3 mm	114	52
Pulmonary, total:		220	
Heart		140	
Total blood volume:		1000. ml	8.3% body weight

(Data from Lawson, 1962).

the pressures are always low, are little influenced by the pulse pressure, and may show large excursions due to muscular activity (Randall, 1970a; Satchell, 1971). No doubt there is a significant capacitance in the venous system of fish, but again no data are available on volume changes under different conditions.

Dual Circuit systems

Beginning with the crocodilian reptiles and including birds and mammals, there are two complete circulatory systems, in effect, pulmonary and systemic. Although both must carry exactly the same total flow, their characteristics are quite different. Of a total blood volume of 8.3% body weight, the distribution of blood within the various portions of the circulatory system of a dog is given in Table 4.1.

The pulmonary circuit is a low-pressure circulation, driven at fairly high volume by the right ventricle. Peak systolic pressure is only about 25 mm Hg and the average pressure about 13 mm Hg (Rushmer, 1965). Since we know that the finest exchange vessels in the lung possess a large total surface area, around 100 140 m^2 in man, and a correspondingly large cross-sectional area, the low volume of blood contained in them (Table 4.1) indicates that they are very short. Nearly two-thirds of the blood volume in the pulmonary circuit is contained in the veins (Lawson, 1962), and since the veins are highly compliant they exert a large capacitance. Observations on man and other mammals indicate that large changes in cardiac output and the corresponding pulmonary flow can occur with only small changes in pressure, which means that the pulmonary vessels must be quite compliant, dilating to accommodate increased flow. The pulmonary circuit on the whole exhibits very little capacity for control.

The systemic circuit, on the other hand, must accommodate widely different types of tissue. Skeletal muscle has a demand for blood flow that may vary severalfold in a very short time during and after exercise. Kidneys require a disparately high flow relative to their metabolic needs, whereas blood flow to brain varies little with exercise, thought, or other disturbance. Blood flow to the skin is extremely temperature-sensitive. The systemic circulation, then, must incorporate rapid and effective means for adjustment of pressure and flow both overall and at a local tissue level.

Blood Vessels

Although the general function of the blood vessels is to distribute blood flow to the various organs and tissues and then to return it to the heart, the various vessels function rather differently according to their place in the circulation. Blood vessels may be conveniently divided, in order from the heart, into elastic arteries, muscular arteries, arterioles, capillaries, venules, and veins (Fig. 4.16). Although the gradation from one type to another is gradual, their structure and function are sufficiently different to warrant the classification.

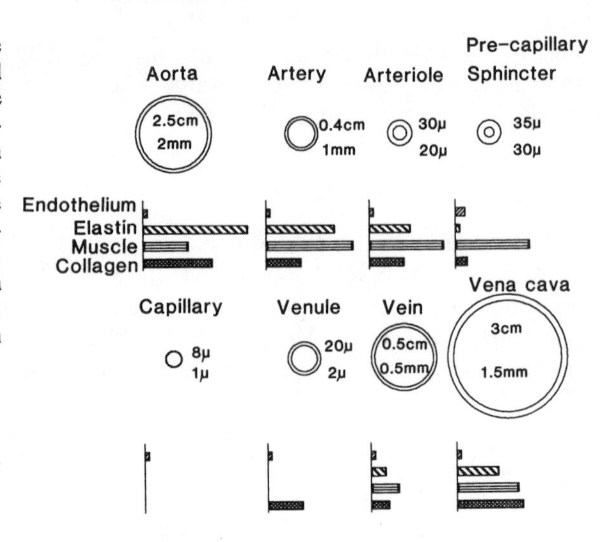

Figure 4.16: Size and structure of the various classes of blood vessels in a large mammal. The upper number indicates a typical lumen diameter, the lower a typical wall thickness. The lengths of the bars indicate the relative thickness of the endothelium, elastin layers, smooth muscle, and collagen bearing connective tissue. (Modified and redrawn from Burton, 1954.)

As blood is ejected from the heart at high pressure and velocity, the elastic arteries, principally the aorta, expand, thereby damping the pressure somewhat. The energy stored by wall distension during systole is released during diastole, resulting in no net work, but a damping or redistribution of the pressure pulse over the cardiac cycle. The walls of the elastic arteries (Fig. 4.17) consist of an endothelial layer; the *tunica intima*, or inner layer; the *tunica media*, or middle

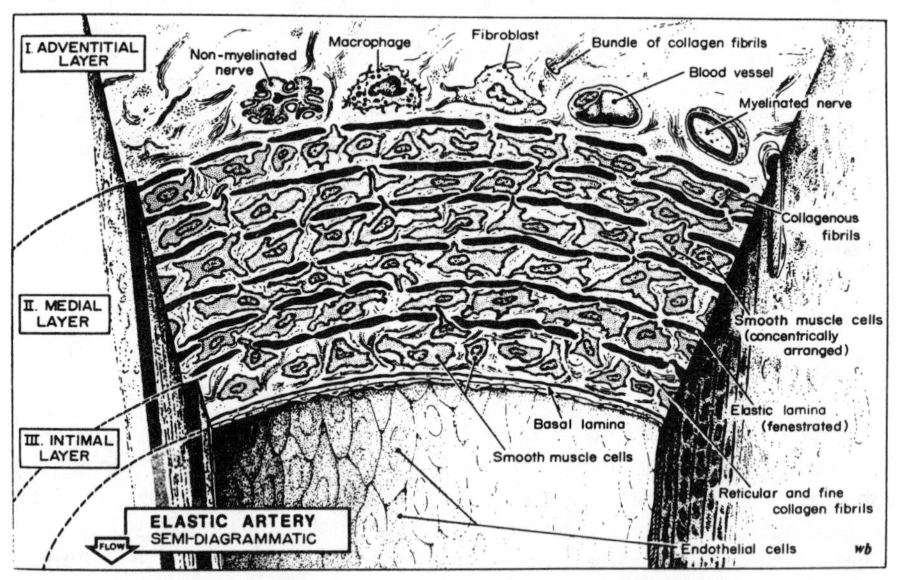

Figure 4.17: Wall structure of a major elastic artery in man. (Drawing by Wm. L. Brudon; reprinted from Rhodin, 1980, by permission, ©Amer. Physiol. Society.)

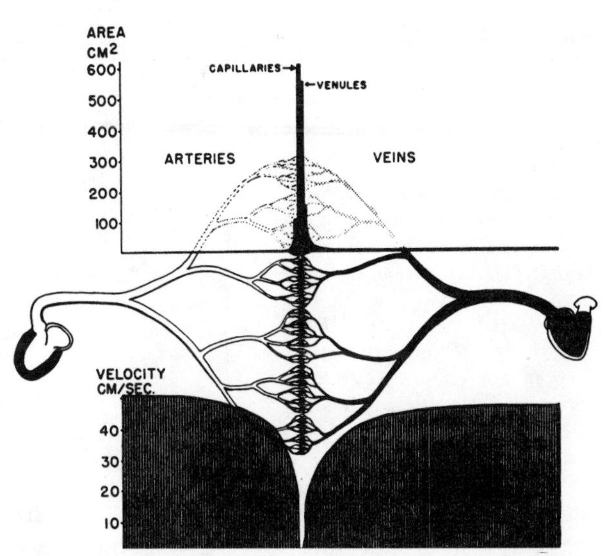

Figure 4.18: Relation between cross sectional area and the velocity of flow in the circulatory system of a 13 kg dog. Note the tremendous area in the arterioles, capillaries, and venules. The velocity of blood flow is inversely proportional to the cross-sectional area, so that blood flows through the capillaries at about 1/100 the speed with which it flows in the large arteries and veins. (From Rushmer, 1965, by permission, ©W. B. Saunders.)

layer; and the *tunica adventitia*, the adventitial layer. The endothelial layer serves as the main barrier to fluid loss but imparts little mechanical strength. The next two layers consist of a few collagen fibers but much more elastin, arranged in layers, interspersed with smooth muscle layers. Finally, the outermost layer contains large amounts of collagen. Elastin, an easily stretched protein, accounts for the bulk of arterial elasticity at pressures from zero to moderate physiological values. Collagen, a much less distensible material, begins to contribute more to the elasticity at higher physiological pressures, but the bulk of the collagen in the adventitia is probably not stretched except in pathological conditions. These properties have been well studied in vertebrate vessels, but one study has found similar mechanical properties in arteries from an octopus (Shadwick & Gosline, 1981).

The elastic vessels are large enough to offer little resistance to flow, so the mean arterial pressure changes little from one end of the aorta to the other. The systolic pressure wave is transmitted rapidly down the length of the aorta, but with one peculiar feature: The peak pulse pressure in many vertebrates (including man) *increases* with distance from the heart. A number of explanations have been offered for this phenomenon, including frequency-dependent wave propagation, pressure wave reflection, and varying vessel wall stiffness.

The muscular arteries differ from the elastic arteries mainly in having less collagen, more smooth muscle (Fig. 4.16), and a different helical winding pattern of the smooth muscle layers (Rhodin, 1980). Most of the pressure drop occurs in the small arteries and the arterioles, and it is at this level that the primary control of blood pressure and flow is achieved through regulation of smooth wall muscle and sphincter tone. Through repeated branching, the total cross-sectional area of the circulation rises even as the mean vessel diameter falls. The distribution of cross-sectional area and flow velocity for the systemic circulation is

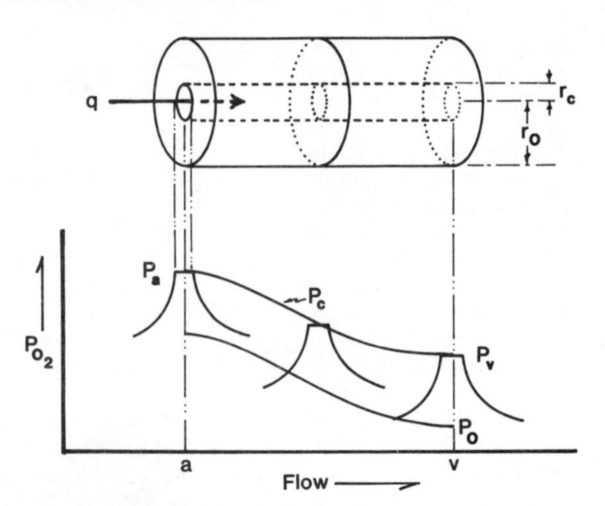

Figure 4.19: The Krogh cylinder model for tissue Po_2, modified from Piiper (1982) to show the nonlinear profile along the capillary length. The oxygen profile outward from the axis of the capillary is shown at the arterial end (a), venous end (v) and along the capillary transit (c). The profile is flat across the radius of the capillary (r_c), and falls off toward the outside (r_o). Blood flow (q), oxygen content and dissociation characteristics, cylinder radius, and diffusion coefficient influence the profiles.

shown in Fig. 4.18, with pressure and flow velocity at various points in the circulation inversely proportional to the cross-sectional area.

Progressing back the venous side of the circulation, properties of the various veins are similar to the arteries, except that the proportion of smooth muscle is lower in even the largest veins (Fig. 4.16), and the relatively thinner walls lend greater distensibility to the venous circulation. Only limited capacitance exists in the arterial system to accommodate rapid vasodilatation in any particular tissue, but the venous reservoirs constitute the major storage (capacitance) site, as evidenced by their high blood volume (Table 4.1). As simple a maneuver as sitting down will affect the pressure and volume distribution of blood in the major veins.

GAS EXCHANGE IN SMALL VESSELS

The thickness of the major arteries and arterioles, their small surface-to-volume ratio, and the large diffusion distances combine to make gas exchange in them insignificant. It is only in the true capillaries that the proper conditions exist for significant diffusive gas exchange. In 1919 Krogh published a simple model for predicting the oxygen gradients along the length of a capillary and laterally from the capillary wall to increasing distances in a tissue cylinder served by that capillary (Krogh, 1919b). The model is a combination of approaches explored in earlier chapters, and yields results similar to those shown in Fig. 4.19. Like other gas exchange models (see Chapter 7), this one is sensitive to a number of variables including the total flow rate, cylinder radius, oxygen content of blood entering the capillary, capillary length, etc. Although capillary geometry in vivo is seldom so neat and convenient (Weibel, 1984), average tissue values for oxygen tension under normal circumstances come out very close to the venous (end) capillary value.

Chapter 5

Transport of Gases by Blood

When animals become too large to depend upon diffusion for their gas exchange needs and a circulatory system develops (see Chapters 3 and 7), the simplest and perhaps the most primitive way to transport gases in the blood is by dissolution alone. A diffusion gradient in the gas exchange organ will cause oxygen, for example, to move from the external environment to the blood until the partial pressures are equal. When this blood is brought by the circulation into contact with actively respiring tissues, oxygen will move down a diffusion gradient from the blood into the tissues. The movements of carbon dioxide are similar but in the opposite direction for normal aerobic respiration.

The quantities of gases that can be transported in this manner are limited by the solubility coefficients of the gases (see TEMPTABL program; Chapter 1). Thus at 20°C and at the ionic strength of seawater, 1 L of blood equilibrated with air can hold 0.25 mM (5.5 ml) of oxygen in solution. At the same partial pressure (159 torr) 1 L of seawater can hold 7.1 mM (159 ml) of carbon dioxide, about 29 times more than the oxygen capacity. As animals become larger and their metabolic rates increase, the limitations of gas transport by solution become serious, as the following example shows.

If we assume that a 1 kg animal has a metabolic rate of 2.2 mM (~50 ml) O_2 per hour, that the blood completely equilibrates with air-saturated water in the gas exchanger organ, and that the blood can give up 80% of its oxygen in tissues, we can calculate the circulatory rate (cardiac output, $\dot{Q}$) needed as:

$$\dot{Q} = \dot{M}O_2/(80\%C_{O_2}) = 2.2/(0.8 \times 0.25) = 11 \text{ L hr}^{-1} \qquad \text{(Eq. 5.1)}$$

or 189 ml min^{-1}, where C_{O_2} is the oxygen content of saturated blood. This cardiac output is a very high, and when we consider that metabolic rates can be more than 100-fold higher and that higher temperatures will further reduce the oxygen capacity of the blood solution, it is easy to see that some additional oxygen capacity is highly desirable. The elimination of CO_2 is a lesser problem due to its higher solubility.

BLOOD PIGMENTS

In nearly all of the more complex animals compounds are present in blood that bind oxygen reversibly, increasing the effective oxygen-carrying capacity. All of these compounds (except deoxygenated hemocyanin) are highly colored, leading to the term *blood pigments* or *respiratory pigments*. The presence of blood pigments has been known since antiquity, but their function in binding gases in the blood was not demonstrated until 1837 when Magnus designed his apparatus for measurement of both oxygen and carbon dioxide (Perkins, 1964). The first oxygen equilibrium curves for hemoglobin were published by Paul Bert in 1878,

Table 5.1: The occurrence of various oxygen-carrying pigments in animals.

Pigment	Environment	Occurrence	Comments
Hemoglobin	Intracellular (RBCs)	Nemerteans Annelids Molluscs (bivalves) Phoronida Echiurida Echinoderms Hemichordates Chordates	Small; monomers to octamers
	Extracellular	Annelids Molluscs Arthropods	Highly variable
Chlorocruorin	Extracellular	Annelids	Formyl substitution on protoporphyrin
Hemerythrin	Intracellular (RBCs)	Brachipods Annelids Sipunculids Priapulids	Nonporphyrin iron
Hemocyanin	Extracellular	Molluscs Arthropods	Probably separate origin in these two

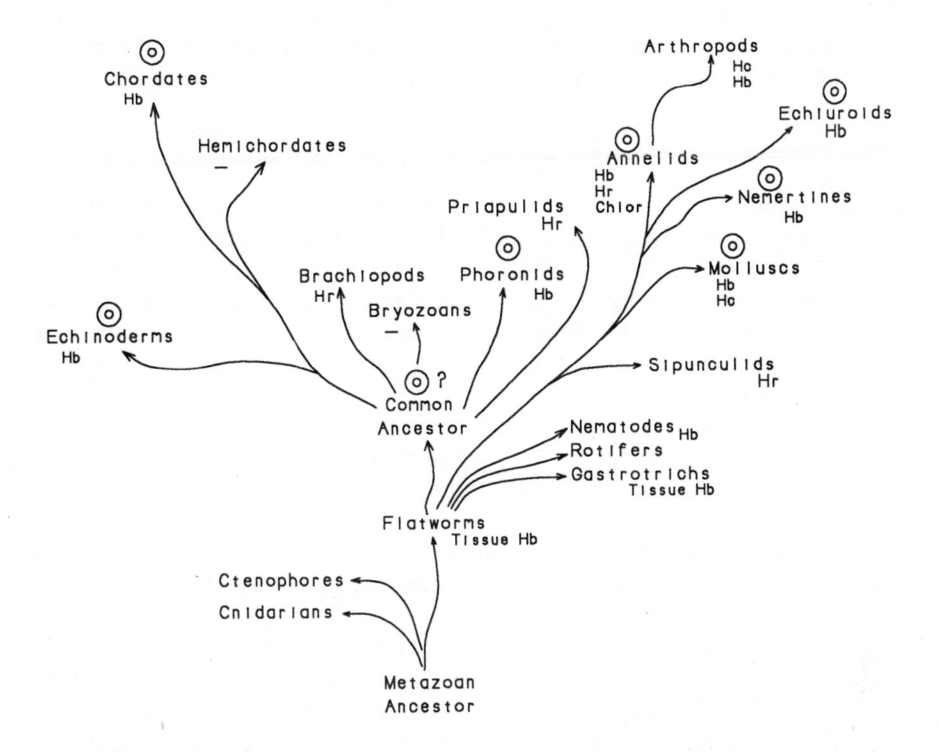

Figure 5.1: Phylogenetic distribution of blood respiratory pigments. Groups with erythrocytes are indicated by the circle logo, and pigments found in the group are listed next to the name. The phylogenetic tree is adapted from Barnes (1980), and the pigment distribution adapted from Mangum (1985) and Mangum et al. (1987).

but their sigmoid (rather than hyperbolic) shapes were not demonstrated until Bohr's studies at around the turn of the century. The first quarter of this century was a very active one for physiological chemistry, and by 1925 or '30 equilibrium curves for blood of many species under widely differing conditions had been determined (Perkins, 1964).

Although all of the known respiratory pigments are metalloproteins that bind oxygen loosely to the metal, there is no generally agreed-upon scheme for their classification. Older textbooks divided them into hemoglobin (Hb), hemerythrin (Hr), chlorocruorin and hemocyanin (Hc), but based upon the similarity between chlorocruorins and other heme proteins they are now generally lumped together. Based upon the protein structure associated with the oxygen-metal binding moiety, however, only the presently known hemerythrins constitute a coherent class (Mangum, 1985) [The coherence may be illusory; expanded knowledge of more Hr's in the future may reveal more heterogeneity (C. P. Mangum, personal communication).] Classes of blood pigments based upon similarities of protein structure are shown in Table 5.1, along with the major phylogenetic groups in which the pigments are found. Occurrence of the various pigments is also shown on a diagram of phylogeny of major phyla, emphasizing the complexity of the pattern of evolution of the various pigments (Fig. 5.1).

Hemoglobin and Heme Proteins

The heme proteins all have iron bound in the center of a porphyrin group (Fig. 5.2), an ancient chemical structure found in single-celled organisms and preserved in oil and coal. The iron is in the ferrous (II) form in active pigment; oxidation to the ferric (III) form produces methemoglobin, which does not bind oxygen reversibly. Oxygen is bound reversibly to the iron by a partial transfer of one electron to the oxygen, producing a ferric state for the iron and a superoxide form(OO^-) of oxygen.

Chlorocruorin differs from other heme proteins only in the substitution at one of the four pyrrole rings of a single formyl group for one of the two vinyl side chains (Fig. 5.2). This substitution shifts the optical spectrum – dilute solutions are a vivid green, more concentrated ones a wine red – but the change appears to have little functional significance. In *Serpula vermicularis* chlorocruorin and protoheme coexist, perhaps even incorporated into the same polymeric molecule (Terwilliger, 1978).

The proteins associated with the porphyrin functional group are quite diverse, at least at the level of quaternary structure, and particularly when comparing the intra- and extracellular heme proteins. Intracellular hemoglobins in the most primitive animal groups are mostly low-molecular-weight proteins not much different from those found in vertebrates (Mangum, 1985). These monomers may, however, be aggregated to a diverse degree in different animal groups, occurring to at least octameric states. In a given individual there may even be different polymers coexisting, and in some cases the aggregation state may alternate with oxygenation and deoxygenation. In one annelid, *Glycera dibranchiata*, more than

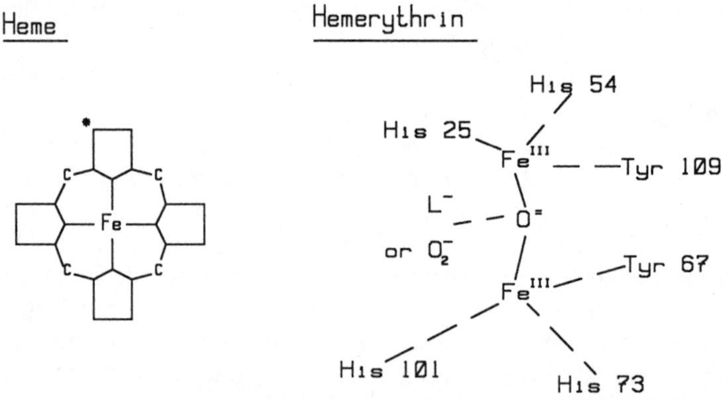

Figure 5.2: Chemical structure of the iron-containing groups that bind oxygen in the heme group of hemoglobin (upper) and in hemerythrin (lower). The asterisk indicates the site of formyl substitution to form chlorocruorin. The iron in heme is in the ferrous (II) state. (Hemerythrin adapted from Klippenstein, 1980.)

one hemoglobin assemblage exists within an individual erythrocyte, contrary to earlier suggestions that the blood might contain a mixture of populations of cells with uniform hemoglobin (Mangum et al., 1989).

Extracellular hemoglobins, on the other hand, are extremely diverse, both in their subunit structure and aggregation state (Terwilliger, 1980). Annelid and insect heme proteins have a subunit of about 15,000 daltons which shares some characteristics with myoglobin. These subunits may be found as monomers or dimers in insects but in aggregates of up to 3 million daltons in annelids. The basic subunit structure of chlorocruorin is similar to that of the large annelid aggregates. The extracellular hemoglobins of molluscs and crustaceans, by contrast, have a very different structure. The subunit may have more than one protoheme, responds differently to aggregation conditions, and may exist in polymers up to 12 million daltons (Mangum, 1985).

Hemerythrin

Hemerythrins represent somewhat of an evolutionary mystery. Similar molecules occur in four phyla: annelids, sipunculids, priapulids and brachiopods. This distribution suggests that the origin of hemerythrin was in a common ancestor, but that is inconsistent with current ideas about invertebrate evolution. Mangum (1985) has summarized our knowledge of hemerythrins as follows:

> [Hemerythrins] are ... polymers of loosely linked subunits each with a single O_2 binding site. The monomers...are 113 – 118 amino acid polypeptides with a molecular weight of ~ 13.5 – 13.9 × 10^3 dalton, and the most frequent multiple is an octamer of about 110 × 10^3 dalton.... Within a phylum such as the sipunculids, the homologies of primary structure are striking.... Why they should be so much more alike than the intracellular [hemoglobins] of the annelids is far from clear. The O_2 binding site, an iron dimer, is located in the middle of four virtually parallel α-helices. This unique conformation, known as the hemerythrin fold, is believed to be a general feature....

The exact details of oxygen binding to the hemerythrin are not completely worked out, but the model of Fig. 5.2 shows the predominant view that the dioxygen binds as a peroxo compound of Fe-III, with the iron held in a fold of the protein chain (Klippenstein, 1980).

Hemocyanin

The only blood pigment to contain a metal other than iron is hemocyanin, which contains a pair of copper atoms bound directly to the protein portion. The best evidence at present suggests that the copper atom pair normally exists in the cupric (Cu^{2+}) state with an endogenous protein bridge. Oxygen is apparently bound as a peroxo group, O_2^{2-}, by partial transfer of electrons from the copper atoms (Ghidalia, 1985).

The protein portion of hemocyanin is actually a glycoprotein, with the percentage of carbohydrate ranging from 1 to 3% in crustacea (Ghidalia, 1985) and reaching much higher values in various molluscs (Mangum, 1985). The monomeric crustacean protein has a molecular weight averaging about 70,000 to 80,000 daltons comprising about 600 to 660 amino acid residues; the circulating form is typically a mixture of hexamers and dodecamers. In molluscs the predominant form is a polymer ranging from 3 million to 9 million daltons in size, made up of much larger subunits ($4 - 5 \times 10^5$ daltons) with multiple active sites per subunit. These fundamental differences in nearly every aspect provide strong support for an independent evolution of hemocyanin in the two groups. Although molluscs and arthropods are both thought to have evolved from annelids or their precursors, the link may not be very direct.

Myoglobin

A final class of heme proteins consists of the myoglobins (Mb), dark red pigments found in muscle cells of vertebrates. Myoglobin has the same heme structure as hemoglobin but is coupled to a 150 amino acid residue protein with a molecular weight of about 17,450 daltons and consists only of the monomer, i.e., one heme per protein chain. Myoglobin has an extremely high oxygen affinity (see below), and its principal function appears to be to facilitate diffusion of oxygen within the muscle cells (Kreuzer, 1970). It may also serve as a temporary oxygen store during muscular contraction. Interestingly, no hemocyanins or hemerythrins with myoglobin-like function have evolved in other animal groups.

Intracellular versus Dissolved Pigments

The occurrence of respiratory pigments in cells (RBC's or erythrocytes) in the higher vertebrates has often been described as an evolutionary innovation allowing greater blood oxygen capacity to support higher metabolic rates. The advantages cited for intracellular pigments include a lower colloid osmotic pressure for equivalent amounts of protein, the convenience of intracellular control of allosteric modulators (see below), and a lower viscosity compared to an equimolar protein solution. Mangum (1985) discussed these various arguments and pointed out that the erythrocyte appears to have evolved several hundred million years before it acquired the properties used to explain its origin, so some other explanation must be sought. In particular, it is curious why some animal groups have lost the erythrocyte in favor of dissolved extracellular pigments.

The contribution of the respiratory pigment to colloid osmotic pressure in those animals without erythrocytes is actually quite small owing to the high degree of polymerization generally seen (Mangum & Johansen, 1975). Osmotic pressure is proportional to the number of osmotically active particles in solution, so if the pigments are carried as very large aggregates, a considerable oxygen capacity may be achieved without contributing substantially to osmotic pressure. In fact many invertebrate pigments can be visualized in electron microscopic images as very large barrel-shaped particles (Mangum, 1976; Terwilliger et al., 1978; van Bruggen, 1983).

The viscosity argument is not particularly convincing either, since packaging of pigment in RBC's (or pink blood cells, PBC's) confers an advantage only in small radius vessels (the Fahraeus-Lindqvist effect). In the invertebrate groups in which corpuscles first appear, the circulatory systems have few small radius vessels and are dominated by larger-bore vessels in which corpuscles actually confer a viscosity disadvantage (Snyder, 1973). Cells containing highly polymerized pigments would not be very deformable, so it is not surprising that in animals with open circulatory systems or with predominantly large-bore vessels the extracellular pigments tend to predominate.

Finally the convenience of intracellular control of allosteric modulators of pigment affinity does not argue strongly for RBC's since the vast majority of animals (invertebrates) having intracellular pigments do not possess the modulating systems found in vertebrates (Mangum, 1985). A further argument against this explanation of intracellular pigments has been provided by research showing modulation of noncellular pigment affinity (viz., some crustacean hemocyanins) by a variety of organic factors (Truchot, 1980; Graham et al., 1983; Mangum, 1983, 1985; Bridges et al., 1984).

The answer to the question of why vertebrate blood pigments were originally packaged within RBC's may indeed be, as Mangum (1985) concluded, that it is simply inherited. The loss of the nucleus in erythrocytes of mammals may indeed enhance the Fahraeus-Lindqvist effect, and the evolution of allosteric modulator systems can be regarded as a late refinement in the evolutionary life history of the RBC.

OXYGEN EQUILIBRIUM CURVES OF HEMOGLOBIN

In order for a respiratory pigment to be physiologically useful, it must bind oxygen when the partial pressure is relatively high and release it at partial pressures suitable for tissue metabolism. The reaction is thus one of dynamic equilibrium:

$$Hb + O_2 \longleftrightarrow HbO_2 \qquad \text{(Eq. 5.2)}$$

If the relation followed simple first-order kinetics the curve would be hyperbolic in shape when the percent saturation of Hb is plotted as a function of oxygen partial pressure and would look like the data for dogfish in Fig. 5.3. In most animals, however, the curves are sigmoid, like the one shown for trout (Fig. 5.3). This curve results from *cooperativity* among the four heme groups in the Hb

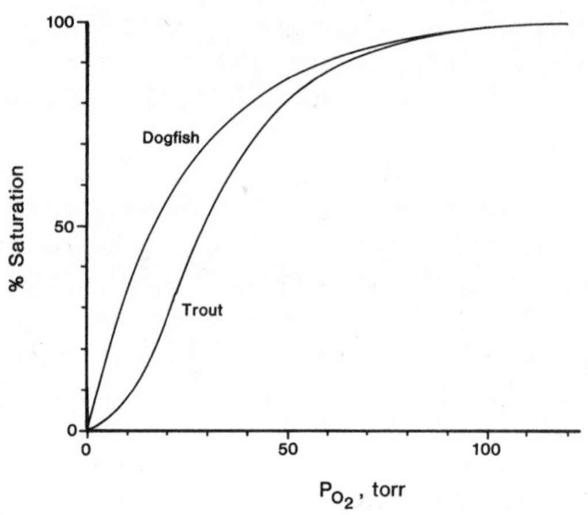

Figure 5.3: Oxygen dissociation curves of hemoglobin from the dogfish (Lenfant & Johansen, 1966) and the rainbow trout (Cameron, 1971a). The trout hemoglobin shows significant cooperativity, resulting in a sigmoid curve, whereas the dogfish does not.

tetramer; i.e., as oxygen is bound to the first heme, it increases the affinity of the others for oxygen. A convenient way of approximating these sigmoid oxygen curves is given by the *Hill equation*:

$$Y = (P/P_{50})^n/\{1 + (P/P_{50})^n\} \quad \text{(Eq. 5.3)}$$

where P is the partial pressure of oxygen, P_{50} is the partial pressure which produces half saturation, Y is the fraction of pigment saturated, and n is a fitted

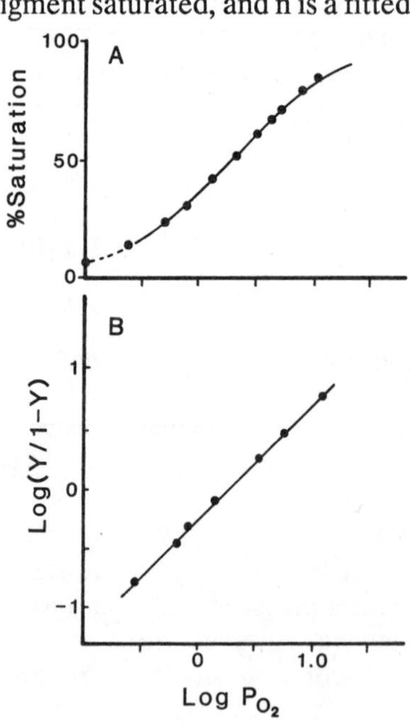

Figure 5.4: Two alternate means of plotting oxygen equilibrium curves. The ordinate for both panels is logarithmic, but the abscissa for the upper panel (A) is fractional saturation, and for the lower panel (B) as shown in the formula. The lower plots are known as Hill plots. The slope is an index of cooperativity, with n = 1 showing no cooperativity. The data are for the monomer of *Limulus* hemocyanin, for which n = 1 and cooperativity is therefore nil. Between about 25 and 75% saturation (log(Y/1-Y) = 0 when saturation = 50%) there is a linear portion whose slope is equal to the cooperativity. Hill plots are typically linear only over a restricted range near 50% saturation. (Re-drawn from Bonaventura & Bonaventura, 1980, by permission, ©Amer. Soc. of Zoologists.)

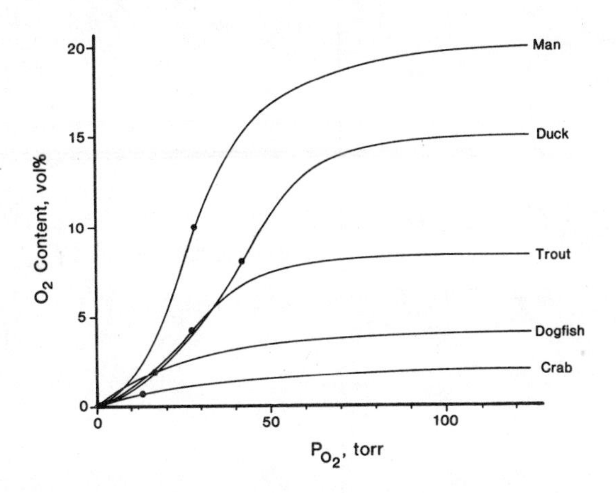

Figure 5.5: When equilibrium curves for various species are plotted as actual oxygen content, rather than percent saturation, the differences in both affinity and capacity are emphasized. (Data from the following sources: man, Altman & Dittmer 1974; duck, Scheipers et al., 1975; trout, Cameron, 1971a; dogfish, Lenfant & Johansen, 1966; crab, Young, 1972a.)

parameter. When $\log[Y/(1 - Y)]$ is plotted against log P the relation is linear over a limited range, and the slope at the P_{50} value ($Y = 0.5$) is given by **n**. The value of **n** is taken as an index of cooperativity (Fig. 5.4)(see Program HILLPLOT, accompanying disk). At the extremes the curve deviates, usually with lower slope.

When the Hill coefficient (**n**) is 1 for a pigment, there is no cooperativity among binding groups; higher values increase the sigmoidicity of the curves and indicate increasing cooperativity. For sharks and some invertebrates the value of **n** is near 1, for mammals the value of **n** lies between 2.4 and 2.9 (Prosser, 1973), and for the large invertebrate hemoglobins **n** values may exceed 5.

The parameters most often used to refer to a particular hemoglobin's properties are its Hill coefficient (**n**), its P_{50} value (partial pressure for half saturation), and its *affinity*. The latter term is not precisely defined but is inversely proportional to the P_{50} value for the pigment.

The concentration of hemoglobin in the blood (the oxygen carrying capacity) as well as its chemical properties naturally influence blood oxygen binding curves. When the abscissae are expressed in terms of the actual oxygen bound to hemoglobin rather than percent saturation these differences are strikingly illustrated (Fig. 5.5). Evolutionary modification of both the kind and quantity of hemoglobin has produced a very wide array of functional properties. Attempts have been made to correlate the quantity and characteristics of various species' pigments to their habits and general level of metabolic activity (Prosser, 1973). The correlation is not very good, however, and is complicated by many other factors such as size (Schmidt-Nielsen & Larimer, 1958).

Influence of The Hemoglobin Environment

Temperature

For nearly all hemoglobins increasing temperature decreases oxygen affinity (i.e., raises the P_{50})(Fig. 5.6). In most animals it is probably a functional ad-

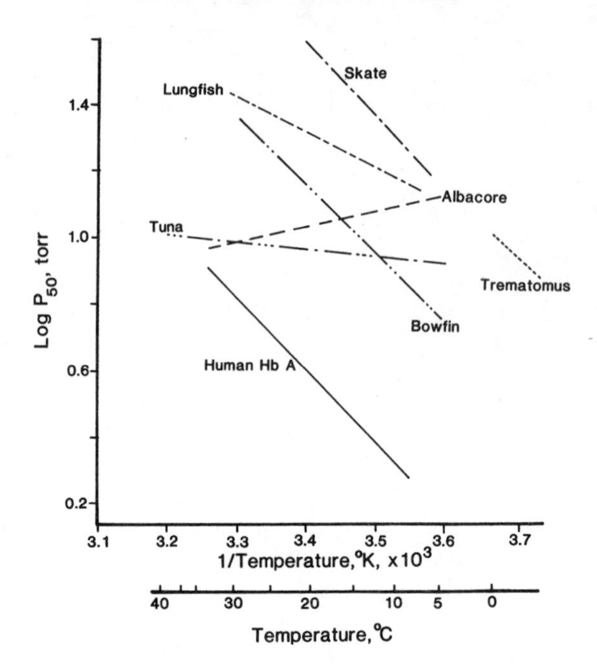

Figure 5.6: Influence of temperature upon hemoglobin-O_2 equilibrium curves. Adapted from Johansen & Lenfant (1972), with additional data on albacore from Cech et al., 1984. The almost flat or even reversed temperature slope of the tuna and albacore blood is thought to be related to rapid changes in temperature in deep muscle (see text).

vantage, since at higher temperatures tissue respiration rates are higher and steeper oxygen gradients probably assist in unloading. There are some cases, however, in which large variations in affinity with temperature may be disadvantageous. As blood circulates from the periphery to the core in heat exchange *rete* of the warm-bodied tuna its temperature may change 10°C or more (Carey & Teal, 1969; Carey et al., 1971). If oxygen affinity changed significantly with temperature, the heat exchange *rete* would also function as an oxygen exchanger, excluding oxygen from the tissues (see Chapter 7). According to some reports, the hemoglobins from tuna have little temperature sensitivity or even a reverse temperature effect (Cech et al., 1984)(Fig. 5.6), although more recent work on another species has shown more conventional properties (Jones et al., 1986). This case is special, however, and most animals' hemoglobins have greatly reduced affinity at higher temperatures.

Carbon Dioxide and pH

The oxygen affinity of most hemoglobins is strongly reduced by increases in the CO_2 concentration of the blood (Fig. 5.7). This shift of the oxygen equilibrium curve to the right is called the *Bohr shift* (Bohr et al., 1904). All vertebrates and most invertebrates show a normal, or rightward, Bohr shift at higher CO_2 con-

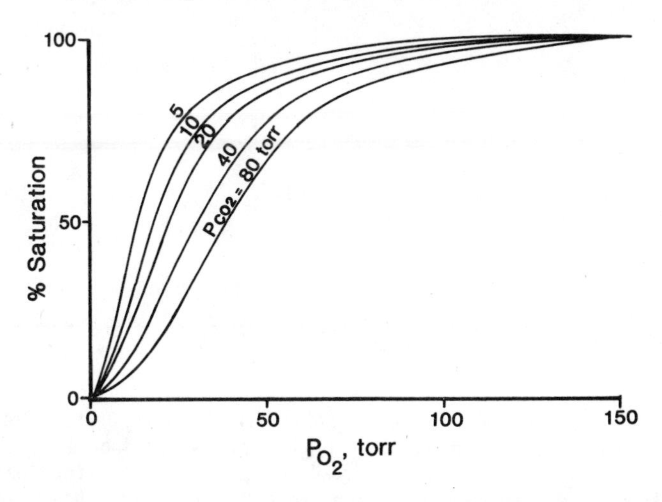

Figure 5.7: Shift of the Hb-oxygen equilibrium curve to the right with increasing P_{CO_2} or decreasing pH is called the Bohr shift. Data of Bohr et al., 1904, for a dog.

centrations, but in some invertebrates (Mangum, 1976) and at extreme pH values in some other animals there is a reverse Bohr shift.

The affinity of the heme site for oxygen in any particular Hb is a function of the molecular shape (tertiary and quaternary structure) and charge configuration near the active binding sites on the heme groups. This charge environment is determined to a large degree by the ionization of α-amino and imidazole groups on certain amino acid residues, and the ionization state is in turn a function of the local pH. Since pH is decreased by increases in CO_2, there is a direct link between not only the overall affinity of the Hb for oxygen but also the cooperativity and Hill constant. Further evidence for this effect of pH on the binding site environment is that CO_2 also affects the binding of carbon monoxide to the heme groups in a manner similar to its effect on oxygen.

In addition to its indirect effect on Hb-O_2 binding through pH, CO_2 also has a direct effect by forming carbamino bonds directly to hemoglobin. The CO_2 binds to uncharged terminal amino groups by the reaction:

$$R\text{–}NH_2 + CO_2 \longleftrightarrow R\text{–}NHCOO^- + H^+ \qquad \text{(Eq. 5.4)}$$

In mammals including man about 2 to 6 vol% CO_2 (1 to 3 mM L^{-1}) is bound to hemoglobin in the carbamino form (Ruch & Patton, 1965), depending upon the oxygenation state. In lower vertebrates the amount of carbamino binding appears to be less (Albers, 1970), but for the invertebrate Hb's there is insufficient information to allow any general statement.

Increases in CO_2 may affect the maximum oxygen binding at high partial pressure as well as the affinity and equilibrium curve (Fig. 5.8). This reduction in the oxygen capacity is known as the *Root shift* (Root, 1931) and is particularly marked in various fishes. Some argue that the Root effect is no more than a pronounced

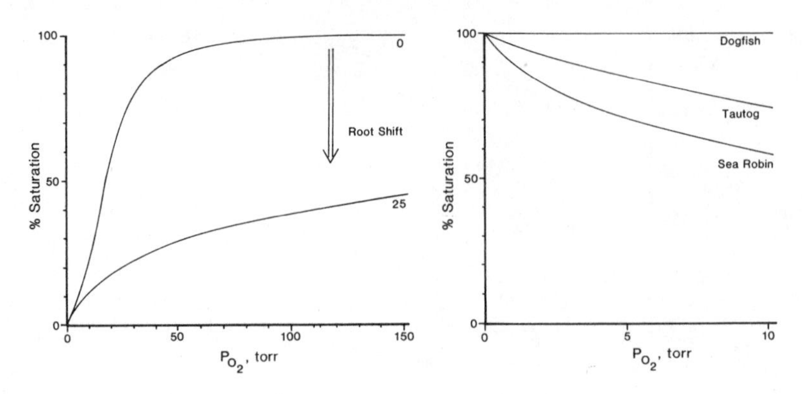

Figure 5.8: Decrease in maximal oxygen binding of Hb at high P_{O_2} is called the Root shift (left panel; P_{CO_2} = 0 and 25 torr, as marked). The original studies were done at very high P_{CO_2}'s; when the data are reexamined over a physiologically realistic range of P_{CO_2} (right panel), the effect is much smaller. For many species there is no Root effect. (Left panel data: tautog; all from Root et al., 1939.)

Bohr shift, and that at very high P_{O_2} values the capacity would be unaffected. Early work on the Root shift was also done using very high partial pressures of CO_2, which exaggerates the importance of the Root effect in the physiological range of P_{CO_2}. Whatever view is correct, the reduction in capacity at physiological oxygen tensions does seem to be real, and is functionally significant in the swimbladder (Forster & Steen, 1969)(see Chapter 7).

Hemoglobin as an Acid

The binding of oxygen to the heme group causes release of H^+ ions from the hemoglobin itself, lowering the pH of its microenvironment. IN other words, Hb itself is a weak acid with dissociable protons, and its acidic strength increases with oxygenation. One immediate effect is that as oxygen is loaded onto Hb and the pH drops, the loss of CO_2 (in lungs, e.g.) is facilitated by the acidification. The converse is true when oxygen is unloaded in tissues, raising pH and promoting CO_2 buffering by the blood. This dependence of buffering or acid strength upon oxygenation is known as the *Haldane effect* (see Chapter 6).

Ionic Strength and Specific Ions

Since Hb is ionized to a certain extent and has weakly dissociated groups, it is affected by solution ionic strength much as any other aqueous solution (Hills, 1973). Increasing ionic strength (μ) reduces pK' by about $0.5(\mu)^{0.5}$, so at higher ionic strength Hb is a stronger acid and its binding affinity is affected. The general effect of increased inorganic ions is to increase oxygen affinity. Ionic strength effects may be somewhat more direct in invertebrates' extracellular Hb's (Mangum, 1973, 1976), since the ionic strength of blood in most of these animals follows that of the environment. For animals with intracellular Hb's the changes may be much less if the cells' internal environments are regulated.

Aside from the general effects of ionic strength, various specific ions affect Hb binding by direct chemical action. Calcium and magnesium increase the oxygen affinity of Hb, as do various organic compounds. Organic modulators of Hb function are known only in the vertebrates and were first discovered by Benesch and Benesch (1967) and Chanutin and Curnish (1967). Working independently, these two groups found that increasing concentrations of intra-erythrocytic organic phosphate compounds, especially 2,3-diphosphoglycerate (2,3-DPG), decrease the affinity (i.e., increase the P_{50}) of the blood of man and other mammals. Although subsequent work has shown that many changes in 2,3-DPG levels reported in the late 1960shave little functional significance, 2,3-DPG modulation appears to be important in the adaptation to high altitude (Lenfant et al., 1971; Bouverot, 1985) and possibly in enhancing fetal-maternal gas exchange (Bauer, 1974). In the fish, with their nucleated erythrocytes and quite different metabolism, 2,3-DPG does not appear to be important (Isaacks & Harkness, 1980), but ATP and other phosphorylated intermediates in the erythrocyte do seem to be involved in regulating responses to hypoxia (Wood & Johansen, 1972) and to seasonal temperature acclimation (Mauer, 1974; Mauer & Cameron, unpublished data). Interestingly, Cl^- binds to the same site on vertebrates' Hb as do the phosphates, but the physiological significance is not clear. In the invertebrates there are no known allosteric modulators of Hb function, but the subject has been so little studied that one cannot rule out the possibility of some being found in the future.

Oxygen Equilibrium In Vivo

In most studies of Hb function variables such as gas partial pressures, ionic strength and pH are held constant with only one variable changing at a time. In vivo, of course, this is not the case: during a typical tissue transit the partial pressure of oxygen declines in a complex manner; CO_2 tension rises in a somewhat

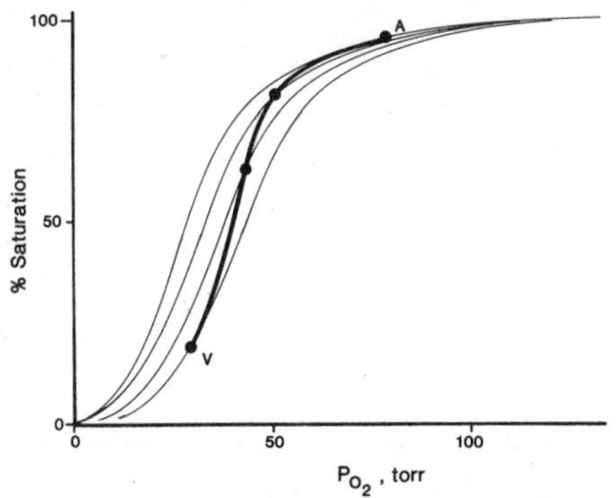

Figure 5.9: Equilibrium curve in vivo is constructed from a family of curves of constant P_{CO_2}. That is, as arterial blood (A) passes through the capillaries, its oxygen tension declines, and CO_2 tension rises, shifting the values from one curve to another, as shown by the dots. The in vivo curve is therefore steeper than in vitro curves determined at constant P_{CO_2}.

reciprocal way; lactate concentration sometimes rises; inorganic ions may change slightly; and pH changes according to the changes in all of the other parameters. A hypothetical in vivo equilibrium curve is shown in Fig. 5.9. Such curves can be approximated rather easily by equilibrating two halves of a blood sample to arterial and venous conditions, then constructing the curve by the mixing technique (Haab et al., 1960; Edwards & Martin, 1966).

OXYGEN EQUILIBRIUM CURVES OF HEMOCYANIN

The color change in the blood of various invertebrates as it passes through gills was first noted by Bert (1867) and later by Fredericq (1878), who coined the term hemocyanin. Winterstein (1909a) confirmed the oxygen transport function of this pigment by direct measurement of oxygen content. Although there were many studies of hemocyanin between 1900 and 1930, the picture of its role in respiration of invertebrates was much less clear than that of hemoglobin in vertebrate respiration (Redfield, 1934). Part of the confusion was no doubt due to the much greater diversity of hemocyanins, particular among various invertebrate phyla (see above), and the fact that hemocyanin oxygen capacity is so much lower than the hemoglobin oxygen capacity in man.

The oxygen equilibrium curves of hemocyanin are similar to those for hemoglobin in that they represent a loose bond of oxygen to the pigment, one that is fully reversible within the physiological range of oxygen partial pressure (Redmond, 1955). In some species the content–partial pressure curves are nearly hyperbolic, but the majority of animals containing hemocyanin have distinctly

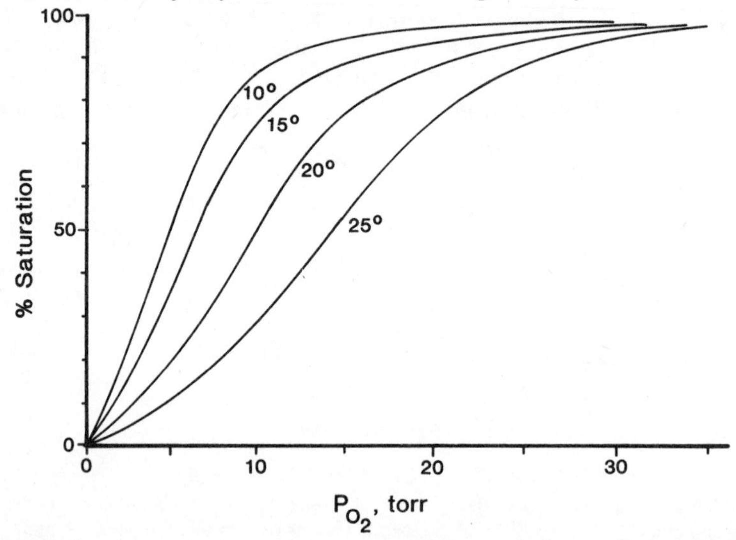

Figure 5.10: Effects of temperature on hemocyanin affinity are similar to effects on hemoglobin. P_{50} for hemocyanin from a spiny lobster, *Panulirus interruptus*, increases about 2.5-fold between 10 and 25°C. (Data from Redmond, 1955).

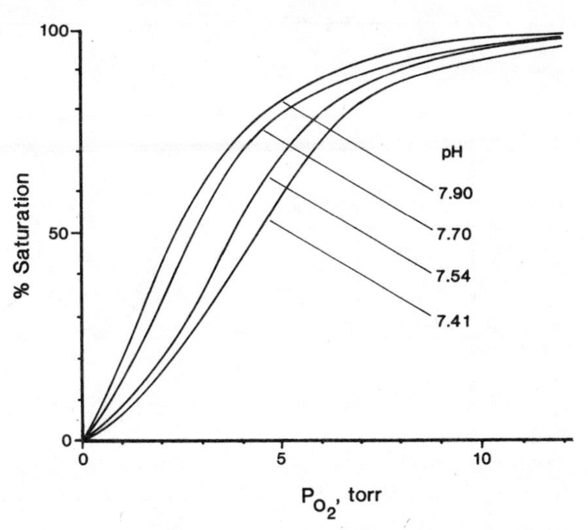

Figure 5.11: Hemocyanin may also have a large Bohr shift, as shown here for the mangrove crab *Cardisoma guanhumi*. The effect is essentially the same whether pH is decreased by increasing P_{CO_2} or by titration at constant P_{CO_2}. (Data from Redmond, 1962.)

sigmoid equilibrium curves with varying degrees of cooperativity among binding sites (Fig. 5.4).

Influence of the Hemocyanin Environment

Temperature

As with hemoglobin, the oxygen affinity of hemocyanin decreases with increasing temperature (Fig. 5.10). At the higher temperatures Redmond's (1955) curves for a spiny lobster (*Panuliris interruptus*) also become somewhat less sigmoid, indicating decreased cooperativity (**n**), but in a more recent study of the blue crab (*Callinectes sapidus*) Young (1972a) could show only a slight, nonsignificant decrease in **n** between 20° and 36°C. From plots of the $\log P_{50}$ against 1/T°K the heat of reaction ΔH° may be calculated, giving an index of temperature sensitivity. For the blue crab Young (1972a) found an average of 13.3 kcal/mole, but species vary considerably in their temperature sensitivity with ΔH° values as low as 8 kcal/mole (Young, 1972b).

Carbon Dioxide and pH

Most hemocyanins exhibit a substantial Bohr shift that can be demonstrated by manipulating the pH either with buffers or indirectly with CO_2 (Fig. 5.11). The slope of the Bohr shift varies considerably among species, with values commonly as high as ($\Delta\log P_{50}/\Delta pH = -0.8$ (Mangum, 1985). In many species of mollusc and the horseshoe crab (*Limulus*) there is a reverse Bohr shift (Fig. 5.12) that is usually nonlinear and may even turn into a "normal" Bohr shift at higher pH values.

The question of whether there is a direct effect of molecular CO_2 or all Bohr effects result indirectly from pH change is still not completely settled (Burnett & Infantino, 1984). Evidence for direct CO_2 effects on several crustacean Hc's

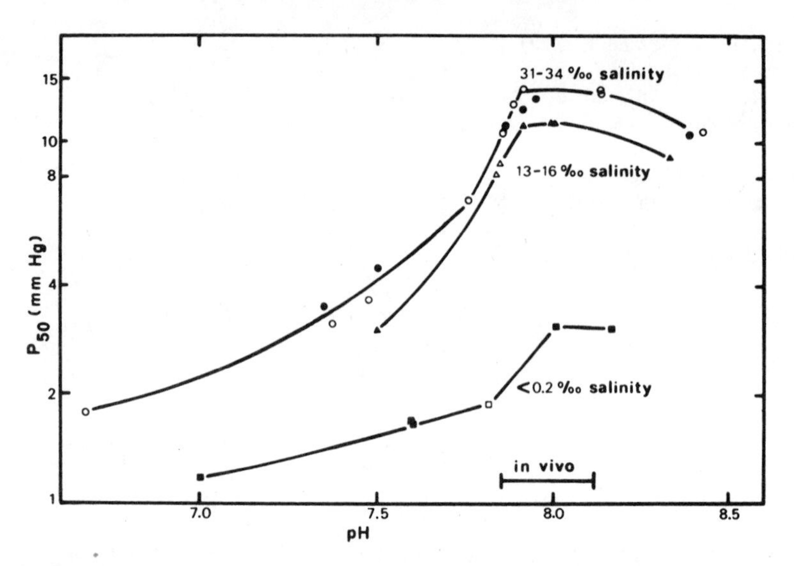

Figure 5.12: The Bohr shift in some invertebrate hemocyanins is reversed, as shown here for the whelk *Busycon* over the pH range of about 6.5 to 7.9. At higher pH values the Bohr shift is normal. (From Mangum & Lykkeboe, 1979, by permission, ©Alan R. Liss, Inc.)

has been presented by Truchot (1973a) and more recently by Mangum and Burnett (1986), but the mechanism for the effect has not been clearly delineated, nor is there an effect in all the species examined. Lykkeboe et al. (1980) offered clear evidence for (carbamino) CO_2 binding to Hc from a cephalopod mollusc, the magnitude of which was comparable to that found in mammals. Their calculations also showed how the pH of arterial blood may actually be *lower* than that of venous blood, due to the stoichiometry of H^+ binding to Hc. I have frequently observed a small decrease in postbranchial blood of the blue crab, which may result from similar equilibria. Analysis of direct CO_2 effects may be complicated by simultaneous anion effects, possibly a HCO_3^- effect on the protein structure similar to that caused by Cl^-. There is no functional equivalent of the Root shift known for Hc.

Ionic Effects

The effects of ions upon the oxygen binding characteristics of hemocyanin must be considered in three categories: specific cation effects, specific anion effects, and general ionic strength effects. Data of Mangum and Burnett (1986) show very little or no effect of NaCl alone, indicating that most ionic effects are the result of specific ion binding. Ionic strength (μ) will exert an effect on ionizable sites (Chapter 1), but it apparently has only insignificant effect on functional properties.

Divalent cations, especially Ca^{2+} and Mg^{2+}, have been known to increase the affinity of hemocyanin for some time (Truchot, 1975a; Mangum, 1980; Diefenbach & Mangum, 1983; Mangum, 1985)(Fig. 5.13). At least part of this effect is

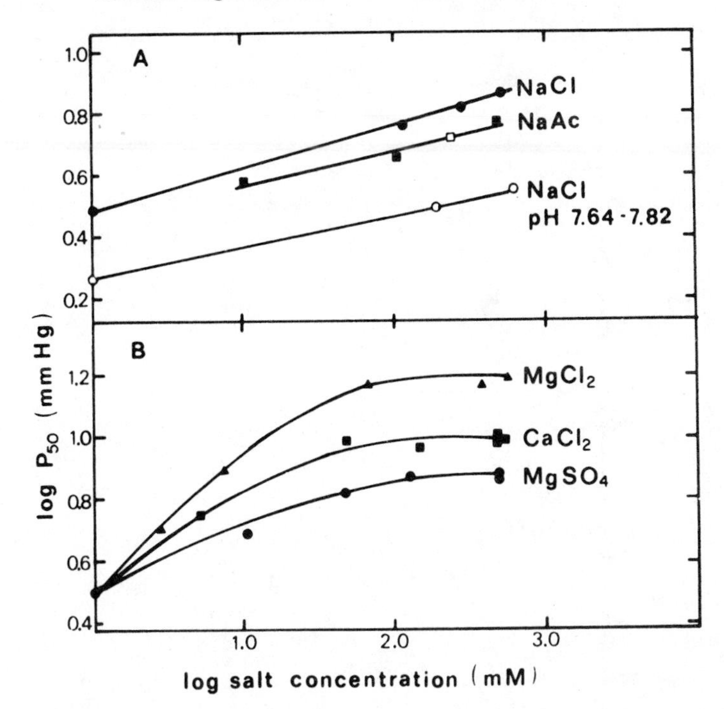

Figure 5.13: Specific ions also affect the P_{50} of hemocyanin. The pH was 7.96 to 8.10 except as indicated in the upper panel. (From Mangum & Lykkeboe, 1979, by permission, ©Alan R. Liss, Inc.)

due to the tendency of Hc subunits to dissociate in the absence of Ca^{2+} and, to a lesser extent, Mg^{2+}; not only does the oxygen equilibrium curve shift to the left as these ions are removed, the cooperativity is lost and the curves become increasingly hyperbolic rather than sigmoid. The physiological significance of these ionic effects is not always clear, since the majority of animals with Hc are osmoconformers at least part of the time, so their ionic concentrations are not very closely regulated.

A final group of ionic effectors has been elucidated recently by Truchot (1980) and others (e.g., Graham et al., 1983). Lactate ion appears to have a specific effect on Hc, increasing Hc affinity as its concentration rises (Fig. 5.14). Since lactate is a major product of anaerobic metabolism during hypoxia, it has been tempting to speculate that this is an adaptive response to increase Hc loading gradients at the gills. Physiological evidence that such a response is helpful is, however, sketchy at present (Mangum, 1985). In some respects lactate could be considered an allosteric modulator, but since it is not produced by the blood and does not appear to be regulated in any similar manner to control Hc affinity, the analogy is perhaps not a good one. Another possible modulator of hemocyanin affinity is uric acid, but it is not at all clear how important it might be physiologically (Morris et al., 1985).

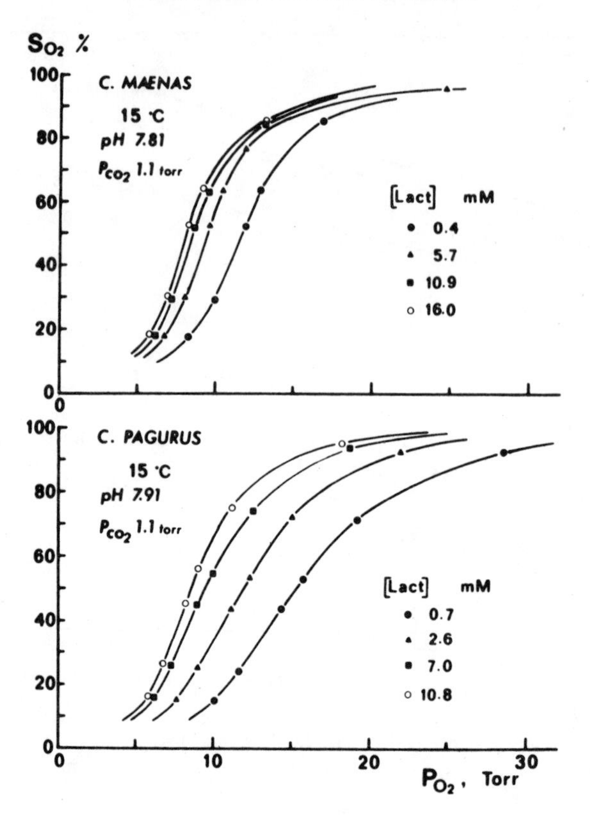

Figure 5.14: Specific effect of lactate ion on hemocyanin equilibrium curves for two crab species. (From Truchot, 1980, by permission, ©Alan R. Liss, Inc.)

Oxygen Equilibrium of Hemocyanin In Vivo

As blood containing Hc circulates from gas exchanger to respiring tissue and back again, there will of course be a cyclic variation in the oxygen content and partial pressure. Concomitant changes in the partial pressure of CO_2, however, may or may not cause an *in vivo* hysteresis like that shown in Fig. 5.9 for Hb. Although CO_2 is produced in tissues at a rate comparable to the oxygen consumption, its uncatalyzed hydration rate is very slow. It is not yet clear if there is any significant reaction of the CO_2 to form HCO_3^- and H^+ ions during the in vivo circulation time (Cameron, 1979)(see Chapter 9). Carbonic anhydrase is absent from the blood but may be present on cell membranes facing the blood (Henry, 1988). Since the effects of CO_2 appear to be mostly indirect through its effect on pH, lack of any significant reaction would mean that the Bohr shift would not occur. A resolution of this question must await better studies of circulation times, reaction rates and physiological catalysis of the hydration/dehydration reactions in vivo.

SPECIAL FUNCTIONS OF BLOOD CELLS

The Chloride Shift

Besides carrying the respiratory pigments, blood cells have several specialized functions related to respiration. In vertebrate erythrocytes one of the most important (and earliest studied) is the "chloride shift," once known generally as the Hamburger shift (Hamburger, 1918). The significance of this process for CO_2 transport is discussed more fully in Chapter 6, but since it provides an effective shuttle of H^+ ions in and out of the blood cell in response to changes in CO_2, the chloride shift is an important part of the oxygen loading and unloading process. When CO_2 is added to blood in the tissues it rapidly diffuses into the blood cells across the blood cell membrane, which is highly permeable to the uncharged molecular CO_2. Once inside the cell, the carbonic anhydrase present there facilitates extremely rapid equilibration with HCO_3^- and H^+ ions. Some of the H^+ ions bind to Hb, causing a Bohr shift. There is a dynamic steady state maintained between intra- and extracellular [HCO_3^-] via a chloride exchanger protein in the membrane, so the increased intracellular [HCO_3^-] increases the movement of HCO_3^- out in a tightly linked exchange for Cl^-. In the lungs when CO_2 decreases, the process is reversed, leading to net movement of Cl^- out of the erythrocyte, a reduction of intracellular H^+, and a reversed Bohr shift, which assists in oxygen loading. This chloride shift mechanism is known for every vertebrate erythrocyte that has been examined (Lückner, 1939; Piiper, 1964, 1969; Cameron, 1978a; Obaid et al., 1979); it has not been described for any invertebrate blood cell, but it may be only because no one has looked for it.

Erythrocyte Metabolism

The supposed advantages and disadvantages of containment of the blood pigments in RBC's have been discussed, but the cells also perform other functions related to gas exchange. Metabolic production of allosteric affinity modulators has also been discussed, but Hb-containing cells also contain an important enzymatic system (methemoglobin reductase) for maintaining the heme group in its reduced (II) state (Jaffe, 1964). Methemoglobin normally comprises less than 1% of the Hb present in man except in disease states (Beutler, 1968), but in several fish species values as high as 3 to 17% have been found (Cameron, 1971b). A similar reductase system has been described for the Hr-containing cells from a sipunculid (Manwell, 1977) but so far for no other invertebrate. It seems likely that they are universally present, however, since the heme group is thermodynamically metastable and an eventual accumulation of the nonfunctional met form would seem otherwise unavoidable.

Energy metabolism of various vertebrate and invertebrate erythrocytes is apparently quite different. Whereas non-nucleated mammalian erythrocytes rely heavily on the pentose phosphate shunt pathway, the nucleated erythrocytes of lower vertebrates and invertebrates have mitochondria, some endoplasmic reticulum, and enzymes characteristic of the Embden-Meyerhof and TCA cycles (Mangum & Mauro, 1985). There is also the intriguing question of whether some of these erythrocytes may actually manufacture their respiratory pigments; some invertebrates' erythrocytes clearly do (Hoffman & Mangum, 1970; Shafie *et al.*, 1976).

Finally there is the conspicuous dichotomy among vertebrate groups between nucleated vs. non-nucleated erythrocytes. The functional advantages of the loss of the nucleus may be related to the enhanced deformability of the non-nucleated cells in small vessels and at higher flow rates. Non-nucleated cell have a life span measured only in weeks, however, whereas nucleated cells may live from 18 months to as long as the animal (Altland & Brace, 1962).

Chapter 6

Acid-Base Regulation

Respiration serves more than the general metabolic functions of obtaining oxygen and eliminating carbon dioxide. Both gases participate in various chemical reactions that affect the acid-base status; CO_2 is particularly important as a variable in acid-base regulation. The respiratory system is usually driven primarily to meet the needs of obtaining oxygen, but not always. Under many circumstances the control of respiratory processes is altered to maintain a particular acid-base status. In this chapter the basic principles of physiological acid-base chemistry are reviewed only briefly. More complete treatments have been published by Albers (1970), Hills (1973), Davenport (1974), Truchot (1983, 1987), and Heisler (1986a).

ACID-BASE CHEMISTRY

Acids, Bases, and Buffers

The most useful general definition of acids and bases for the purposes of physiological chemistry is that of Brönsted and Lowry, in which an *acid* is defined as a *proton donor* and a *base* as a *proton acceptor*. Any acid-base reaction always involves a conjugate acid-base pair, consisting of the proton donor and a corresponding proton acceptor. In dilute aqueous solution, a generalized acid (HA) will tend to donate a proton to water:

$$H_2O + HA \longleftrightarrow H_3O^+ + A^- \qquad \text{(Eq. 6.1)}$$

in which case the water acts as the proton acceptor. Donation of a proton is a general acid characteristic but the tendency to do so (i.e. acid strength) in aqueous solution varies widely. Those acids with only a slight tendency to dissociate are called *weak acids*, whereas those that tend to be completely dissociated in water are called *strong acids*. This tendency to dissociate can be quantified by the equation:

$$K = [H^+][A^-]/[HA][H_2O] \qquad \text{(Eq. 6.2)}$$

where the square brackets have their usual meaning of concentration in moles per liter. Usually the water concentration is considered to be constant and is incorporated into K, leaving only:

$$K = [H^+][A^-]/[HA] \qquad \text{(Eq. 6.3)}$$

For nonideal solutions the concentration and ionic strength will affect the reactivity of each species; if activities are used instead of concentrations in Eq. 6.3 then the value for the *apparent* dissociation constant is denoted as K′. Thus for a weak acid such as acetic acid, CH_3COOH, the conjugate base is the anion CH_3COO^-, the K′ value is 1.74×10^{-5}, and the negative log of K′, called the pK′, is 4.76. Acids with pK′ values close to the acidic end of the pH scale are strong acids, and bases with high pK′ values are strong bases. The logarithmic form of Eq. 6.3 is:

$$pK = \log\{[HA]/[H^+][A^-]\} \qquad \text{(Eq. 6.4)}$$

Note that under this definition of acids and bases, NH_4OH is neither an acid nor a base, since its dissociation into NH_4^+ and OH^- is similar to the dissociation of an electrolyte such as NaCl and involves no transfer of protons. NH_3 is a base because it has a strong tendency to accept a proton from water, forming NH_4^+, which in turn is its conjugate acid. More general theories of acid-base chemistry are required for non-aqueous systems, but the Brönsted-Lowry definition will suffice for the present purposes.

If a strong acid is titrated over the pH range with OH^-, at first there is almost no pH change, since added OH^- simply combines with excess H^+ in the solution to form water. At the point where the added OH^- almost equals the original acid concentration, there is a rapid transition from excess H^+ to excess OH^-, followed by a further flat portion with nearly constant high pH (Fig. 6.1). These "stair-step" titration curves are typical of strong acids, but weak acids have the rather different titration curves shown in Fig. 6.1. As base is added to these solutions, the H^+ ions reacting with the added OH^- are partially replaced by further dissociation of the original acid in equilibrium in the solution. Equal additions of OH^-

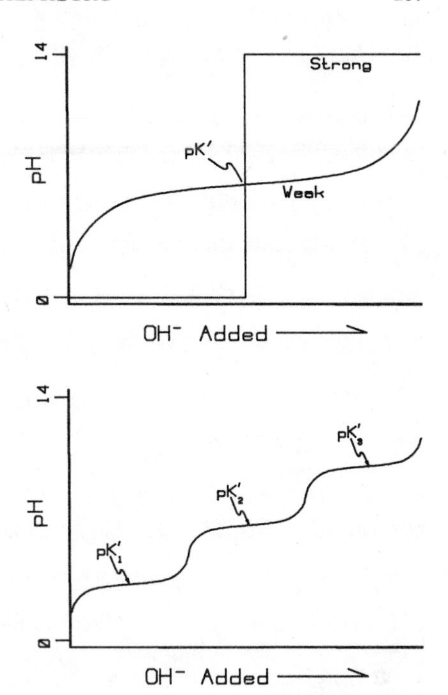

Figure 6.1: (Top) Titration curve of a strong and a weak acid showing the buffering action of the latter. The pK′ falls at the inflection point of the titration curve, and the flatter slope at that point corresponds to maximum buffering capacity. (Bottom) Titration of a complex acid, showing three pK′ values from different groups. The curve could be for an inorganic acid such as H_3PO_4 or an amino acid such as histidine, with the first pK′ from the α–carboxyl group, the second from the imidazole moiety, and the third from the α–amino group.

cause greater shifts of the ratios as the pK′ value is approached and the slope of the titration curve reaches a minimum at the pK′ value. If we consider the flatter slope as a resistance to pH change with added base, or a *buffer capacity*, we see that the buffer capacity is maximal at the pK′ value. The symbol β is often used to denote buffer capacity, and is defined as the change in base per unit pH change, or:

$$\beta = \Delta[\text{base}]/\Delta\text{pH} \qquad \text{(Eq. 6.5)}$$

This point is important in that it means that a particular conjugate acid-base pair has significant buffer action only near its pK′ value. At pH 10, for example, a solution of acetic acid is almost completely dissociated and has virtually no capacity to generate further protons in response to an addition of base.

Many acids contain two or more dissociable protons and actually behave as a series of acids. Phosphoric acid, for example, has three separate acid-base dissociation reactions:

$$H_3PO_4 \longleftrightarrow H_2PO_4^- + H^+ \qquad pK' = 2.14$$
$$H_2PO_4^- \longleftrightarrow HPO_4^{2-} + H^+ \qquad pK' = 7.20$$
$$HPO_4^- \longleftrightarrow PO_4^{3-} + H^+ \qquad pK' = 12.4 \qquad \text{(Eq. 6.6)}$$

and a titration curve of this acid over the whole pH range will show three distinct plateaus at the pK′ values for each dissociation step (Fig. 6.1).

The Carbonic Acid Buffer System

Carbonic acid, H_2CO_3, is a weak acid with a pK′ of about 3.8. In water at around physiological pH values it is strongly dissociated according to:

$$H_2CO_3 \longleftrightarrow HCO_3^- + H^+ \qquad \text{(Eq. 6.7)}$$

The bicarbonate ion, HCO_3^-, is also a weak acid, further dissociating:

$$HCO_3^- \longleftrightarrow CO_3^{2-} + H^+ \qquad \text{(Eq. 6.8)}$$

with a pK value of about 10.2. At physiological pH, which seldom exceeds 8.0, less than 1% of the bicarbonate will form carbonate (CO_3^{2-}).

From the discussion of buffers given above, the question of how carbonic acid is a useful physiological buffer immediately arises: Since its pK′ value is far from the physiological range (3.8 versus 7.0 to 8.0) it should have only very weak buffer action in that range. The answer lies in the further equilibrium in water between gaseous dissolved CO_2 and carbonic acid:

$$CO_2(\text{gas}) + H_2O \longleftrightarrow H_2CO_3 \qquad \text{(Eq. 6.9)}$$

and the fact that there is a large effective reservoir of the molecular CO_2. In other words, as pH changes in the physiological range, H_2CO_3 would normally have little further capacity to donate protons, since less than 1% of the total present would remain undissociated. Consider the compound reaction and the equilibrium distribution of forms shown in Fig. 6.2. As H^+ is removed from the equilibrium system (e.g., by adding OH^-), the small amount of H_2CO_3 present would not have much capacity to generate more HCO_3^- and H^+. With a large reservoir of the gaseous CO_2 present, however, new H_2CO_3 is formed to maintain the equilibrium. The overall system, then, acts as an effective buffer even though the primary species alone, H_2CO_3, would not.

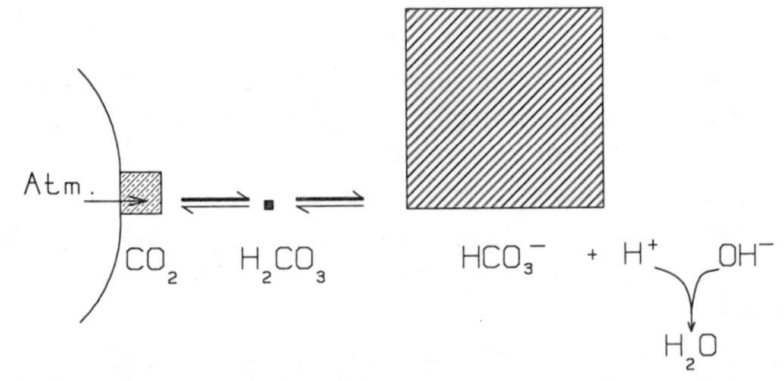

Figure 6.2: Carbonic acid equilibria. The size of each compartment indicating the approximate abundance of different forms at physiological pH. The atmosphere is an infinite reservoir; carbonic acid is present in such low concentrations that only a small dot would appear. As the system is titrated with OH^-, H_2CO_3 could not supply any significant amount of H^+ for buffering without the large reservoir of atmospheric and dissolved CO_2 gas with which it is in equilibrium. These secondary equilibria account for the carbonic acid system's being a better physiological buffer than would be predicted from its pK′.

Following the general analysis scheme for acid-base reactions, the various K′ and pK′ values can be calculated:

$$L = [CO_2]/[H_2CO_3] \quad \text{(Eq. 6.10)}$$

$$K_1 = [H^+][HCO_3^-]/[H_2CO_3] \quad \text{(Eq. 6.11)}$$

and

$$K_2 = [H^+][CO_3^{2-}]/[HCO_3^-] \quad \text{(Eq. 6.12)}$$

The first two steps may be combined into:

$$[H^+] = (K_1/L)([CO_2]/[HCO_3^-]) \quad \text{(Eq. 6.13)}$$

and the constant K_1/L redefined as K_1', or the first apparent dissociation constant for carbonic acid. Its value is around 6.4 at physiological temperatures, emphasizing the point that the system acts as a better physiological buffer than would be predicted from the characteristics of carbonic acid alone.

The rather complex equation set for the carbonic acid system contains several quantities that are not directly analyzable in the laboratory. The variables usually measured with greatest ease are the pH, the partial pressure of CO_2 (PCO_2), and the total CO_2 content. The latter is denoted in this book as C_T and is defined as the sum of all forms of dissolved plus chemically combined CO_2:

$$C_T = \alpha PCO_2 + [HCO_3^-] + [CO_3^{2-}] \quad \text{(Eq. 6.14)}$$

where α is the solubility coefficient and the concentration of H_2CO_3 is usually ignored. Beginning with Henderson (1909) and Hasselbalch (1917) various

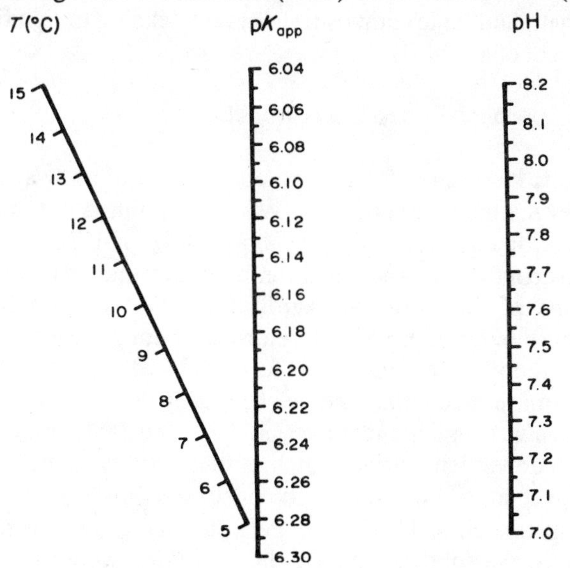

Figure 6.3: pK′ nomogram for rainbow trout true plasma between 5° and 15 °C. By drawing a straight line from the ambient temperature to the measured pH, the pK′ can be estimated from the intersection with the center line. (From Boutilier *et al.*, 1984, by permission, ©Academic Press.)

methods of simplifying the equation set have been employed, most of which generate an equation similar to what is usually called the Henderson-Hasselbalch equation:

$$pH = pK' + \log[(C_T/\alpha P_{CO_2}) - 1] \qquad \text{(Eq. 6.15)}$$

When the concentration of CO_3^{2-} is small enough to be ignored, then the term $(C_T - \alpha P_{CO_2})$ is approximately equal to $[HCO_3^-]$ so:

$$pH = pK' + \log([HCO_3^-]/\alpha P_{CO_2}) \qquad \text{(Eq. 6.16)}$$

In Eqs. 6.15 and 6.16 the new value of pK′ is no longer a constant but a complex function of pH, temperature, ionic strength, and specific ion concentrations. No simple mathematically explicit solution for the complete equation set can be generated, so these approximations and simplifications are a matter of convenience and necessity. Various authors have published tables, equations and nomograms such as the one shown in Fig. 6.3 for computing the correct value of pK′ as a function of other variables (Sigaard-Anderson, 1974; Boutilier et al., 1984; etc.).

The utility of the Henderson-Hasselbalch equation has been demonstrated for a long time, and whole systems of graphical and numerical acid-base analysis have been developed from it (Davenport, 1974). For human clinical physiology the risk of errors of approximation is reduced by the narrow range of temperature and electrolyte composition encountered. In some pathological states, however, and certainly in comparative animal studies the range of temperature, pH, ionic strength and ionic composition is great, leading to significant errors in some cases (Nicol et al., 1983).

An Alternative Approach Based Upon the SID

Recently Stewart has argued for a complete revision of the way in which acid-base status is evaluated (Stewart, 1978, 1981). He pointed out that the solution of complex sets of equations, virtually impossible in 1920, may now be accomplished in seconds on any personal computer and indeed on many hand-held calculators. One of the principal justifications for the simplification of the Henderson-Hasselbalch approach, therefore, is no longer valid.

The second important argument put forward by Stewart is that, of the many parameters of solution equations, some are clearly dependent variables, others independent. Only the independent variables, furthermore, may be considered causative. An independent variable such as the quantity of Na^+ ion present in solution may be changed only by addition or subtraction of ion from outside the solution. Variables such as $[HCO_3^-]$, on the other hand, are determined by other conditions within the solution and are therefore dependent variables. He objected to the conventional approach to acid-base analysis on the basis that most of the measurements are of dependent variables, and expressing the relation between them for any set of solution conditions cannot reveal anything about causation (Stewart, 1981).

Proceeding on these premises, Stewart constructs a set of eight equations:

1. Dissolution of CO_2

$$[CO_{2(dissolved)}] = \alpha \times PCO_2 \qquad \text{(Eq. 6.17)}$$

2. Equilibrium with carbonic acid

$$[H_2CO_3] = K_H \times PCO_2 \qquad \text{(Eq. 6.18)}$$

3. Water dissociation

$$[H^+] \times [OH^-] = K_w' \qquad \text{(Eq. 6.19)}$$

4. Weak acid dissociation

$$[H^+] \times [A^-] = K_A \times [HA] \qquad \text{(Eq. 6.20)}$$

5. Weak acid conservation

$$[HA] + [A^-] = [A_{Tot}] \qquad \text{(Eq. 6.21)}$$

6. HCO_3^- formation

$$[H^+] \times [HCO_3^-] = K_C \times PCO_2 \qquad \text{(Eq. 6.22)}$$

7. Carbonate formation

$$[H^+] \times [CO_3^{2-}] = K_3 \times [HCO_3^-] \qquad \text{(Eq. 6.23)}$$

8. Electrical neutrality

$$[SID] + [H^+] - [HCO_3^-] - [A^-] - [CO_3^{2-}] - [OH^-] = 0 \qquad \text{(Eq. 6.24)}$$

with the following definitions:

SID The difference between all strong (i.e., completely dissociated) electrolyte cations and anions.

α The Bunsen solubility coefficient for CO_2 gas dissolved in the solution.

PCO_2 The partial pressure of CO_2.

K_w' The equilibrium constant for water dissociation, $[H^+] \times [OH^-]$, with the concentration of water incorporated into K, hence the prime notation.

K_A The equilibrium constant for weak acid in the system.

A_{Tot} The total concentration of weak acid.

K_C An equilibrium constant for bicarbonate formation from dissolved CO_2; equal to $K_H \times$ the equilibrium constant for carbonic acid dissociation.

K_H An equilibrium constant for partial pressure of carbon dioxide and carbonic acid, incorporating the solubility coefficient and water concentration; equivalent to $\alpha \times L$ in Albers' (1970) notation.

HA, A^- Notation for unspecified weak acid in nondissociated and dissociated form.

K_3 The second dissociation constant for carbonic acid, i.e., for formation of carbonate ion from bicarbonate. Equal to Albers' K_2.

The procedure outlined by Stewart for solving this system is to substitute appropriately into the equation for electrical neutrality (Eq. 6.9), deriving an expression containing only one dependent variable, $[H^+]$, the three independent variables (SID, A_{Tot}, and PCO_2) and the various constants:

$$\begin{aligned}[H^+]^4 &+ \{K_A + [SID]\}[H^+]^3 \\ &+ \{K_A([SID]\ [A_{Tot}])\ (K_C \times PCO_2 + K_w')\}[H^+]^2 \\ &- \{K_A(K_C \times PCO_2 + K_w') + K_3 \times K_C \times PCO_2\}[H^+] \\ &- K_A \times K_3 \times K_C \times PCO_2 = 0 \qquad \text{(Eq. 6.25)}\end{aligned}$$

This equation can be solved by a short computer routine employing iterative approximation, and the resulting value for $[H^+]$ can then be employed in Eqs. 2 through 8 to solve for the other dependent variables.

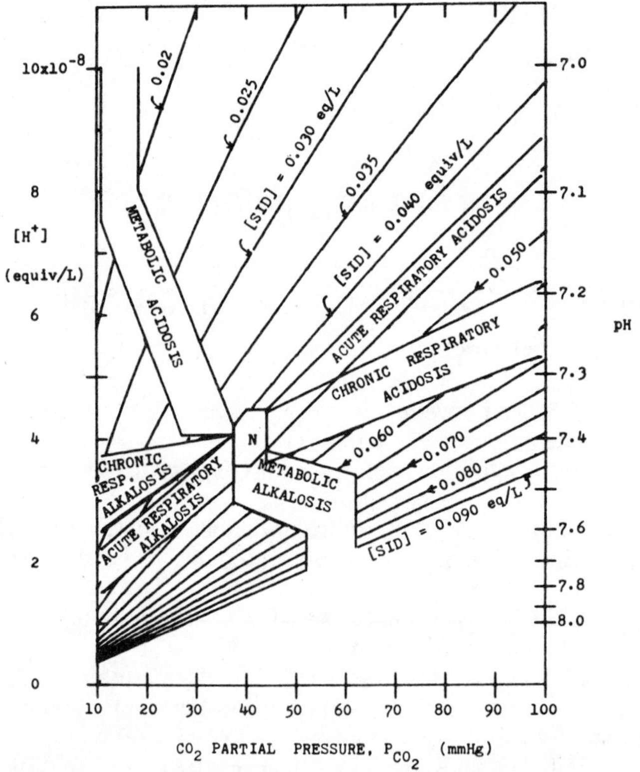

Figure 6.4: Acid–base summary diagram constructed from Stewart's equation set (see text). The normal range for humans is designated by N, and various kinds of disturbance are indicated by zones on the diagram. The isolines are for constant SID, and a constant value for A_{tot} is assumed. (From Stewart, 1981, by permission, ©Elsevier North Holland, Inc.)

Using this system, Stewart (1981) analyzed a number of common acid-base situations and generated graphs of the relations between variables. In his summary graph (1981)(Fig. 6.4) he presented [H^+] (dependent, y-axis) as a function of P_{CO_2} (independent, x-axis) at a range of [SID] values (independent, isolines) at an (assumed) constant value for [A_{Tot}]. The graph seems unfamiliar, since the conventional approach is to plot pH as the x-axis and [HCO_3^-] as the y-axis with P_{CO_2} isolines (see Fig. 6.13 and below). [SID] is not normally presented, and [A_{Tot}] enters into the analysis only indirectly as it affects pK′ and the buffer line. The conventionally derived pH–HCO_3^- diagram employs the same variables as those given by Stewart (1981) for generation of Fig. 6.4. The difference may be more apparent than real, however; if Fig. 1 is rotated 120° counterclockwise, each of the zones representing different sorts of acid-base disturbance are brought into roughly the same orientation as in the conventional diagrams. Had Stewart elected to plot [SID] on the y-axis, pH on the x-axis, and P_{CO_2} as isolines, the graphs would seem much more familiar.

As a consequence of reexamination of ideas following Stewart's publications, there is considerable confusion in acid-base physiology. In the conventional system, pH and either P_{CO_2} or total CO_2 measurements were sufficient to describe the system, employing the highly derived pK′ nomograms worked out for humans and other animals. In Stewart's system, SID, A_{Tot}, and P_{CO_2} must all be measured and 4 constants known (K_w, K_C, K_A, and K_3). Not only is one not certain what to measure, it is not abundantly clear how the measurements of SID and A_{Tot} are to be made. In most of his analyses Stewart assumes that A_{Tot} remains constant at 25 mEq L^{-1} and does not consider ion activity, only concentration. It seems important at this juncture to attempt to (1) clarify the theoretical relationship between these two different approaches, (2) generalize the equations to cover situations in which ionic strength and A_{Tot} vary significantly, and (3) evaluate the utility of each approach in some experimental situations chosen to emphasize extreme variations.

Other Physiological Buffers

Proteins and Amino Acids

Physiological buffers other than carbonic acid form an important part of the overall acid-base regulating system but are generally regulated by various physiological mechanisms for purposes other than acid-base maintenance. The most important extracellular buffer type is protein. The amino acids that make up proteins contain both weak acid and weak base moieties with pK′ values for the various groups given in Table 6.1. As discussed above, only those compounds with pK′ values near the physiological range may be expected to exert significant buffer action, so the physiologically important amino acid residues of proteins are almost exclusively histidine (Fig. 6.5) and cysteine with very minor contributions from N-terminal amino groups, tyrosine and glutamic acid. The importance

Table 6.1: pK′ values for ionizable groups of amino acids and related compounds.

Group	pK′
α–COOH	1.3–3.3
α–NH3	8.6–10.6
Aspartic acid, β–COOH	3.86
Glutamic acid, γ–COOH	4.25
Histidine, imidazole	6.95
Cystine, SH group	8.33
Tyrosine, phenol group	10.07
Lysine, ϵ–amino	10.53
Arginine, guanidinium	12.48

(Data for imidazole from Reeves, 1972; other data from Lehninger, 1975.)

of cysteine is reduced further, since sulfide bridges between adjacent cysteine residues render the ionizable group inactive in most proteins.

From Table 6.1 and the foregoing discussion of ionizable acid and base groups in proteins, one may suppose that the titration curves of proteins might have the following characteristics: a sigmoid portion (inflection) around pH 2 to 3, representing the α-carboxyl dissociation; one to perhaps several inflections around neutral pH, representing R-group (mostly histidyl) dissociation; and a final inflection around pH 9 or 10, representing the α-amino group dissociation (Fig. 6.1). The actual titration curves of complex proteins tend instead to be fairly smooth curves with no obvious inflection points over the mid-range of pH values (5 to 9). The curves are straight because the pK′ for any ionizable R-group in the protein is affected by neighboring residues and by the spatial geometry of the protein. Thus histidyl residues in myoglobin and hemoglobin have pK′ values ranging from nearly 6 to over 10, depending on their position in the protein (Botelho et al., 1978; Matthew et al., 1979). A consequence of the spreading of pK′ values is that the net ionization, or net charge, of a particular protein is highly dependent upon the pH of the solution.

Phosphate Compounds

Due to the multistep dissociation of phosphoric acid, inorganic phosphate salts will exert significant buffer action in the physiological pH range (Fig. 6.1). The

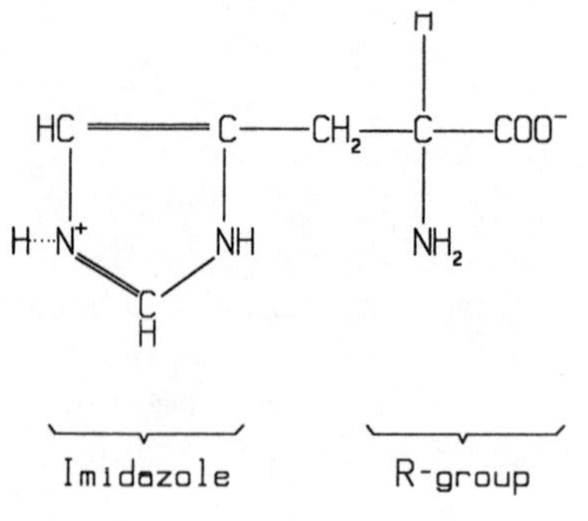

Figure 6.5: The structure of histidine, showing the imidazole moiety and the R-group. Substitution at different positions in either histidine or histidyl residues of proteins can have a considerable effect on the ionizing potential of the imidazole moiety.

total phosphate concentration in the blood of most animals is low, generally around 1 to 4 mEq L^{-1}, but intracellular concentrations may be higher and stores of inorganic phosphates may be replenished from other sources including organic compounds and bone. In muscle, for example, the hydrolysis of inorganic phosphate from creatine phosphate can be extremely rapid, so even with freeze clamping it is difficult to estimate the in vivo concentrations of each (Ellington, 1983). Excess dietary phosphorus is removed principally by the kidney, which can cause some problems when large amounts of divalent cations (Ca^{2+} and Mg^{2+}) are also being excreted. $CaHPO_4 \cdot 2H_2O$ and $MgHPO_4 \cdot 3H_2O$ have low solubility products and tend to precipitate in the urinary system. The low pH of urine undoubtedly helps to prevent precipitation, but at least in marine fish such precipitates seem fairly common (Hickman & Trump, 1969).

Ammonia

In the physiological pH range ammonia is about 99% combined with water and ionized to NH_4^+. The pK′ for the reaction lies between 9 and 10 and the actual concentration in small, so the buffer contribution of ammonia is normally unimportant. In animals on high protein diets, however, both excess ammonia and H^+ must be excreted and HCO_3^- retained. Ammonia diffusing as a gas across the kidney tubule combines with secreted H^+, forming NH_4^+ and thereby reducing the H^+ gradient against which the kidney must work (Pitts, 1973). Other than unusual environmental situations that present high ammonia concentrations to aquatic animals, the situation in the kidney is probably the only example of ammonia acting as an important buffer.

PATTERNS OF pH REGULATION

Effects of Temperature Change

Physicochemical Parameters

As temperature increases, we have already discussed how the solubility of gases declines in aqueous solutions (see Chapter 1; Program TEMPTABL). The solubility of CO_2 in water is about one-half as much at 37°C as it is at 12°C.

Temperature also changes the equilibrium constants for the reactions important to acid-base regulation. One common way to express the change in pK′ with temperature is by giving the value for ΔH°. This quantity, the heat of enthalpy, is a measure of the change in heat energy of an equilibrium system when temperature is changed. It can be calculated from measurements of pK′ at any two temperatures using the van t'Hoff equation:

$$\Delta pK' = \Delta H°(T_2 - T_1)/(2.303RT_2T_1) \qquad \text{(Eq. 6.26)}$$

where R is the gas constant, 1.98719 cal $°K^{-1}$ $mole^{-1}$, and ΔpK′ is simply the difference between the pK′ values at temperatures T_1 and T_2 (absolute or degrees

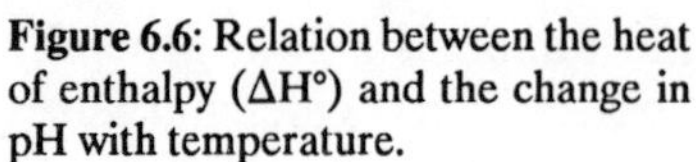

Figure 6.6: Relation between the heat of enthalpy ($\Delta H°$) and the change in pH with temperature.

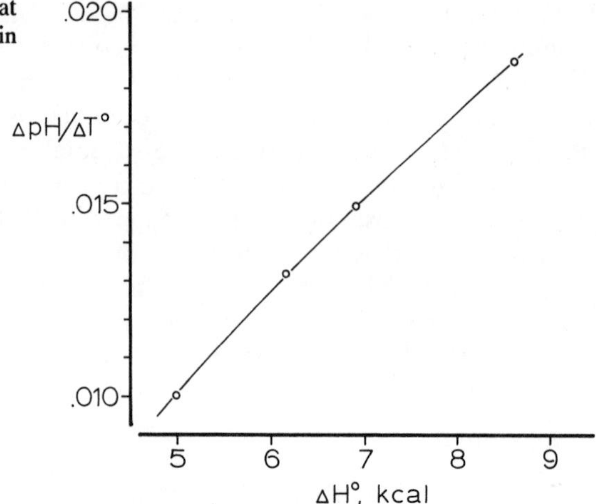

Kelvin). Once the value of $\Delta H°$ is known, it may then be used to predict the (purely passive) change in pK′ over any temperature range. Water itself has a $\Delta H°$ value of about 7300. For the carbonic acid equilibrium of Eqs. 6.13 and 6.16, the value for $\Delta H°$ is 2240 (Edsall & Wyman, 1958; Edsall, 1969) and the value for $\Delta pK/\Delta T°$ is about –0.006. That is, the pH-temperature slope for the carbonic acid system is relatively flat. For the phosphate buffer system it is even flatter, with $\Delta H°$ of about 1000 and a $\Delta pH/\Delta T°$ of –0.003. Histidine and various related compounds have $\Delta H°$ values in the 6000 to 9000 range, however, giving them much steeper pH-temperature slopes with values of $\Delta pH/\Delta T°$ in the range of –0.016 to –0.024.

Experimental evidence shows that the pK′ values for histidine residues of proteins are influenced by their local charge environment (see above)(Botelho et al., 1978) and apparently the $\Delta H°$ values will also vary (Roberts et al., 1969). Various imidazole compounds which differ only in the R-group have quite different $\Delta H°$ values: 4-methyl imidazole has a $\Delta H°$ value of 8600 cal mole^{-1} compared to the 6900 for free histidine and 7700 for free imidazole (Fig. 6.6). Neighboring amino acid residues in proteins apparently have similar effects on imidazole in proteins. Greater knowledge of how $\Delta H°$ values vary within proteins would be an important contribution to understanding acid-base regulation, as we shall see below.

Acid-Base Status

For nearly all cold-blooded animals studied a common pattern of acid-base response to temperature has emerged. The inverse relation between blood pH and temperature in a variety of animals is shown in Fig. 6.7. This sort of temperature response was first observed in work on the alligator by Austin et al. (1927) and elaborated for the turtle by Robin (1962). A large number of poikilothermic animals, vertebrate and invertebrates, have now been found to have pH-tempera-

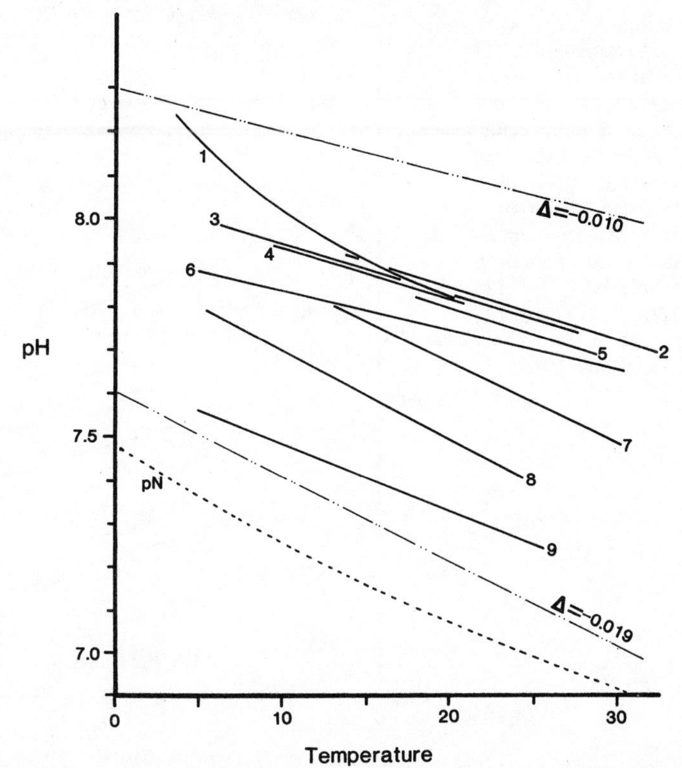

Figure 6.7: Acid-base status of the blood versus temperature for a wide variety of poikilotherms, including sipunculids (8: *Phascolopsis gouldi*; Mangum & Shick, 1972), annelids (2: lugworm, *Arenicola marina*; Toulmond, 1977), marine crabs (3: blue crab, *Callinectes sapidus*, Cameron & Batterton, 1978a), terrestrial crabs (7: Pacific mangrove crab, *Cardisoma carnifex*, McMahon & Burggren, 1981), elasmobranchs (4: spotted dogfish, *Scyliorhinus stellaris*, Heisler et al., 1976), freshwater teleosts (1: rainbow trout, *Salmo gairdneri*, Randall & Cameron, 1973; and 2: channel catfish, *Ictalurus punctatus*, Cameron & Kormanik, 1982a), marine teleosts (5: silver seatrout, *Cynoscion arenarius*, Cameron, 1978c), and a reptile (6: red-eared slider turtle, *Chrysemys picta*, Glass et al., 1985). The reference lines are also shown for pH–temperature slopes of –0.010 and –0.019 and for the "pH of neutrality," or $^1/_2pK_w$.

ture slopes varying from –0.008 to –0.021 (reviews by Cameron, 1984, 1986; Heisler, 1986a; Truchot, 1987) with the average at around –0.016 to –0.019 pH/°C.

Since these relations closely parallel the change in the pH of neutrality[1] with temperature (Fig. 6.7), there has been some discussion of whether the important physiological variable is pH or $[H^+]$ per se or perhaps the $[H^+]/[OH^-]$ ratio (Austin et al., 1927; Rahn, 1966). The probable answer is neither; rather, the relation between pH and temperature has to do with the maintenance of protein function. This idea was discussed by various authors between 1925 and 1935 (Stadie et al., 1925; Austin et al., 1927) and has been brought up more recently by Reeves (1972, 1976). The terminology differs: during the 1920s they wrote in terms of constancy of "base bound to protein," whereas Reeves discussed "alpha im-

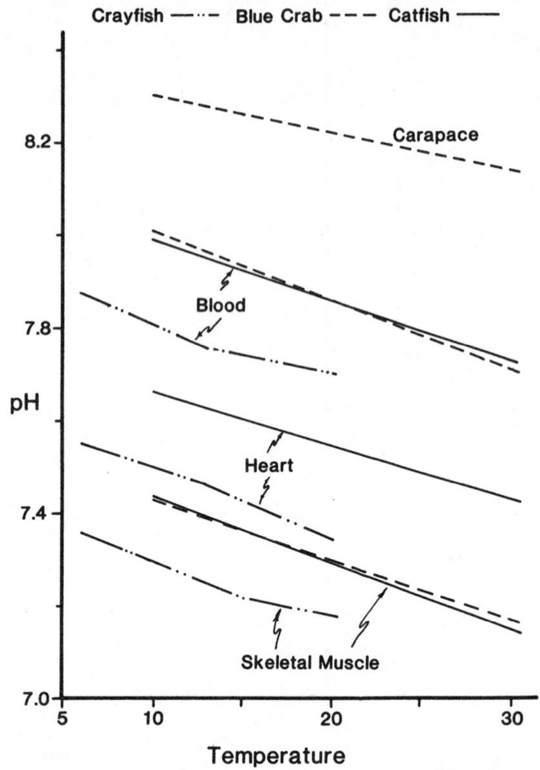

Figure 6.8: Relation between intracellular and extracellular pH and temperature for three animals. The tissues are all about 0.3 to 0.5 pH units more acidic than blood, but the carapace fluid compartment in the crab is more alkaline. Consistent differences among tissue types occur in all three animals. Crayfish data from Gaillard & Malan, 1985; blue crab data from Wood & Cameron, 1985; and catfish data from Cameron & Kormanik, 1982a.

idazole" (with alpha defined as a fractional index of dissociation). When the underlying models and equations are examined, the ideas are similar; i.e., that as temperature changes the blood pH must be managed in such a way as to maintain the net charge configuration of proteins (Wilson, 1977; Yancey & Somero, 1978; Somero, 1981). A change in pH equal to the change in pK means that the net charge of a protein will not change.

Intracellular versus Extracellular pH

For most tissues examined the intracellular pH (pH_i) is about 0.5 units below that of the blood or extracellular fluid. Since the negative inside potential of cells and the selective permeability properties of the cell membrane lead to predictions of higher inside H^+ activity, there has been a half-century-long controversy as to whether the observed pH difference was simply a result of passive equilibration or active H^+ transport. Experimental evidence on a variety of tissue and cell preparations has now firmly established the nonequilibrium nature of pH_i and has also elucidated Na^+/H^+ exchanges and other ion transport mechanisms responsible for its maintenance (Thomas, 1977, 1988; Roos & Boron, 1981; Schlue & Thomas, 1985). Since CO_2 is freely diffusible across cell membranes, the lower intracellular pH must be accompanied by lower intracellular [HCO_3^-], implying a relatively low permeability of the cell membrane to

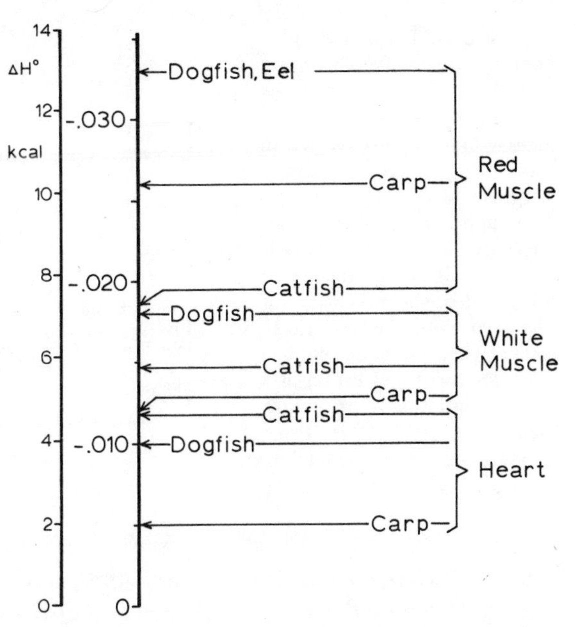

Figure 6.9: Differences in pH_i from different fish tissues. Catfish data from Cameron & Kormanik, 1982a; dogfish data from Heisler et al., 1976; carp data from Heisler, 1980; and eel data from Walsh & Moon, 1982.

HCO_3^-. The intracellular contents are also much better buffered by higher protein and phosphate concentrations.

Despite these differences, the pattern of intracellular pH change with temperature in poikilotherms is very similar to the extracellular pattern (Fig. 6.8; Malan et al., 1976; Gaillard & Malan, 1985; Heisler, 1986a; Cameron, 1986). If a weak acid indicator such as 5,5-dimethyl-2,4-oxazolidinedione (DMO)(Waddell & Bates, 1969) is used to estimate the average whole body pH_i, the temperature slope is similar to that of blood. Interesting differences, however, have been observed from different tissues: The temperature slope for red muscle is consistently the largest, for white muscle intermediate, and for heart muscle the least (Fig. 6.9). Whether this difference represents a difference in predominant proteins in these tissues or has some other basis is purely a matter for speculation at present. We do know, however, that the maintenance of an appropriate intracellular pH-temperature relation is critical to the functioning of intracellular enzymes (Somero, 1986)(Fig. 6.10). Earlier comparative biochemical studies in which a constant pH was used over a range of temperatures sometimes yielded inappropriate conclusions about enzyme regulation (Wilson, 1977; Somero, 1981).

Mechanisms of pH Adjustment with Temperature

Ventilation

The changes in blood pH in vivo are similar to changes observed by Rosenthal (1948) when he warmed or cooled samples of human blood in a closed system. As temperature was decreased in this constant content system the pH rose

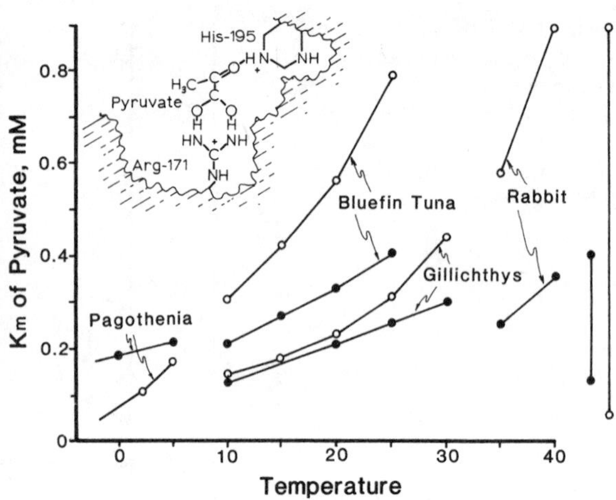

Figure 6.10: Effects of pH–temperature regulation on enzyme function. The open circles show how the affinity of skeletal muscle LDH would change with a constant pH, and open circles show the affinity at physiologically appropriate pH. The affinity of the enzyme system is better regulated with the in vivo pH regimen. The vertical lines at the right show the total range with constant pH (open circles) compared with pH varied in a physiologically appropriate manner (closed circles). (From Somero, 1981, by permission, ©Elsevier North Holland, Inc.)

and the P_{CO_2} fell. Animals are open systems, in the sense that they are free to exchange CO_2 with the atmospheric pool. In air-breathing vertebrates such as the frog, decreases in temperature cause a relative increase in ventilation such that the P_{CO_2} drops; these animals regulate their open systems in a way that mimics Rosenthal's in vitro closed system (Jackson, 1971; Jackson et al., 1974). (Ventilation may actually decrease as temperature falls, but not as sharply as the rate of metabolic CO_2 production.) In water breathers, however, the situation is quite different. Due to limitations of the oxygen supply in aquatic environments, ventilation cannot be reduced significantly without the animal becoming hypoxic. Ventilation is therefore not a mechanism available for primary adjustments of acid-base status. What happens instead is that as temperature decreases the $[HCO_3^-]$ concentration (and total CO_2) increases, producing the same inverse pH-temperature relation with smaller changes in P_{CO_2} (Fig. 6.11; Randall & Cameron, 1973; Cameron & Batterton, 1978a; Cameron, 1978c). The changes in P_{CO_2} that do occur are due more to the decrease in solubility and change in pK′ than to active adjustment of ventilation.

Since in the air-breather blood P_{CO_2} rises at higher temperature through the regulation of ventilation, little or no further adjustment in the intracellular compartment is required to maintain the proper pH-temperature relationship. In other words, the intracellular compartment in air-breathers can be considered a closed system, much like the in vitro systems studied by Rosenthal (1948). These systems require no mass transport of ionic species to regulate acid-base status. In certain situations this probably represents a considerable energy saving. Blood passing from the core to the extremities in an animal such as a moose undergoes considerable cooling, then rewarms as it returns to the core. The studies of Rosenthal (1948) and Reeves (1972, 1976) showed that an appropriate pH_i will

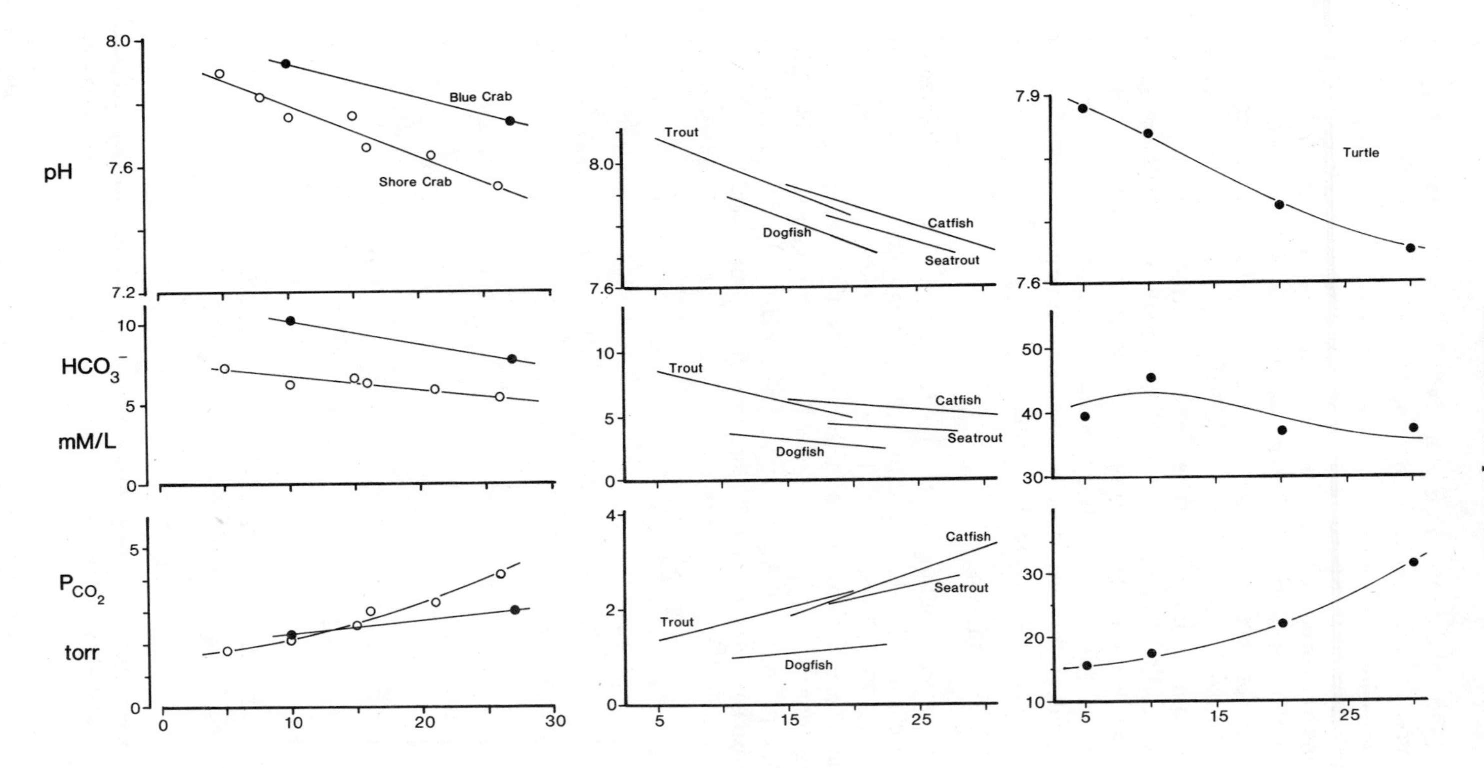

Figure 6.11: Acid-base parameters as a function of temperature in invertebrates, fish and a reptile. Blue crab (*Callinectes sapidus*) data from Cameron and Batterton, 1978a; shore crab (*Carcinus maenas*) data from Truchot, 1973b; rainbow trout (*Salmo gairdneri*) data from Randall & Cameron, 1973; dogfish (*Scyliorhinus stellaris*) data from Heisler et al., 1976; channel catfish (*Ictalurus punctatus*) data from Cameron & Kormanik, 1982a; seatrout (*Cynoscion nebulosus*) data from Cameron, 1978c; and turtle (*Chrysemys picta bellii*) data from Glass et al., 1985.

be maintained throughout this temperature cycle solely as the result of passive reequilibration of the chemical buffer system.

In the aquatic animals, on the other hand, the pattern of P_{CO_2} change shows that closed system behavior is not being mimicked through the regulation of ventilation, leading us to expect considerable ionic exchange as temperature varies. Measurements of net acid-base transfers between animal and environment have now been made for several fish, bearing out the prediction (Heisler, 1978; Cameron & Kormanik, 1982a). There need to be transfers not only between animal and environment but also between intra- and extracellular compartments in order to maintain the proper acid-base regimes in both. A further consequence of the difference between P_{CO_2} regulation through ventilation and HCO_3^- regulation through ion transfers is that temperature adjustments take longer in the aquatic animals and are accompanied by sometimes large transient acid-base disturbances (Cameron, 1976; Heisler, 1986a).

Ion Transport Mechanisms

Kidney: In the air-breathing vertebrates the importance of the kidney in responding to chronic acid-base disturbance has long been appreciated (see reviews by Hills, 1973; Pitts, 1973). Through the differential regulation of Na^+/H^+ and Cl^-/HCO_3^- exchanges in the tubules, the vertebrate kidney is capable of excreting either acid or base loads. Renal acidification is possible down to a final urine pH of about 4.5 during acidosis, and substantial quantities of HCO_3^- may also be excreted in an alkaline urine during alkalosis.

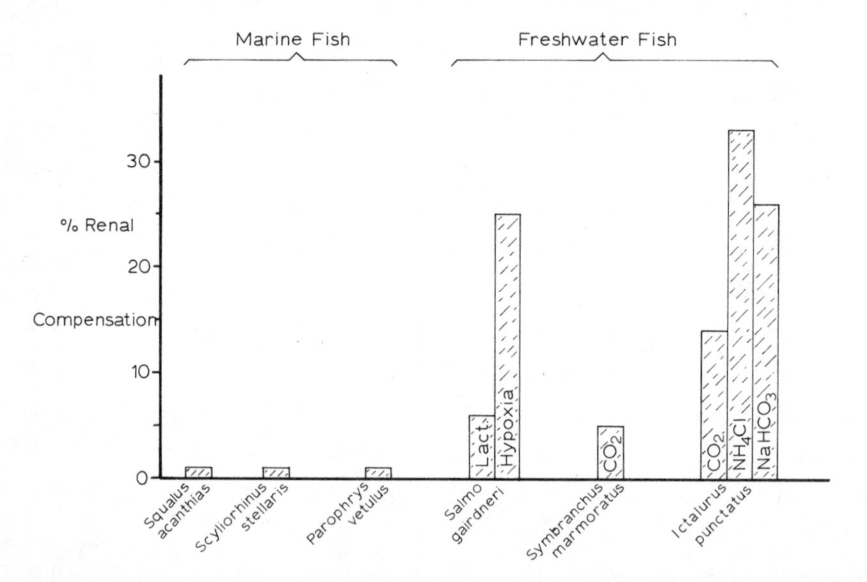

Figure 6.12: Proportion of compensation attributable to the kidney in acid-base compensation in several marine and freshwater fish. The type of acid-base disturbance is indicated on each bar. (From Cameron, 1984).

In aquatic animals, however, the kidney and its various invertebrate analogues appear to have only a small role in acid-base regulation (Fig. 6.12). In freshwater fish the rate of urine flow is fairly large in order to maintain water balance, and up to about 15% of acid or base loads can be excreted via this route (Cameron, 1980; Cameron & Kormanik, 1982b). In the marine teleosts and elasmobranchs, however, total urine flow is small, the kidneys may be reduced to varying degrees (Smith, 1961; Hickman and Trump, 1969), and the kidney is generally insignificant in acid-base regulation (Hodler et al., 1955; Cross et al., 1969; Wood & Caldwell, 1978; McDonald et al., 1982; Heisler, 1986a). The antennal glands of crustaceans serve some of the same functions as the vertebrate kidney: Regulation of water balance and of divalent cations is accomplished mainly through filtration and selective resorption. Their role in compensation of acid-base disturbance is also negligible, however (Cameron & Batterton, 1978b; Truchot, 1979). There is a report of the coxal glands of the horseshoe crab (an arachnid) responding to acid-base disturbance (Towle et al., 1982), but an actual net H^+ excretion rate could not be calculated.

Gills: In August Krogh's (1937,1938) studies of ionic balance in freshwater animals, he proposed that Na^+ and Cl^- could be taken up independently from the environment by linked electroneutral exchanges in the gills or skin. Without hard evidence, he proposed that Cl^- was transported inward in exchange for HCO_3^- and Na^+ for either H^+ or NH_4^+. Since then a great deal has been learned about these active ion exchanges (Maetz & Garcia-Romeu, 1964; Evans, 1975, 1984), but the acid-base importance of them was not recognized until the work of DeRenzis and Maetz (1973), who showed that ionic disturbance led to acid-base disturbance, and Cameron (1976) who showed that acid-base disturbance led to changes in the ion flux rates. Since skin is a relatively small portion of the external surface area and has quite low ion permeability, the bulk of this active regulation occurs across the gill surface, apparently through most of the respiratory epithelial cells (Girard & Payan, 1977). Though the requirements of ionic balance and acid-base regulation might at first seem to be contradictory, Stewart's analysis (see above) makes it clear that the two are in fact one and the same problem, i.e., of maintaining the proper balance of all electrolytes in solution, including H^+.

Although the gills also show a number of other ionic exchange mechanisms, a general model for the respiratory and acid-base interactions can be drawn (Fig. 6.13). The Na^+/H^+ exchange appears to be an apical process, sensitive to amiloride, a specific Na^+ site blocker (Cuthbert & Maetz, 1972; Kirschner et al., 1973). The transport may actually involve two steps, the first of which is the Na^+/H^+ translocation across the apical surface and the second of which may be a Na^+/K^+ exchange at the basolateral surface. Presumably the K^+ subsequently diffuses back out the basolateral side. The Cl^-/HCO_3^- exchange may also be apical, although there is less evidence for it. It is inhibited by thiocyanate (SCN^-) and by various stilbene derivatives such as SITS and DIDS (Epstein et al., 1973). Since the epithelial cell is substantially negative inside, the Cl^- may simply dif-

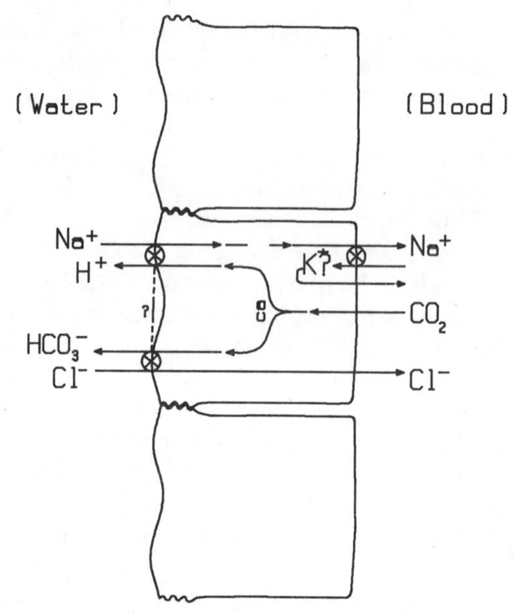

Figure 6.13: Model of ion exchange and transport processes in the fish gill. Modulation of the relative rates of these processes is thought to effect acid-base compensation (see text).

fuse down its electrochemical gradient to the serosal side, balancing the passive rediffusion of K^+. Modulation of the one exchange relative to the other provides a net acid-base flux to the inside, thereby acting to regulate the internal environment. We still lack the kind of information we would like to have in order to understand the quantitative kinetics of these transport processes, but the experimental evidence that they are the primary acid-base regulating mechanisms in aquatic animals in general is overwhelming.

An interesting related question is the importance and function of carbonic anhydrase (CA) to this overall process. Although aquatic vertebrates lack CA in their plasma and invertebrates have none at all in blood, there is a large amount of CA present in the gills of all aquatic animals that have been studied (Maren, 1967). Since HCO_3^- permeability is generally low, there is also a question as to how the HCO_3^- that is exchanged for Cl^- reaches the apical cell surface. One possibility is that dissolved CO_2 provides the source by diffusing into the cell, hydrating therein to HCO_3^- and H^+, and then supplying the counterions for both the Na^+ and Cl^- exchanges (Henry & Cameron, 1983). This process seems to provide a satisfactory role for the intra-epithelial CA, since it both speeds up the hydration reaction by several thousand-fold and facilitates the diffusion of CO_2. Without it, local gradients might develop in the region of the cell membranes, which would inhibit the transport processes. There are persistent suggestions, however, that carbonic anhydrase might be positioned in the outer cell membranes in such a way as to enable direct dehydration of external HCO_3^-. The most recent immunocytochemical evidence, however, still does not show definitively if HCO_3^- must first cross the membrane before reacting with CA (Rahim et al., 1988).

We have tended to think of these two exchanges as independent since the work of Krogh (1937, 1938), but more recent work indicates that in at least some tissues there is a doubly linked $Na^+/H^+/Cl^-/HCO_3^-$ co-transport (Schlue & Thomas, 1985). Whether this coupled exchange also occurs in gills, or there is only an indirect coupling of the two exchanges via internal pH, as Shaw (1964) suggested, remains to be discovered.

Other Organ Systems. Not all aquatic animals have either impermeable skins or large gills, and in some of these cases the skin may be important in carrying out acid-base relevant exchanges. The frog skin, for example, has been the subject of epithelial transport studies for decades (Biber & Mullen, 1976; Ehrenfeld & Garcia-Romeu, 1977). Various fishes, particularly complete or partial air-breathers, have much reduced gills and generally smooth, scaleless skins through which a considerable portion of their respiration may take place (Johansen, 1972; Romer, 1972). Some salamanders have neither gills nor lungs and rely completely on the skin (Moalli et al., 1981). In these cases it is reasonable to suppose that the primary organ of acid-base regulation is also the skin.

Finally, the gut may be important in some animals, though its specific function in acid-base regulation has received almost no attention. Marine fish drink habitually to balance their osmotic water loss (Smith, 1961), and marine crustaceans may drink considerable quantities of water just after moulting (Towle & Mangum, 1985; personal observation).

Intracellular to Extracellular Ion Transfers

For a given change in the acid-base status of the blood and extracellular fluids with temperature, whether transport of ions between intra- and extracellular compartments is required is a matter of the composition of the intracellular buffer system and the nature of the extracellular change (see above). For example, if the extracellular acid-base status changes between 20° and 10° as shown in Figs. 6.7, 6.8 and 6.11, there is a combination of bicarbonate, phosphate, and protein (imidazole) buffers that will produce the parallel change intracellularly with no net ionic transfer (Reeves, 1972, 1976). As we have seen, however, different tissues appear to have characteristic pH-temperature slopes that do not match the predominantly imidazole buffer pool's chemistry (Fig. 6.9). For a number of aquatic animals, measurements have now been made of intercompartmental transfers in response to temperature change. For the dogfish, most of the animal-to-environment transfer originates in the intracellular compartment, and the same appears to be true of the freshwater catfish (Cameron & Kormanik, 1982a). The mechanisms for these transfers appear to be similar to the ionic exchanges occurring across the gills, i.e., active H^+ exchange for Na^+ and HCO_3^- exchange for Cl^- (Thomas, 1977; Roos & Boron, 1981). During temperature changes the actual amount of change in ion concentrations in either compartment is small. Two consequences are that (1) it is difficult to make direct measurements against the rather large ionic background, and (2) the effect of ionic changes intracellularly is expected to be small. For the case of larger acid-base disturbances of different origin, the changes may be larger (see below).

The pH–HCO_3^- Diagram as an Analytical Tool

With the conventional approach to acid-base analysis using the Henderson-Hasselbalch equation (Eq. 6.15) the three principal variables of interest are pH, C_T, and P_{CO_2}. In order to represent all three on a two-dimensional graph, the usual approach is to plot pH as the x-axis, C_T (or $[HCO_3^-]$) as the y-axis, and to represent various partial pressures of CO_2 as isolines or isopleths. If P_{CO_2} is held constant at any particular value, an exponential curve will describe the possible combinations of pH and C_T (from Eq. 6.16). Setting P_{CO_2} equal to a second value generates another curve (isoline) and so on. A typical pH-HCO_3^- diagram, also called a "Davenport" diagram (stemming from Davenport's popular text on acid-

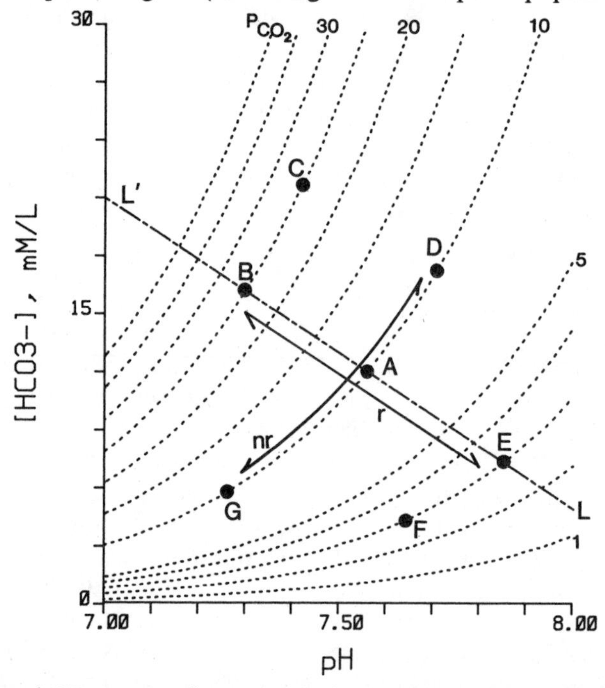

Figure 6.14: Sample "Davenport" diagram. pH is given on the abscissa, and the $[HCO_3^-]$ on the ordinate, and P_{CO_2} isolines are plotted as dashed lines. The isolines are derived from the values for pK′ and solubility appropriate to the temperature and ionic strength. Point A represents a normal resting individual, with a true plasma buffer line shown as L–L′. Purely "respiratory" disturbances cause shifts in status in the direction indicated by the line marked r and non-respiratory (i.e., metabolic) disturbances by the line marked nr. B indicates an acute respiratory acidosis, and C a partially compensated one. D and G represent pure nonrespiratory alkalosis and acidosis, respectively. E represents a respiratory alkalosis, and F a partially compensated respiratory alkalosis or mixed respiratory alkalosis and nonrespiratory acidosis. Similar diagrams can be generated by the programs DAVGM and DAVPLOT, on the accompanying disk.

base physiology), is shown in Fig. 6.14. The two constants required for the generation of the plot, α and pK′, are obtained from tables at appropriate values of temperature, ionic strength, and sometimes other factors as well (Sigaard-Anderson, 1974; Boutilier et al., 1984; Heisler, 1984). A "nonbicarbonate buffer line" is determined by titrating a sample of blood in vitro with CO_2 and determining a series of pH/[HCO_3^-] points. The slope of the line obtained is called the buffer slope and is denoted by β. Since these diagrams are tedious to construct by hand calculation, two computer programs are included on the accompanying diskette. The first, DAVGM, is provided as BASIC source code and in compiled version. It will run on any IBM-compatible computer and generates a screen plot that can be printed on a dot matrix printer. The second, DAVPLOT, is a PASCAL program that generates a plot on either a Houston Instruments DMP-40 or Hewlett-Packard 7475 plotter. (It should work with little or no modification on any Houston Instruments plotter.)

On the example of Fig. 6.14, typical values for a teleost fish at 25°C are shown. The point A marks normal values, and the nonbicarbonate buffer line (L–L′) is drawn through the control data. By referring to the control point, the buffer line, and the P_{CO_2} isopleths, a change in any direction from the control point can be interpreted as resulting from respiratory and nonrespiratory (i.e., "metabolic") causes.[2] A change in status from A to B, for example, would represent a pure respiratory acidosis caused by increasing blood P_{CO_2}. The change is respiratory since it involves change only in P_{CO_2}, i.e., the variables follow the same relation as they did when blood was titrated in vitro with CO_2. It is also an acidosis because it moves the pH to the left along the x-axis, toward more acidic pH. A subsequent shift of the blood variables to point C would represent a non-respiratory, or "metabolic" alkalosis with respect to B and can be seen as a compensatory process since the pH tends to return toward the control point. Change from A to G represent purely nonrespiratory acidosis, and from A to D a non-respiratory alkalosis. Point E represents a respiratory alkalosis.

Acid-base disturbances are seldom purely respiratory or nonrespiratory, and in these cases the pH-HCO_3^- diagram can be used for at least an approximate quantitative analysis. In the case of a mixed acidosis, e.g., C relative to A, if lines are drawn horizontally from A, B, and C to the Y-axis, the amount that [HCO_3^-] increases due to the change in P_{CO_2} can be compared to the amount due to the nonrespiratory compensation as a measure of the quantitative contribution of the two processes. Partitioning of the change A–F into a respiratory alkalosis and a nonrespiratory (compensatory) acidosis could also be performed in the same way. The principal pitfall with this sort of analysis is that it assumes that the buffer slope, β, remains constant under all conditions. Frequently fluid shifts attend acid-base disturbance, altering, for example, the protein concentration of blood, which would in turn change the value of β. In many cases, however, this error is not large, and the diagrammatic approach suffices.

Unfortunately, temperature comparisons are not possible using the pH-HCO_3^- diagram approach, since the parameters α and pK′ are both highly temperature-dependent, shifting locations of all the isolines considerably. Perhaps a four-dimensional plot in three dimensions could be worked out with some of the newer desktop computers, but for the moment different approaches must be taken.

Responses to Hypercapnia

Hypercapnia (sometimes "hypercarbia") simply means an elevation of P_{CO_2} above values normal for the animal. It may be caused by increased retention of metabolically produced CO_2 or by an increase in the ambient partial pressure, but in either case it has a rapid and ubiquitous effect on the body fluids. The immediate result of hypercapnia is a decrease in blood and intracellular pH, which in air-breathing animals such as man is quickly compensated by an increase in the ventilation minute volume, thereby tending to restore the normal P_{CO_2}. A primary driving force in the ventilatory control system of air-breathing vertebrates is in fact the arterial P_{CO_2} (see Chapter 8). In the air-breather increasing ventilation normally has little effect on oxygen transport, since the blood pigment is already fully saturated in the lungs and remains so with greater ventilation.

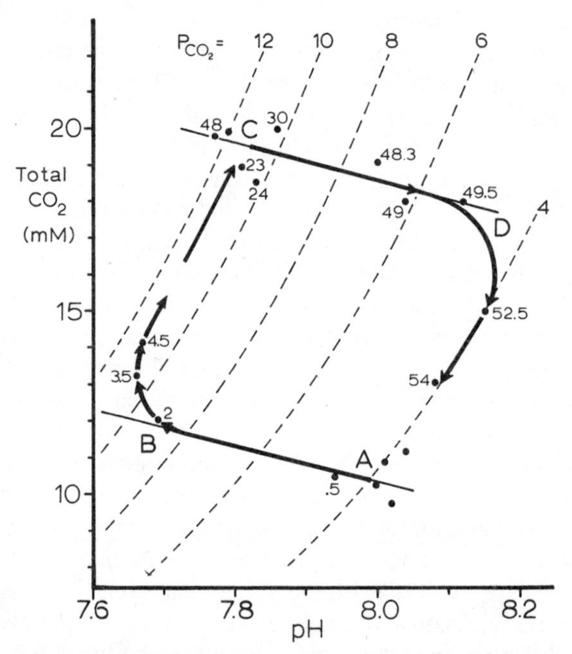

Figure 6.15: Responses of the blue crab to hypercapnia, plotted on a Davenport diagram such as the one in Fig. 6.14. A shows the control resting values, B the status after 2 hours of hypercapnia. The partially compensated animals stabilize at C, shift quickly to D when the hypercapnia is stopped, and eventually return to A. Numbers next to various points give the time in hours after onset of hypercapnia or recovery treatments. (Redrawn from Cameron, 1978b.)

If hypercapnia is sustained in man or other air-breathing vertebrates, two further processes come into play to bring about a long-term (chronic) return to normal. Fluid and ion shifts between intra- and extracellular compartments tend to buffer the pH change through the greater buffer capacity of the intracellular compartment. Inverse changes in H^+ and Na^+ can be measured during the first hours of hypercapnia, e.g., indicating a process similar to temperature adjustment in poikilotherms. These shifts come into play within the first hours, but by 3 to 4 hours the action of the kidney becomes increasingly significant. By 1 to 3 days in man the kidney has excreted an acid load equivalent to the load imposed by the CO_2 titration of hypercapnia (Hills, 1973), and there appears to be an increase in the number of membrane transport sites in the kidney tubules (Talor et al., 1987). Maintenance of the increased arterial P_{CO_2} over a long period is possible at normal pH by the achievement of a new steady state [HCO_3^-]. The overall response, then, is a complex sequence of immediate ventilatory response, intermediate-term amelioration by fluid shifts, and a chronic response of the kidney to restore normal acid-base status.

In aquatic animals the sequence of responses and the time course of acid-base changes are similar, although the organ system responsible for the chronic adjustment is the gills rather than the kidney, and ventilation may not play such an important role. In the blue crab, for example, an initial respiratory acidosis is followed within hours by progressive non-respiratory compensation (Fig. 6.15). The compensation occurs via Na^+/H^+ and Cl^-/HCO_3^- exchanges in the gills (Cameron, 1976, 1978b) but reaches a steady state at about 60% pH compensation (Siesjö, 1971) at 24 hours. After removal of the external hypercapnia, the P_{CO_2} is restored to normal very quickly, leaving a nonrespiratory alkalosis as a result of the elevated blood [HCO_3^-]. Reversal of the original ionic compensation in the gills restores the normal acid-base status within a further 24 hrs (Fig. 6.15). Similar data have been published for a variety of fish (Claiborne & Heisler, 1984; review by Heisler, 1986a) and for several crustaceans (Truchot, 1983).

A common feature of hypercapnic compensation in aquatic animals is that it is incomplete, 60 to 80% pH compensation appearing to be typical (Cameron, 1978a; Truchot, 1979; Heisler, 1986a; Cameron & Iwama, 1987). The idea of an upper bicarbonate threshold around 30 mM L^{-1} has been proposed as the major limitation of hypercapnic response in poikilotherms (Claiborne & Heisler, 1984; Heisler, 1986a; Boutilier & Heisler, 1988). It seems unlikely, however, because if the P_{CO_2} is increased further additional compensation occurs to values for [HCO_3^-] that would have provided 100% compensation at the lower P_{CO_2} (Cameron & Iwama, 1987)(Fig. 6.16). The answer to this puzzle may revolve around the ionic imbalance created during hypercapnic compensation; i.e., the Na–Cl difference is increasing in direct proportion to the increasing [HCO_3^-], mostly by decreases in [Cl^-] (Claiborne & Heisler, 1984; Cameron & Iwama, 1987). It may be that a kind of dynamic steady state is reached at each level of hypercapnia that represents a compromise between the need for ionic balance and the need for pH homeostasis.

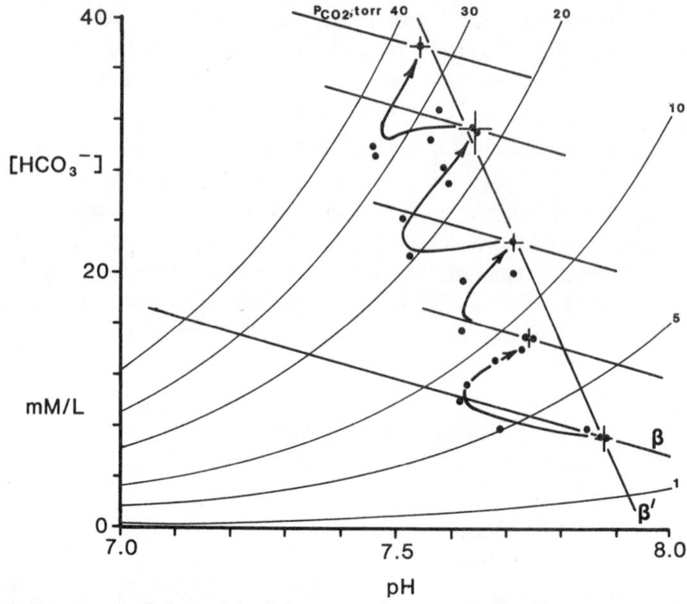

Figure 6.16: Responses of channel catfish to stepwise, progressive hypercapnia. The fish were held at each level for 24 to 48 hours. Note that although the pH compensation was incomplete at each level, higher concentrations of HCO_3^- were possible at increased P_{CO_2}. (Redrawn from Cameron & Iwama, 1987.)

A final point of interest with respect to hypercapnia and the water-breathers is that there does not appear to be any direct relation between P_{CO_2} and ventilatory drive as there is in the air-breathers. In some studies there is a ventilatory increase in response to a step increase in P_{CO_2}, but in nearly all cases the change is secondary to either stimulation of activity (*i.e.* a behavioral response) or to hypoxia from Bohr or Root shifts (Smith & Jones, 1982). Aquatic crabs also do not respond directly to CO_2 (Batterton & Cameron, 1978), but interestingly enough some air-breathing crabs do increase ventilation linearly in response to changes in P_{CO_2} (Cameron, 1975b; Smatresk & Cameron, 1981). This finding poses the interesting evolutionary question of whether the CO_2 response has evolved independently in the air-breathing crabs, or is present but suppressed in the aquatic ones and thus common to vertebrate and invertebrate lines.

Responses to Hypoxia and Exercise

Hypoxia per se has little direct effect on acid-base state, but indirect effects arise from the increased ventilation (in order to maintain O_2 supply) and in severe cases from increased metabolic production of acidic end-products such as lactic acid. Hypoxia can be brought about either by greatly increased metabolic O_2 demand (exercise) or by a limitation of environmental supply. During exercise the metabolic production of CO_2 is also increased, and the acid-base disturbance

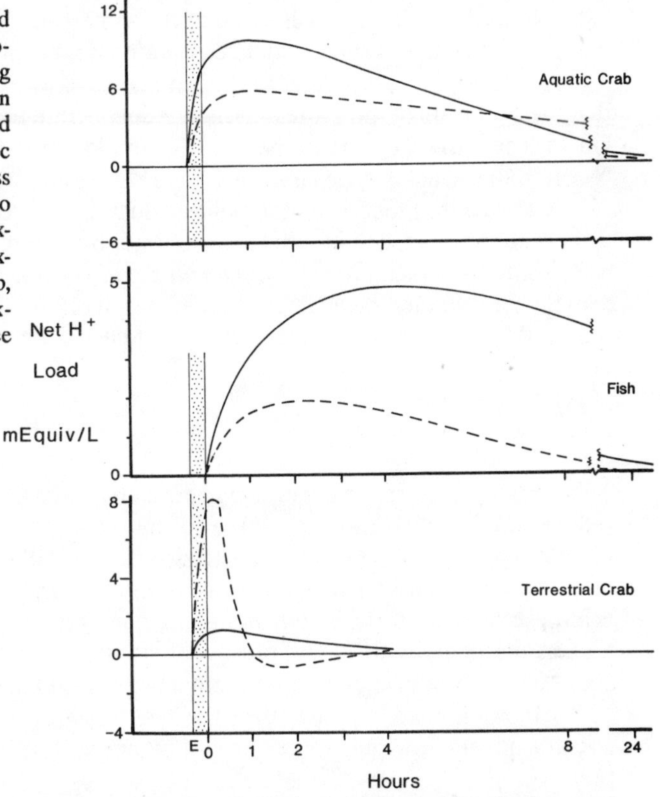

Figure 6.17: Net H^+ load and lactate load (mEq L^{-1}) appearing in the blood following periods of severe exercise in an aquatic crab, a fish, and a land crab. For the two aquatic animals, the H^+ load is less than the lactate load due to shuttling of the H^+ to the external environment via ion exchanges. For the land crab, however, the H^+ load exceeded the lactate load. See text for discussion.

results from a combination of respiratory alkalosis from hyperventilation and respiratory acidosis from increased CO_2 production. The acidosis and alkalosis are likely to nearly balance each other, but it is common to see a net alkalosis in both crustaceans (McDonald et al., 1979; Booth et al., 1984) and man (Dejours, 1975).

When the exercise is severe enough to generate significant amounts of anaerobic end-products, a nonrespiratory acidosis develops. In the air-breathing vertebrates it can be compensated for by moment-to-moment changes in ventilation, much as in the case with hypercapnia. In the aquatic animals, however, a more common pattern is for the H^+ ions to appear in the blood much more rapidly than the lactate$^-$, and for the H^+ ions to be temporarily "shuttled" to the environment via branchial ion exchanges. Thus the measured net H^+ load in the blood at any time after exercise tends to be much less than the lactate load (Fig. 6.17)(Wood et al., 1977; McDonald et al., 1979; Holeton et al., 1983; Holeton & Heisler, 1983). An exception to this general pattern was found in the semiterrestrial crab *Gecarcinus lateralis* exercised in air, in which the H^+ load actually exceeded the lactate load (Smatresk et al., 1979). This animal cannot "shuttle" the H^+ ions to water, so the more rapidly diffusing H^+ load precedes the lactate ion in the blood.

Ambient hypoxia produces some of the same changes without the complication of increasing CO_2 production, circulation, and ventilation brought about by exercise. One interesting feature of the response to ambient hypoxia in fish is that the perfusion of the white muscle is reduced until the ambient O_2 is restored. During the early recovery phase, muscle circulation increases, causing a lactate "flush" that can actually be large enough to kill the animal (Burggren & Cameron, 1979). The consequences of metabolic acidosis from excessive production of lactic acid are also avoided in some invertebrates by a metabolic shift to other pathways. In the sipunculid worm *Sipunculus nudus* up to 24 hours of anoxia can actually be attended by a slight intracellular alkalosis due to shifts to pathways that lead to alternate end-products and that have different H^+ budgets (Pörtner et al., 1984a,b).

Hyperoxia

The primary response to hyperoxia in both air- and water-breathing animals is a reduction of ventilation in response to the rising partial pressure of O_2 in the blood. In air-breathing vertebrates this reduction rather quickly leads to an internal hypercapnia, which restimulates ventilation and practically eliminates any acid-base effects. In the water-breathers, however, the general pattern of ventilatory insensitivity to P_{CO_2} holds; ventilation remains low and a hypercapnic acidosis develops (Dejours & Beekenkamp, 1977; Dejours et al., 1977; Heisler, 1980). If maintained for a long time the hypercapnia is compensated in the same way as a normoxic hypercapnia but without any compensatory change in ventilation.

ACID-BASE REGULATION AND THE SKELETON

In the vertebrates the chief mineral component of the skeleton is hydroxyapatite, a rather complex crystalline solid with the approximate formula $3Ca_3(PO_4)_2{\cdot}Ca(OH)_2$. Smaller amounts of brushite, a similar calcium phosphate compound, and amorphous $CaCO_3$ also occur. Hydroxyapatite and brushite are virtually insoluble in water in the physiological pH range, so no particular acid-base regulation is required for their formation or maintenance (Cameron, 1985b). In the invertebrates, on the other hand, the most common skeletal material is $CaCO_3$, a material significantly more soluble and highly sensitive to pH changes near the physiological range. In order to form and maintain a skeleton of $CaCO_3$ an alkaline environment must be maintained to prevent its dissolution. In the blue crab we have found that there is a functionally separate fluid pool associated with the carapace that is maintained about 0.5 pH units above that of the blood and that this relation holds over a wide temperature range (Wood & Cameron, 1985). During the postmoult phase, when new $CaCO_3$ is being formed at a high rate, the H^+ ions liberated by its formation are rapidly transported to the external environment in order to maintain the alkaline environment for deposition (Cameron & Wood, 1985).

Perhaps inadvertently, the carbonate skeleton can also act as a buffer store during acute acid-base disturbances in many animals. After exercise in air, the circulating [Ca^{2+}] rises in terrestrial crabs, indicating some buffering of excess H^+ by the skeletal carbonate (Smatresk et al., 1979; DeFur et al., 1980; Henry et al., 1981; Wood & Randall, 1981b). The skeletal carbonates contribute only a small percentage of the total hypercapnic compensation in crabs when breathing water (Cameron, 1985a). Hypoxic acidosis can produce an increase in circulating [Ca^{2+}] in molluscs when the shells are closed (Dugal, 1939). Under various conditions of metabolic and respiratory acidosis, some buffering can also be contributed by dissolution of $CaCO_3$ from mammalian bone (Bettice, 1984; Bushinsky & Lechleider, 1987), but whether this buffering is physiologically significant is not clear.

Footnotes

[1]The pH of neutrality (pN) is defined as the point where pH = pOH for any particular temperature.

[2]The term non-respiratory is preferred over metabolic, since processes such as active ion transport are nonrespiratory but not part of what is usually considered metabolism. The term metabolic is more appropriate in the case of lactacidosis but is not as general.

Chapter 7

Models of Gas Exchange

The process of gas exchange involves such a large number of variables that it is difficult to get an intuitive feel for how the system might behave in response to changes in any one variable. From an experimental standpoint one would like to change each variable independently to arrive at such an understanding, but experimental animals are not so cooperative. During exercise, for example, ventilation increases but so do cardiac output, tissue demand, and a series of other parameters, making it difficult to say anything about the effects of the increased ventilation alone. One of the primary reasons, then, for working with models of gas exchange systems is explore, at least in a theoretical way, the behavior of the total system. It is also of interest to calculate various indices of exchange efficiency (see Chapter 3), and to explore the sensitivity of exchange efficiency to each of the primary variables.

In order for a useful model to be constructed, a considerable amount of experimental data must already be available. Of the roughly 30 extant animal phyla, there are probably only two phyla for which sufficient information is available to make such models of gas exchange: the arthropods and the chordates. (Some of the information is also available for a handful of mollusc species.) For the purpose of illustrating the general procedure, the kinds of conclusions one can draw from such models, and the limitations of them, a model of counter-current exchange in a fish gill is described in detail and a partial model of capillary exchange in the human lung is discussed to show the influence of nonlinearity in equilibrium curves.

A COUNTER-CURRENT MODEL: THE FISH GILL

Description of the Model

The exchanging unit of a fish gill (or a crab or mollusc gill) is the water channel in the gill sieve plus the adjacent lamellar blood channels (Fig. 7.1a). Since the flows of water and blood are in opposite directions, the gill can be represented by the simplified schematic shown in Fig. 7.1b. The water flow over the gills is represented by $\dot{V}_g$, blood flow by $\dot{Q}$, and exchange by $\dot{M}$.

If one assumes that the oxygen equilibrium curves are linear, relatively simple differential equations to describe such a system can be written and solved by or-

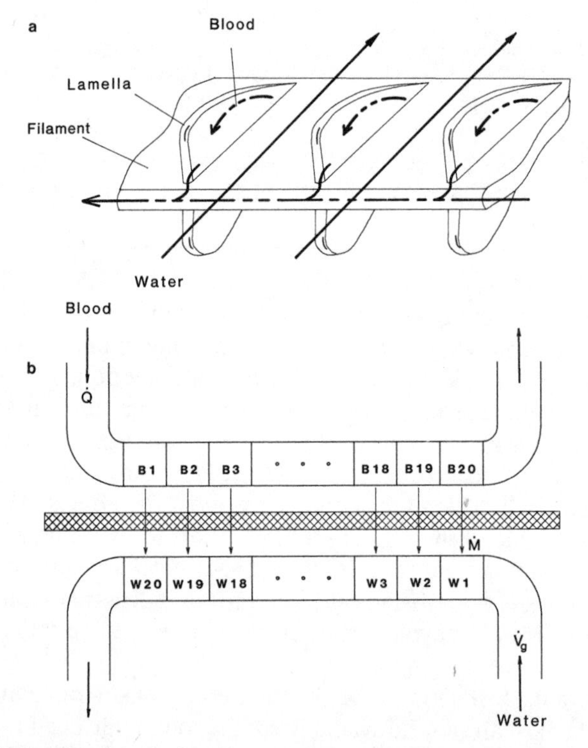

Figure 7.1: (a) Blood and water flow through a teleost gill. A small section of one filament is shown, with secondary lamellae arranged perpendicularly on both sides. The solid arrows show water flow, and the dashed lines and arrows show blood flow orientation. (b) Compartmental model used to simulate gas exchange in the fish gill. See text for discussion.

Table 7.1: Attributes of a 175-gm trout used in the simulation model.

Symbol	Value	Description (Source)
A	300 cm^2	Gill surface area (Hughes, 1966)
$\dot{Q}$	0.053–0.265 ml sec^{-1}	Blood flow (see text)
$\dot{V}$	0.65–3.29 ml sec^{-1}	Water flow (see text)
D	1.98×10^{-5} cm^2 sec^{-1}	Diffusion coeff. (Radford, 1964)
x	8. μm	Diffusion path (Hughes, 1966; see text)
GWV	0.68 ml	Water vol. in 2° lamellae (Hughes, 1966)
GBV	0.148 ml	Blood vol. in 2° lamellae (Hughes, 1966)
α	0.04462 ml L^{-1} $torr^{-1}$	Bunsen solubility coeff. (Cameron, 1985)
S	80. ml L^{-1}	Hb–O_2 binding capacity
N	20	Number of compartments variable
dt	0.001 sec	Integration time step

Assumes half of gill surface area is functional.

dinary integration (Piiper & Scheid, 1984). For a model that incorporates non-linear functions for the relation between oxygen tension and oxygen content, however, mathematically explicit differential equations are difficult to derive and difficult to solve. The approach outlined below gets around these difficulties by using a numerical computer method.

Both the blood and water channels are divided into a number of compartments, and an equation is written to describe the rate of change of oxygen in each compartment. Next the concentrations and flows in the water and blood are set and the fixed parameters of the model system are used to calculate the constants for the equations. Finally the simulation system is "run" until a steady state is reached for each blood and water compartment. When steady state is reached, the individual compartment values provide a "stair-step" approximation of the profiles from end to end of the blood and water channels, and the concentrations leaving the downstream blood and water compartments allow calculation of the overall exchanger performance.

For each compartment, the flux equation will have the form:

$$\text{Flux} = (\text{inflow} \times \text{conc.}) - (\text{outflow} \times \text{conc.}) - \text{diff. loss} \qquad \text{(Eq. 7.1)}$$

or since inflow and outflow from each compartment are the same:

$$\text{Flux} = (\text{flow} \times \Delta\text{conc}) - \text{diff. loss} \qquad \text{(Eq. 7.2)}$$

If we designate the compartments as 1 through N then the general procedure is as follows:

1. Initialize each compartment. The simplest method is to assume a linear profile across the gill at the start.
2. Calculate the change in each compartment during some small time increment *dt*.
3. Reset all compartments' contents to reflect the change during *dt*.
4. If the values are still changing, repeat.

5. If steady state has been reached, stop and report results.

Since the fish that has been studied most intensively is the rainbow trout (see Chapter 10), a simulation model was constructed following the steps given above and using data from various literature sources. The model applies to a 175gm trout at 13°C, with the attributes given in Table 7.1. The value for gill surface area is derived from Hughes' (1966) measurements plus his observation that only about half of the actual secondary lamellar surface is underlaid by blood space; the rest is underlaid by various supportive tissues and does not appear to function in gas exchange. The values for blood and water flow were taken from Cameron and Davis (1970), with the low values representative of resting trout, and the higher values representing a 5-fold increase during maximal exchange conditions. Although Davis (1972) has estimated that only about 20% of the secondary lamellae are perfused at rest, it seemed more appropriate when analyzing the maximum performance of the whole gill system to employ the higher values in most of the calculations. An alternative analysis using only 20% of the surface areas and the lower flow rates produces similar results.

In substituting the arithmetic expression $\Delta C/x$ for the differential term dC/dx of Eq. 1.13, the conductance of the system will be overestimated if x is taken to be simply the thickness of the epithelium separating blood and water. The value of x should be increased to reflect incomplete mixing in both media and the resultant partial pressure gradients that develop (see Chapter 3). Precisely how much to increase x is not known, but the theoretical studies of Scheid et al. (1986) indicated that 1/4 of the thickness of the water channel would be a reasonable amount to add to the epithelial thickness. For the rainbow trout the epithelial thickness is about 3 μm and the water channel width about 20 μm, so $3 + (\frac{1}{4} \times 20) = 8$ μm.

For the model calculations it is necessary to know the total volume of water contained in the water and blood spaces of the secondary lamellae. From data of Hughes (1966; Hughes & Morgan, 1973) the values given in Table 7.1 can be calculated. In dividing the gill into N compartments, each compartment is assigned 1/Nth of the water and blood volumes and 1/Nth of the gill surface area, with the

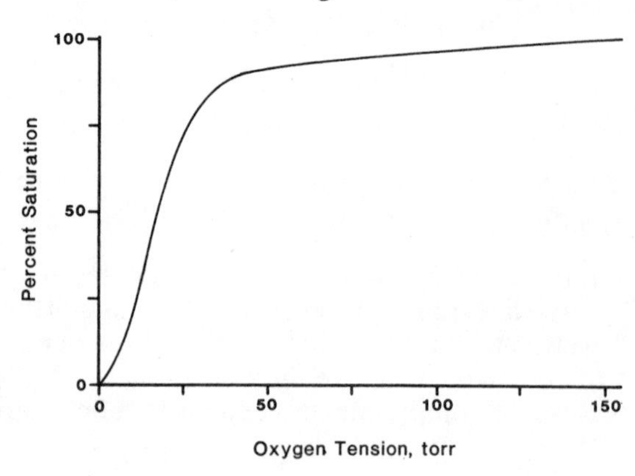

Figure 7.2: Plot of the in vivo oxygen equilibrium curve used in the simulation model.

total blood and water flow going through each compartment.

The relation between partial pressure and content in the water compartments is simply the linear constant α, but for the blood a sigmoid equilibrium curve function was required. Using arterial and venous O_2 and CO_2data for rainbow trout (Cameron & Davis, 1970; Cameron, unpublished data) and in vitro equilibrium curves, an in vivo curve was estimated (Fig. 7.2) and then approximated with a sigmoid mathematical function:

$$\text{Content} = 1000\alpha P + [S/(1 + 80e^{(-0.25P)})] - 1 \qquad \text{(Eq. 7.3)}$$

where content and S (the O_2 capacity of the hemoglobin) are in ml L^{-1}. Since this expression cannot be solved explicitly for P, an iterative approximation routine was employed to calculate P from content.

The complete computer program for the gill simulation model is included on the accompanying disk under the name GILLEX. The Pascal source code (Turbo Pascal version 3.0, ©Borland International) is stored with a .PAS extension, and the compiled object code with a .COM extension. Some brief documentation describing the program's operation is stored as GILLEX.HLP. (See Appendix 1 for notes on the program.)

Assumptions of the Model

As with any model, a number of simplifying assumptions have been made. The model first of all assumes that the rates of blood and water flow are uniform and that there is no shunting of blood or water. Shunts are actually easy to incorporate – one simply reduces the flow through the exchanger by the proportion shunted, then adds that portion to the effluent stream, calculating an average tension and content. Shunting will directly reduce the overall exchanger efficiency.

The model also assumes that the blood channel is of uniform dimensions along its length, and therefore that the blood flow velocity and length of contact time are uniform along the length. Secondary lamellae are actually somewhat wedge-shaped (Hughes & Morgan, 1973), but the model studies of Malte and Weber (1985) indicated that lamellar shape was not very important.

Finally, the model assumes that the blood is completely mixed and that the loading and unloading of hemoglobin are instantaneous. Lack of complete mixing simply increases the effective diffusion pathlength (x), which can be easily modelled, and the reaction of O_2 with Hb is so fast as to be nearly instantaneous (Roughton, 1964).

Definitions of Indices

All indices of flow, performance, and effectiveness are those described in Chapter 3. From the indices of effectiveness of transfer for blood and water (E_w and E_b or $E_{b\text{-mod.}}$) an overall effectiveness is derived, $E_{tr} = 1 - E_b - E_w$. This parameter

Randall's (1970b) definition of transfer factor is:

$$T_{O_2} = \dot{V}_{O_2}/\{½(P_{I_{O_2}} + P_{E_{O_2}}) - ½(P_{a_{O_2}} + P_{v_{O_2}})\} \quad \text{(Eq. 7.4)}$$

which is really just a method of arithmetically averaging the oxygen difference between blood and water, and of calculating the oxygen transferred per arithmetic gradient unit. This index assumes implicitly that the change in oxygen tension in both water and blood is linear from one end of the gill exchange unit to the other.

Results

Influence of *dt* and N

As *dt* gets smaller and N larger, the computer time required to reach steady state increases to the point where the model is not useful. On the other hand, values for N of 10 or less and for *dt* of much more than 0.001 result in too coarse resolution. With N = 20 and *dt* = 0.001, useful simulations can be completed, i.e., run to steady state, in 1 to 3 hours on a 4.77 MHz IBM PC-XT with an 8087 math co-processor. The AT (or other 80286- and 80386-based machines) and

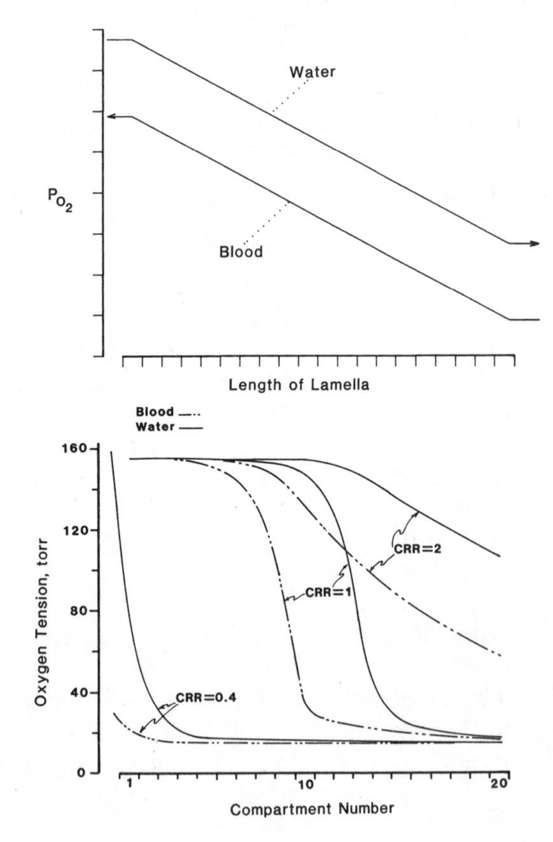

Figure 7.3: (Upper) Oxygen profiles in a conventional linear counter-current gill model. When flows are perfectly matched, the profiles are linear and the gradient between blood and water is constant over the length of the lamellar exchanging area. (Lower) Oxygen tension profiles in blood and water generated by the nonlinear simulation model at CRR = 0.4, CRR = 1.0, and CRR = 2.0.

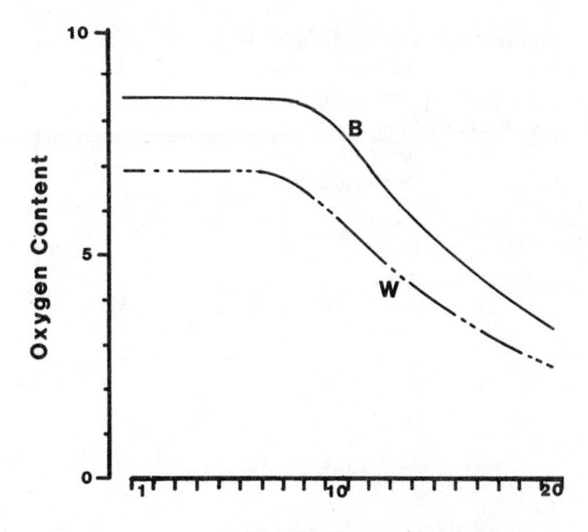

Figure 7.4: Oxygen content profiles from the simulation model under conditions similar to those used to generate the partial pressure profiles of Fig. 7.3. These content profiles correspond to CRR = 1, the middle lines of Fig. 7.3.

machines operating at higher clock speeds can reduce that time considerably. Although this time is slow, in 1973 a similar model written in FORTRAN IV required 20 to 30 hours on a large mini-computer (Data General, Nova series)(Cameron & Polhemus, 1974).

Conventional Counter-Current Profiles

Many authors have presented O_2 profiles across fish gills derived from heat exchanger equations that assume linear solubility functions for both media (Hughes & Shelton, 1962; Hughes, 1964; Piiper & Scheid, 1972; Dejours, 1975; Piiper, 1982; Piiper & Scheid, 1984). In the special case where the capacity-rate ratio (CRR) is 1.0, straight line profiles are obtained (Fig. 7.3a) with a constant PO_2 gradient. When the CRR is not equal to 1, the profiles become curvilinear, but retain the constant gradient feature.

The GILLEX model, with its built-in O_2 equilibrium curve, produces the tension and content profiles shown in Figs. 7.3b and 7.4 at a CRR of 1.0. Not only is the blood profile nonlinear, but so is the water profile. A point-by-point subtraction of the blood tension from the water tension across the gill produces the plot of O_2 gradients shown in Fig. 7.5, a marked departure from the constant gradient found in the conventional counter-current model. Piiper and Scheid (1984) considered nonlinear profiles for blood but not for water (their Fig. 10, p. 248). As the CRR increases or decreases, the profiles shift, as shown in Fig. 7.3b.

Mean O_2 Gradients

A proper calculation of the mean O_2 gradient involves summing the differences across the gill or integrating the curve of Fig. 7.5, which is simple to do as part of the model output (see GILLEX listing). The usual calculation of arithmetic mean O_2 gradient performed from affluent and effluent blood and water tensions produces grossly erroneous results, as shown in Fig. 7.6. Malte and Weber (1985)

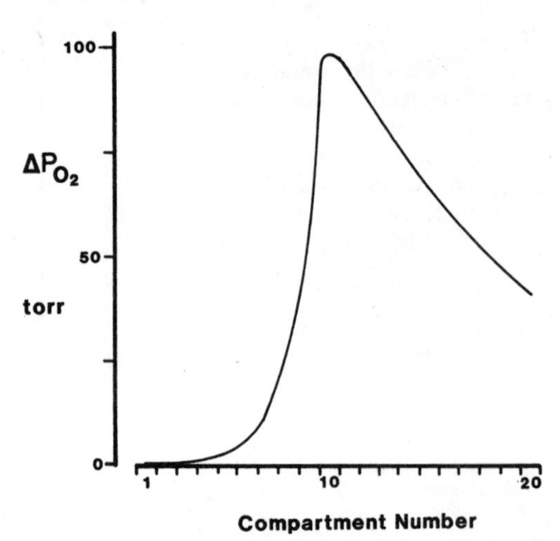

Figure 7.5: The O_2 gradient between blood and water across the length of the exchanging lamella. An arithmetic average will obviously bear little relation to the real mean gradient.

mean O_2 gradient performed from affluent and effluent blood and water tensions produces grossly erroneous results, as shown in Fig. 7.6. Malte and Weber (1985) concluded that the arithmetic mean gradient always overestimates the true mean gradient, but from Fig. 7.6 it is obvious that it depends on the actual values of other parameters such as blood and water flow and the resultant changes in CRR.

Transfer Factor

As defined by Randall (1970b), the transfer factor is simply the oxygen transferred (consumed) divided by the arithmetic mean O_2 gradient. A correct version employs the integrated mean O_2 gradient, and as expected this corrected transfer factor remains constant over a range of CRR from 0.25 to 3.00, whereas the uncorrected formula yields numbers ranging from 0.1 to 10 times the proper value. The transfer factor theoretically depends only upon fixed parameters of the model: A, D, and x, but might change considerably in a real animal as the degree of perfusion of the lamellae or other parameters change.

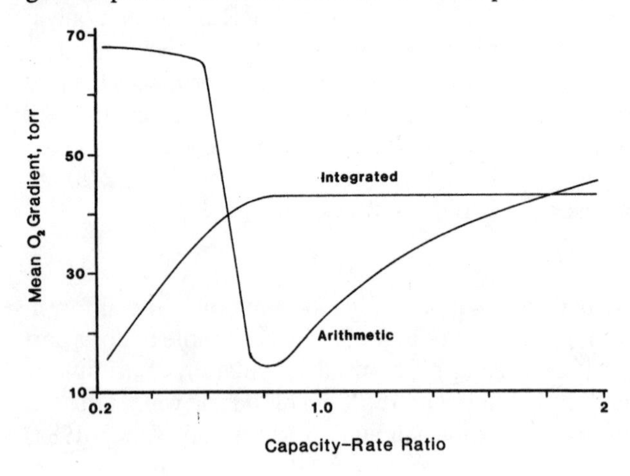

Figure 7.6: Influence of CRR on the integrated and arithmetic mean O_2 gradients, calculated as explained in the text.

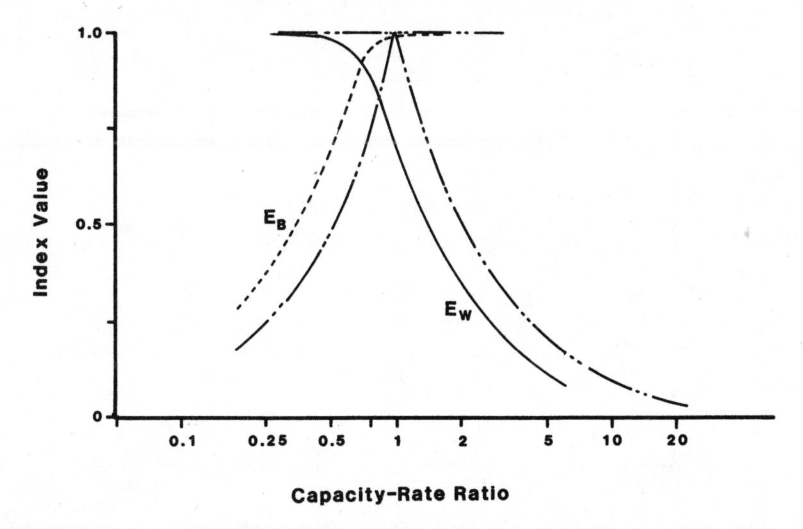

Figure 7.7: Plot of E_{b-mod} (dashed line) and E_w (solid line) obtained from the nonlinear simulation compared to the usual values obtained from linear models (dot/dash lines) for CRR ranging from 0.1 to 20.

Efficiency of Transfer

In the conventional linear counter-current model the efficiency of transfer of oxygen to blood (E_b) describes a hyperbolic curve from CRR values of 0 to 1 and remains constant at 1.0 for all CRR values greater than 1.0 (Fig. 7.7). The efficiency of transfer from water (E_w) has an inverse relation, i.e., constant at 1.0 from CRR values from 0 to 1 and a declining hyperbolic curve at CRR values greater than 1.0. In the present nonlinear model both these relationships change shape and shift relative to the CRR (Fig. 7.7), with the interesting result that there is no single point at which both efficiencies are at 1.0. If we define the optimum operating conditions for the exchanger as those that produce the minimum value of the expression ($1 - E_w - E_{b\text{-mod.}}$) or a minimum of E_{tr} then the optimum falls at a CRR of about 0.72 when the venous oxygen content is 40%.

The Influence of Venous Content

In most previous model studies the venous oxygen content has been assumed to be zero (Hughes & Shelton, 1962) or has been held constant at some typical teleost value (Piiper & Scheid, 1984; Malte & Weber, 1985). The definition of CRR (see Chapter 3) actually assumes a zero oxygen content for venous blood, so it seems worthwhile to examine the effects of changes in venous oxygen content. As shown in Fig. 7.8, the optimum venous saturation is about 20%, with the values for $\{1 - E_w\ E_{b\text{-mod.}}\}$ decreasing at both higher and lower percent saturation. The total O_2 exchanged by the gill decreases as venous content increases (Fig. 7.8) and appears to have a maximum at 10 to 12% saturation. The integrated

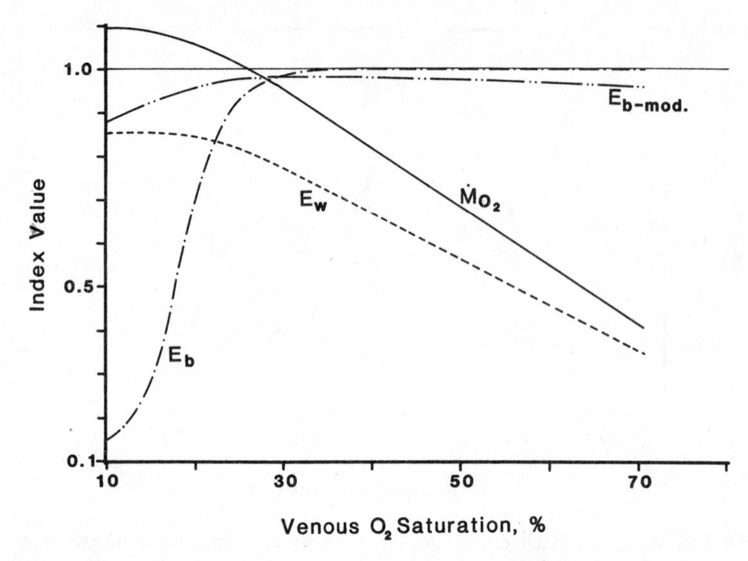

Figure 7.8: Effects of venous saturation on oxygen transferred ($\dot{M}O_2$), $E_{b\text{-}mod}$, E_b, and E_w.

mean oxygen gradient (Fig. 7.9) shows a decline at higher venous contents, exactly corresponding to the decline in oxygen consumption (transfer), whereas the conventional arithmetic mean O_2 gradient provides nonsensical data.

Exchange Potential Ratio; A New Index

The results of varying venous oxygen content and the consequences of a non-linear model suggest a definition of a new index for exchanger performance,

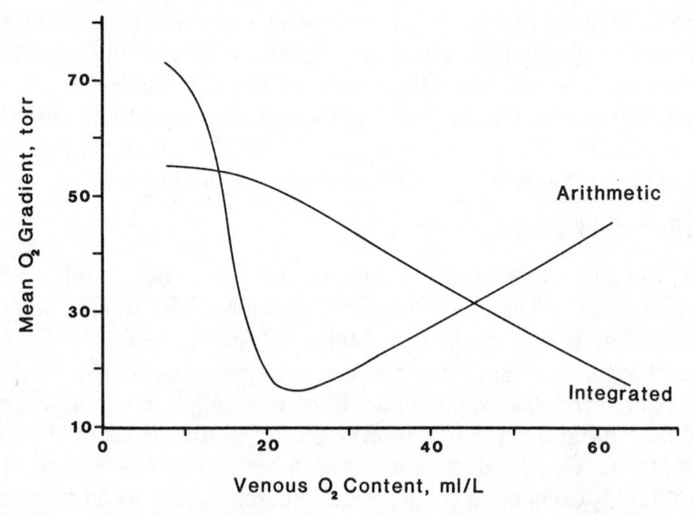

Figure 7.9: Effects of venous saturation on the arithmetic and integrated mean O_2 gradients.

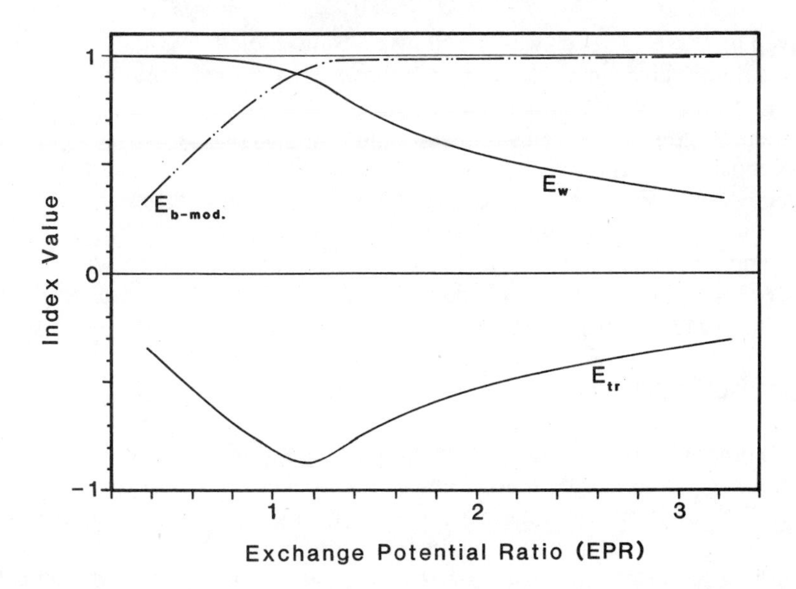

Figure 7.10: Variation in indices as a function of the exchange potential ratio (EPR). See text for definitions and discussion.

which will be called the "exchange potential ratio" (EPR) to avoid confusion with the CRR. This index is the ratio of the maximum potential for the blood to take up oxygen to the maximum potential for the water to give up oxygen, or:

$$EPR = \dot{Q}[\theta(P_I) - Cv] / [\dot{V}_G\alpha(P_I - P_v] \qquad \text{(Eq. 7.4)}$$

where $\theta(P_I)$ is a function giving the oxygen content of the blood when equilibrated to the partial pressure of inspired water. Both the sigmoid equilibrium curves and the effect of variation in venous oxygen content are taken into account in this expression When the various efficiencies are recalculated as a function of the EPR, the optimum exchanger efficiency falls at EPR values of about 1.2, which corresponds to a CRR of about 0.72 for the standard simulation conditions (Fig. 7.10).

Sensitivity Analysis

Although a mathematically rigorous sensitivity analysis cannot be carried out with the numerical approximation method, the same result can be obtained by varying each parameter by the same proportion and measuring the effect on the main output variable, in this case oxygen exchanged (consumed). Starting from the standard conditions of A = 300 cm^2, Cv = 32 ml L^{-1} (36.8% saturated), $\dot{V}_G$ = 3.29 ml sec^{-1}, $\dot{Q}$ = 0.265 and the other parameters

Table 7.2: Sensitivity of net oxygen exchange of the gill model to specific variables when the CRR is 1.

Variable	%/%
$\dot{Q}$	1.105
Cv	–0.65
A, D, or x	<0.02
$\dot{V}_G$	<0.02

as given in Table 7.1, each was varied by 10% and the effects on output studied. The results are given in Table 7.2, which shows that the model output is most sensitive to blood flow and venous saturation, and relatively insensitive to the other parameters. Another way of stating this conclusion is that at a CRR of 1 the fish gill is mostly perfusion-limited, with a lesser limitation resulting from blood capacity; the sensitivity analysis does not point out the greatest limitations to gas exchange, only the sensitivity to changes in various parameters. The sensitivity to A, D, and x is equal, since they are combined into a single parameter which is more or less equivalent to the conductance (cf. Piiper, 1982) or number of transfer units (NTU; Malte & Weber, 1985).

Comparison with the Linear Model

The combination of a strongly nonlinear oxygen equilibrium curve and the fact that normal venous oxygen tension and content fall on the steepest portion of the equilibrium curve makes it mandatory that any model incorporate these effects. The errors generated by assuming a linear model range from small for E_w, particularly at high water flows, to very large in the case of oxygen gradients (Fig. 7.9), E_b, and oxygen profile calculations. This conclusion is similar to that reached by Malte and Weber (1985), although they did not explore some effects such as variations in venous saturation and errors in the mean gradient and transfer factor calculations.

COUNTER-CURRENT MULTIPLIERS

Characteristics

The high efficiency characteristics of counter-current exchangers have been utilized in the course of evolution for a variety of specialized structures collectively known as counter-current multipliers (CCMs). Their structure is similar to the gill just described, except that the same fluid passes on both sides of the exchanger surface, and the outflow of one side is looped around to become the inflow of the other (Fig. 7.11). The unique feature of these structures is that large end-to-end gradients may be built up with only very small gradients maintained at any point along the length of the exchange surface. In Fig. 7.11a an example of a counter-current heat multiplier is shown at steady state. If there is a constant heat input in the loop portion, the warm return stream entering the exchanger will lose some heat to the cooler inflowing stream all along the exchanger. At any given point along the exchanger the temperature gradient is relatively small, but the end-to-end gradient in the example is 30 times larger.

Some general characteristics of an effective CCM are (1) a large exchange surface, (2) small diffusion (or exchange) barrier thickness, and (3) an energy source or sink to drive the multiplication. In biological systems the first two require-

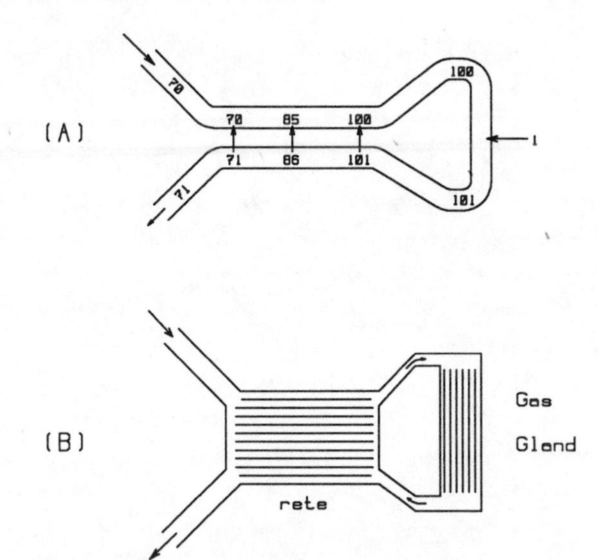

Figure 7.11: (A) Basic features of a countercurrent multiplier. The numbers illustrate that with only a small input in the loop and small gradients from outflow to inflow side, a large end-to-end gradient can be achieved. The example (B) shows a fish swimbladder.

ments are met by dense networks of capillaries, with inflow and outflow interspersed in a regular pattern (Fig. 7.12). The third characteristic varies depending upon what the particular CCM is multiplying.

Occurrence

Counter-current multipliers may be found that multiply (or concentrate) gases, temperature, or salts. Since only gas exchangers concern us here, the others will be described only briefly.

Temperature conservation is important in the extremities of many animals, especially birds and mammals of northern climates. Gulls, for example, can stand on ice in extremely cold weather without losing very much heat. This remarkable ability depends upon a counter-current heat exchanger in the vessels leading to and from the legs (Steen & Steen, 1965). In this case heat from the warm arterial blood flowing out the limb is transferred in an exchange *rete* (= network) to the returning venous blood, so by the time the arterial blood arrives at the foot, it is already cooled from around 40°C to perhaps 4° to 6°C, greatly reducing the gradient for heat loss to the environment. Similar rete are found at the bases of the extremities of many marine mammals.

Just the opposite kind of heat conservation is effected by the heat exchanges of tuna, however. The large lateral vessels give rise to branches to the interior muscle mass, which generates considerable heat from its elevated metabolism. This heat, carried by venous blood, is transferred in a rete system to the inflowing arterial blood, thus keeping much of the heat in the loop and maintaining a high muscle temperature (Carey et al., 1971).

Salt concentration is the object of CCMs found in organs such as the mammalian kidney and the avian salt gland. The general arrangement is like that

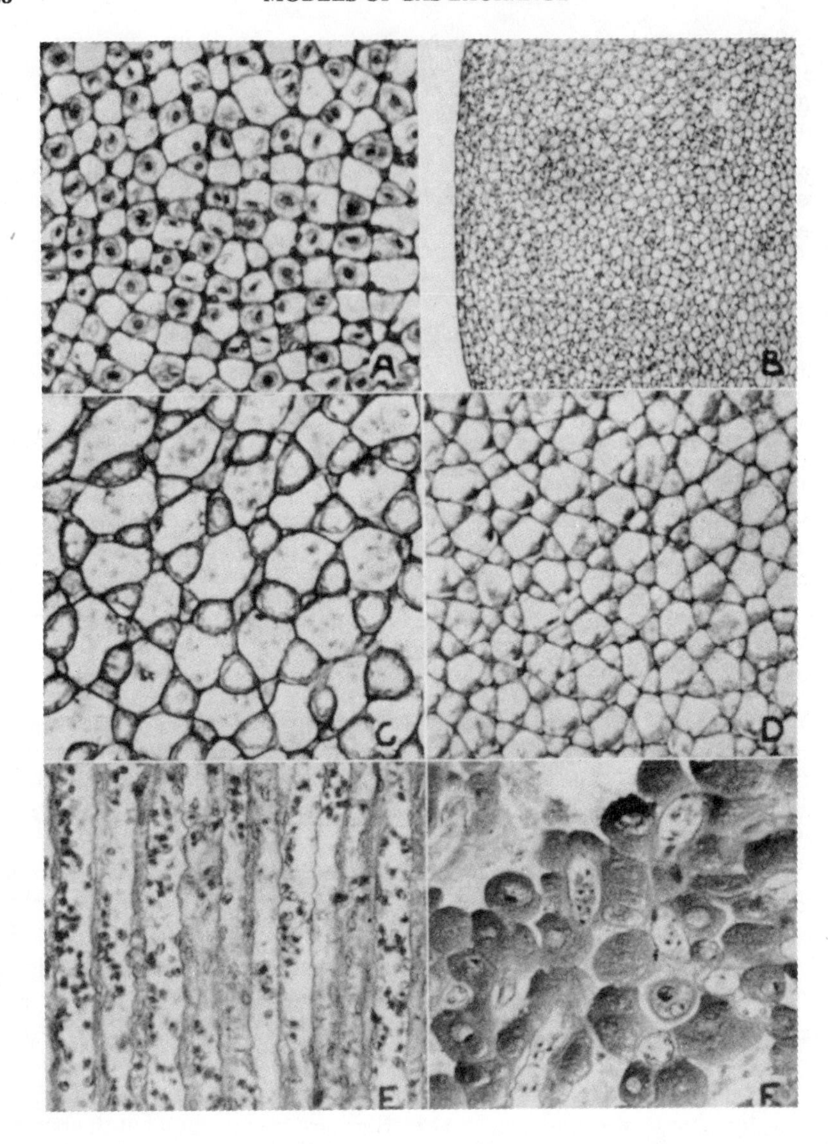

Figure 7.12: Cross sections of the capillaries of some swimbladder retia of various deep sea fish. (A) *Synaphobranchus* sp., ×300; (B) Same as A, ×100; (C) *Coryphaenoides* sp., ×300; (D) *Sebastes* sp., ×300; (E) *Coryphaenoides*, longitudinal section ×300; (F) glandular portion from *Coryphaenoides*, ×225. (From Scholander, 1954, by permission, ©Biol. Bulletin.)

shown in Fig. 7.11, but in this case the energy source is active transport distributed all along the tubules, rather than an input or output in the loop portion (Schmidt-Nielsen, 1964).

CCMs for gas exchange are most remarkable in two quite different structures: the eye and the swimbladder of teleost fish. The vessels to and from the eyes of many fish go through a capillary exchange rete, the choroid rete, which is responsible for maintaining a high oxygen concentration in the aqueous humor of the eye (Wittenberg & Wittenberg, 1962). Similarly, the vessels to and from a small organ known as the gas gland on the swimbladder pass through an exchanger known as the *rete mirabile*. Since the swimbladder is the best studied and involves more than one mechanism for gas concentration and secretion, it will be discussed in some detail.

The Swimbladder Gas Gland

Structure

The basic structure of the swimbladder gas gland in an eel is shown in Fig. 7.11. There is an artery leading to the *rete* section, the *rete* itself, then a secondary circulation leading to an organ on the surface of the gas bladder known as the gas gland. The efferent circulation from the gas gland re-enters the venous capillary side of the *rete* and finally returns to the general venous circulation. The length of the *rete* section is approximately 1 cm in the eel, but in deep sea fishes longer *retia* are the rule (Scholander, 1954). The diffusion distance between the arterial and venous sides of the *retia* capillaries averages 1.5 μm, and the capillaries are packed so as to provide maximal exchange area (Fig. 7.12)(Scholander, 1954).

Function

Gas concentrations from the swimbladders of fish caught at great depth must be at least equal to the prevailing hydrostatic pressure. Since pressure increases approximately 1 atm for each 10 m depth, a fish with a swimbladder living at 1000 m must be able to generate pressures of at least 100 atm in order to accomplish secretion of gas into the swimbladder. Analyses of swimbladder gases consistently show the bulk of the gas to be oxygen, around 90% usually, but there is also a significant amount of nitrogen, up to 10 atm or more (see reviews by Scholander, 1954; Steen, 1963). The challenge in the study of the swimbladder has been to discover mechanisms by which gases can be concentrated to this remarkable degree.

The primary driving force for the concentrating action is the metabolic activity of the gas gland tissue, which produces both CO_2 and lactic acid. Addition of lactic acid to the blood prior to its entering the return portion of the *rete* will reduce the solubility coefficient for gases, producing a "salting out" effect. This effect was recognized early, but various model studies have estimated that it can account for only a small proportion of the total gas concentration (Scholander, 1954; Kuhn & Kuhn, 1961). It is, however, the only satisfactory mechanism to account for the concentration of nitrogen, since nitrogen does not participate in any binding reactions in the blood. That the nitrogen concentration is purely a solubility effect is confirmed by the constant nitrogen–argon ratio in seawater, blood, and swimbladder gas (Scholander, 1954).

At high oxygen pressures, many fish will nevertheless show a reduction in hemoglobin-oxygen binding when CO_2 is increased or pH reduced, i.e., they show a Root shift. A second important mechanism, then, is the release of oxygen from hemoglobin caused by the acidifying action of the gas gland. In an early model, Kuhn and Kuhn (1961) showed that if the lactic acid did not diffuse across the capillaries of the *rete*, quite satisfactory matches between theoretical calculations and observed gas concentrations resulted. When it was later shown that the *rete* is permeable to lactic acid, however, another mechanism was required. Forster and Steen (1969) later demonstrated that the time required for the Root "off-shift" was very short, about 0.2 sec in the eel, but that the time required for the "on shift" was relatively long, about 10 seconds. Thus when the blood is acidified by the gas gland it will quickly release some oxygen, increasing the partial pressure in the return side of the *rete* and generating the gradient necessary to drive gas concentration. As the lactic acid diffuses into the arterial side and the pH difference is reduced, oxygen tension remains high in the returning side because there is not sufficient time for the Root on-shift to occur. A more recent model by Sund (1977) has verified theoretically that the combined action of salting out, the Root shift, and the time hysteresis between on- and off-shifts provides a satisfactory agreement between mathematical models, experimental work, and field observations of depth distribution and gas composition.

Probably the remaining mystery is why some fish appear to have gas glands and concentrating *retia* but do not show an appreciable Root shift (Scholander & van Dam, 1954). The more recent experimental work on swimbladders has been done mostly with the eel and others that do have Root shifts, and the earlier data have largely been forgotten. Perhaps it is time to repeat the cycle of testing models and making some new experimental observations with some of these species.

CAPILLARY EXCHANGE IN THE HUMAN LUNG

A Simple Model

A number of models of gas exchange in the human lung have been constructed, some very elaborate (e.g., Milhorn et al., 1965), but none which adequately take into account the non-linearity of the oxygen equilibrium curve. With a uniform pool model, i.e., one in which the alveolar gas reaches a steady state concentration, the modelling can be very much simpler: One simply defines a "packet" of blood that enters the capillary at time zero and exchanges with this fixed alveolar gas pool for a time equal to the mean residence time for lung capillaries. The time course of oxygen tension and content in this packet, assuming constant flow, will be the same as the distance profile from one end of the capillary to the other.

For the human model an in vivo oxygen equilibrium curve was taken from data of Roughton (1964) and approximated by the sigmoid equation:

$$\text{Content} = (1000\alpha P) + [S/(1 + 32e^{(-0.105P)})] \quad \text{(Eq. 7.5)}$$

over the range of 8 to 100% saturation. The computer routine used in a Pascal computer program LUNGEX to solve this expression is similar to that used in the GILLEX program (LUNGEX is available from the author upon request).

All that is now needed is parameters to satisfy the equation:

$$dQ/dt = AD\alpha(100 - z)/x \quad \text{(Eq. 7.6)}$$

where z is the blood Po_2, 100 is the steady-state alveolar tension (Dejours, 1975), and the other variables have their usual meanings. The blood volume of the human lung is 140 ml (Weibel, 1972) so if we consider a 1 ml (1 cm^3) slug of blood, the appropriate surface area is (500,000/140) = 3571 cm^2 (Weibel, 1972). The harmonic mean diffusion pathlength is 0.4 μm (Weibel, 1972), D for air is 0.189 cm^2 sec^{-1} (Radford, 1964), and the value for plasma oxygen solubility, α, is 2.93×10^{-5} cm^3 O_2 cm^{-3} blood $torr^{-1}$ (Dejours, 1975). The program LUNGEX.PAS was used with a time step of 0.1 msec and a numerical technique to generate the solutions shown in Fig. 7.13.

Results

The profiles of blood oxygen tension and content as a function of time (= distance) in the lung capillary are shown in Fig. 7.13 for venous saturation values ranging from 3 to 70%. At the lowest venous content complete oxygen loading is complete in about 65 msec, compared to a resting capillary contact time of about 750 msec (Roughton, 1964; Piiper, 1969; Crandall & Bidani, 1981). The use of 0.4 μm for the mean harmonic thickness of the lung epithelium, however, probably overestimates the diffusive conductance since it ignores the thickness of the

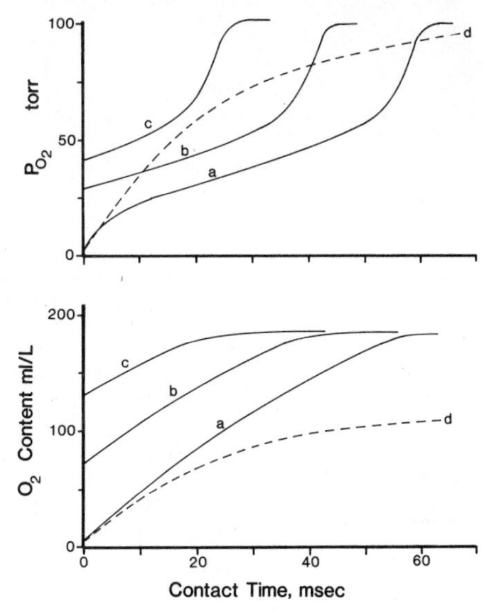

Figure 7.13: Profiles of Po_2 (upper) and oxygen content (lower) with contact time (= distance) in the lung, generated with a computer simulation model incorporating a nonlinear blood oxygen equilibrium curve. The various curves show results for a venous oxygen tension of 3 torr (a), 40 torr (b), and 70 torr (c), as well as results using a linear solubility function with capacity equal to normal human blood (d). See text for further discussion.

blood layer and assumes no gradient at all within the alveolar air. A more realistic figure might be 1 to 2 μm, which would yield oxygen loading times of about 150 to 300 msec. The half-time for O_2 binding to hemoglobin is on the order of 30 to 50 msec, which would add some additional time. Nonetheless, the loading seems comfortably quick in the resting condition but is close to the contact time estimates of 450 – 500 msec during exercise (Roughton, 1964; Piiper, 1969).

The advantages of a sigmoid oxygen curve of fairly high affinity are also illustrated in Fig. 7.13. If the oxygen carrier had a linear binding function it would not only be limited by the alveolar PO_2 but would load more slowly. The effect of the sigmoid curve is to hold the PO_2 lower for a longer period and since the O_2 flux is proportional to the gradient, the pigment loads faster and more completely.

MODELS OF GAS EXCHANGE IN LUNGS

A number of useful models of lung gas exchange have been constructed in which the processes occurring along the length of the lung capillaries are lumped, and the blood and air pools in the lung are considered as uniform compartments. The performance of the exchanger can then be modelled with an input/output approach, and various indices of performance assessed. One example of this level of modelling is illustrated by a general gas exchange model developed by Piiper and Scheid (1972; 1977), and applied to the turtle and other animals by White & Bickler (1987). The model focusses primarily on conductances: that for air convection (ventilation), G_{vent}; for diffusion across the air/blood barrier in the lung, G_{diff}; and for blood perfusion, G_{perf} (Fig. 7.14). Limitation indices (L) are derived

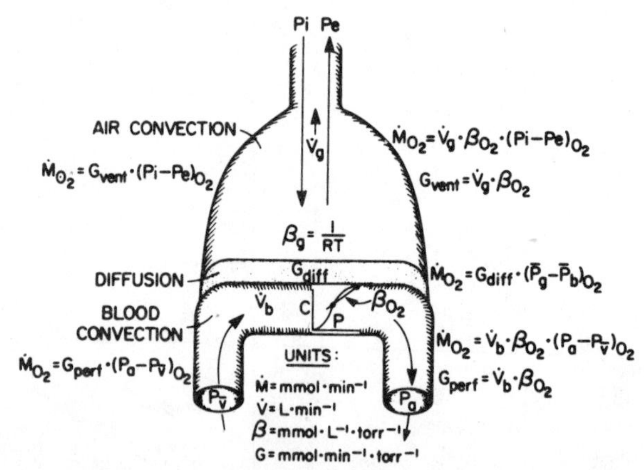

Figure 7.14: Components and quantities used in the model analysis of gas exchange in the turtle. Note that the designation *v* has been used for pulmonary arterial blood, and *a* for pulmonary venous blood. The symbol β is used as a capacitance for the gas phase but represents a nonlinear equilibrium curve for blood. Conductances, calculated after Piiper & Scheid (1972, 1977), are represented by *G*. (Reprinted from White & Bickler, 1987, by permission, ©Amer. Soc. of Zoologists.)

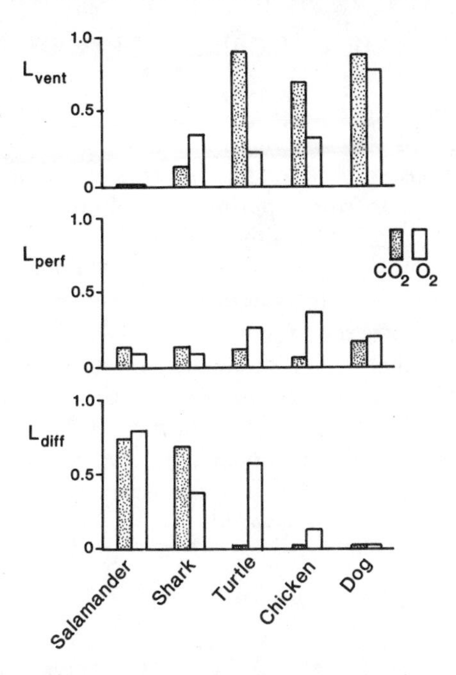

Figure 7.15: Limitation indices for the principal components of gas exchange in various vertebrates. A limitation index (L) value of 0.6 means that gas exchange is reduced 60% from what it would be without limitation by the processes indicated (ventilation, perfusion, diffusion). See text for discussion. (Redrawn using data from White & Bickler, 1987.)

by comparing the performance when a particular factor (e.g., ventilation) is not limiting to actual performance, so a value for L of 50% means that the factor under consideration reduces gas exchange performance by 50%. White and Bickler fitted this model to a variety of animals, including the skin-breathing salamander, a gill-breathing dogfish shark, a bird, the dog, and the turtle (Fig. 7.15).

One interesting conclusion is that in all the lung breathers CO_2 exchange is strongly ventilation-limited, whereas in the gill- and skin-breathers the diffusion limitation is greatest (Fig. 7.15). This conclusion is in accord with the different ways in which these groups regulate pH (see Chapter 6). For oxygen the picture is more complex, with ventilation the dominant limitation in the mammal, ventilation and perfusion about equal in the bird, and diffusion the primary limitation in the turtle, dogfish, and salamander.

This approach has yielded some interesting information and serves as a basis for comparison of either different systems or the same system under different operating conditions. One disadvantage of this modelling approach, however, is that changes in many parameters show up only as changes in the calculated resistances. For example, a change in the regional heterogeneity of lung ventilation or a change in venous saturation would cause changes in the overall performance indices in ways that could not be predicted from the model parameters.

GENERAL APPLICABILITY OF MODELS

Perhaps the most useful aspect of any modelling exercise is that it forces a clear focus on the nature of the system to be modelled. Components must be delineated, the relationship between parameters defined, and actual values for parameters selected either from experiments or literature. Working with such models can provide the sort of intuitive feel for how all the components work together in a way that might take years of experimental work. Conducting any sort of sensitivity analysis helps to focus on the most critical parameters and highlights areas in which work is required.

Rather than ends in themselves, models should probably be viewed as a step in the research process: experiment, model, experiment. That is, experimental observations provide the input data and conceptual framework for constructing and working with a model, and the model in turn will suggest a direction for the next generation of experiments. The model is no more than an abstract tool for understanding and probing the real world.

Chapter 8

Control of Respiration

An animal with constant metabolism living in a constant environment could theoretically have a respiratory system that operated at a fixed rate. Neither the metabolic requirements nor the environment are so simple, of course, so the animal needs the ability to vary the operation of its respiratory system in an appropriate manner. In previous chapters some responses to variation in internal requirements or external conditions have been discussed, but the issue of how the animal "knows" what to do has not yet been addressed. Varying the operation of the respiratory system correctly requires the participation of sensory systems, neural reflexes and information processing, and systems for putting the correct response into effect i.e., a control system. Although the control systems in different animals vary widely, there are common themes that permit a functional approach in the following discussion.

FEEDBACK CONTROL SYSTEMS

Mechanical & Electrical Systems

A simple mechanical system comprised of a faucet at the bottom of a standpipe represents a steady state system with no control. The faucet serves as a regulator, allowing water to flow only in some fixed relation to the size of the orifice in the faucet. From the standpoint of engineering analysis this system has the general elements of a process rate, the water flow, and an effector mechanism, the faucet. It is not, however, a controlled system, since as the water level in the standpipe gradually falls, the rate of flow will decrease (Fig. 8.1).

In order to make the system into a complete controlled system some further elements must be added. In particular, a flow sensor, a set-point reference, an error comparator, and a servo valve would be required in order to create a completed closed-loop, negative feedback control system (Fig. 8.1). Each of these elements has important properties that are relevant to biological as well as physical systems.

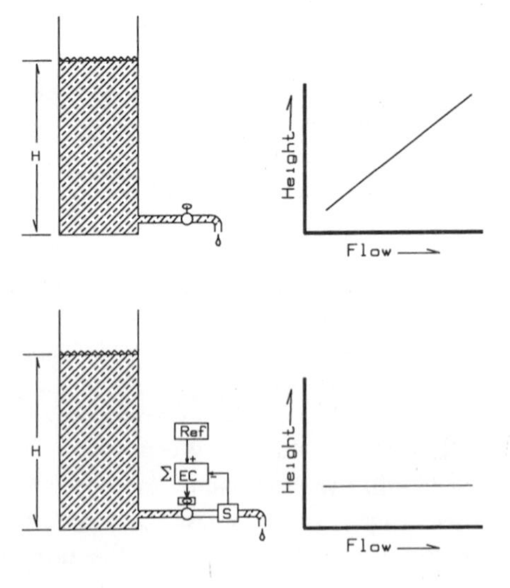

Figure 8.1: Engineering examples of controlled and noncontrolled, steady-state systems. (Upper) At any faucet setting, there will be a linear relationship between reservoir height (pressure) and output flow. (Lower) With addition of flow sensing, a reference signal, a summing error comparator, and a servo valve, the system is controlled and maintains constant output. See text for further explanation.

Sensors (Receptors)

The function of the sensor is the most obvious: In order for any process to be controlled, information in some form must be provided concerning its rate. The sensor is usually a transducer as well, which means that it changes information from one form to another. In this example the bulk flow rate of water might be represented by the magnitude of an electrical voltage, but many other forms of information transduction and coding are imaginable. The sensor provides the information to other elements of the system, usually in a very low-power fashion. The water flow rate might be thousands of gallons per minute, for example, and the electrical signal a 4 to 20 mA current loop.

Set-Point Reference

In addition to information about the present flow rate, information about the desired flow rate must be provided in the form of a set-point reference. For the water flow example, it might simply be a rheostat controlling voltage. Information in this pathway may or may not be of the same form as that from the sensor element but is also usually very low powered.

Error Comparator

The error comparator, a critical element, has the function of comparing the present output of the system to a desired output (the set-point) and generating an error signal that is predictably proportional to the difference (system error). Usually the sensor signal is inverted before being summed with the set-point at the comparator, giving rise to the term "negative feedback." The signal from the error comparator is similarly low-powered and may or may not be of the same form as information from other elements.

Effector Mechanism

Finally an effector is required, i.e., some way to correct the error (if any) and change the rate of the controlled process. In the present example a screw-type valve actuated by a motor would suffice; application of a positive voltage would drive it open, a negative voltage closed. Since the effector often requires considerable power output, there is usually power amplification at this stage and sometimes another change in the form of the information.

Feedforward Elements

In addition to the feedback elements described above, many controlled systems also have feedforward elements, or commanders. In the example of Fig. 8.1 the reference serves as a feedforward element, setting the target or goal for the system, in this case a particular flow rate. Feedback and feedforward elements are combined in the general control model shown in Fig. 8.2. Houk (1988) has also described a third type of control element that he calls an adaptive controller. An adaptive controller is one that acts on a longer time scale to change the characteristics of the feedforward controller or, less often, the feedback controller.

We may now examine how this system might work. A change in the reference from zero to some positive flow rate would exert a feedforward command for flow to begin. Since the feedback elements at first would indicate no flow, the error signal would be large and the valve would begin to open at a rapid rate. As the flow approaches the reference level, the error signal decreases until a stable flow is reached at a zero error signal. If the level in the reservoir were to rise by 20%, the pressure drop across the valve would be increased proportionately, leading to a 20% increase in water flow. The increased water flow would cause a greater electrical signal from the transducer, which in turn would cause a change in the error comparator's output from zero to negative. This inverted signal would then be passed along to the solenoid valve where it would cause a reduction in coil current, partially closing the valve to restore the error signal to zero and the flow to the desired set-point. Thus the system acts in a self-regulating manner to maintain output (flow) at a constant value despite variations in the pressure head.

In the real world things are a bit more complex, particularly when rapid fluctuations in disturbing conditions may be present. If, for example, the pressure head in the reservoir of Fig. 8.1 (lower) fluctuated, and there were a finite time lag between when the valve received a change in control signal and when it actually completed its adjustment, oscillations would very likely result. In order to stabilize such systems the controlling network will usually sense not only the rate of the controlled process but its rate of change. Mathematically the rate of change functions requires the incorporation of first and sometimes second derivatives in

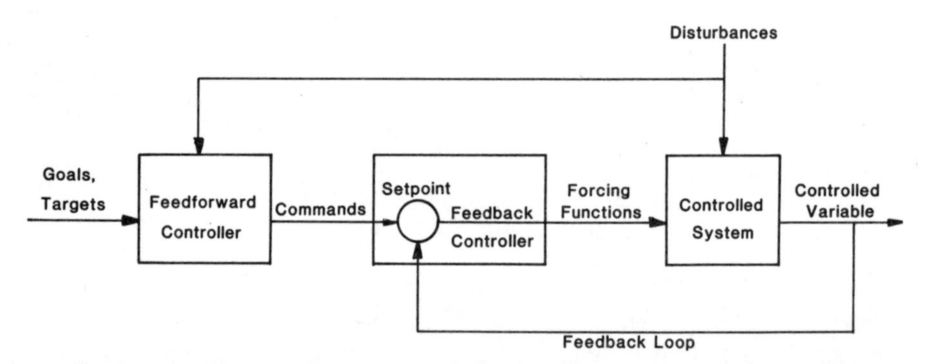

Figure 8.2: Controlled system. The feedforward controller processes information about goals or targets as well as information about potential disturbances into commands that are sent to the feedback controller. The feedback controller generates forcing functions to the controlled system itself, controlling an output variable called the controlled variable. Information about the controlled variable is sent back to the feedback controller via a feedback loop. (Modified from Houk, 1988.)

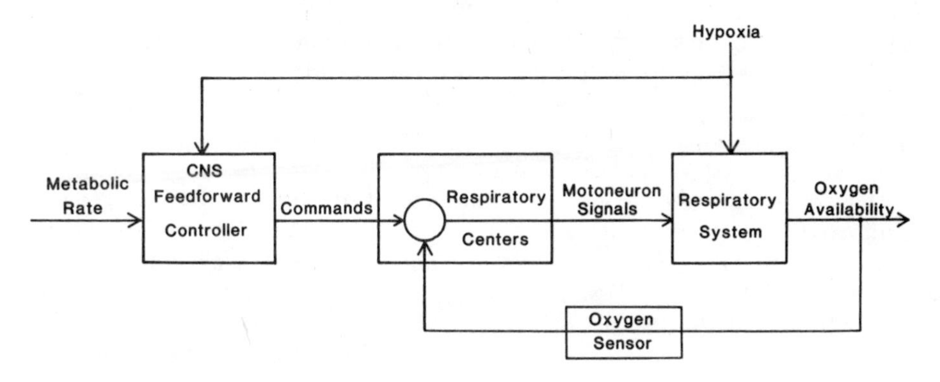

Figure 8.3: The respiratory system as a controlled system. The availability of oxygen (in blood or tissue) is determined by the respiratory system operating under the influence of the forcing functions from the feedback controller. This forcing function is in turn a product of sensory feedback, feedforward command level, and neural processing of this information into the motoneuron signals that constitute the forcing function. External disturbances, such as hypoxia, have a direct effect and may also lead to a change in the feedforward command level by indirect pathways.

the function describing the relation of sensor output to process rate (Milhorn, 1966; Milsum, 1966).

Biological Systems

The engineering approach is well suited to the study of biological control systems, particularly in that it requires identification and characterization of the elements critical to achieving a stable controlled system. We must know what the controlled process is, how it is sensed, what the set-point reference mechanism is, what the effector mechanisms are, and how the loop is closed to effect control. In biological systems, however, it is often difficult to identify and characterize the various elements, so much of our analysis is based upon inference and incomplete information about the control system. As a rule it is the information elements of the system that are less well known, i.e., the sensors and the neural links between sensor, central nervous system, and effector organs. Identifying the critical elements of the control system does, however, serve to focus thinking and experimental effort.

The respiratory system may be represented as a controlled element with the availability of oxygen as the primary output variable (Fig. 8.3). Oxygen sensors provide a feedback loop to the feedback controller, which sums that information with a command signal from the feedforward controller to produce an error signal. This error signal is processed into appropriate motoneuron impulses to drive the breathing apparatus. The feedforward (command) signal is based upon the system goal, which is to provide oxygen adequate for the animal's metabolism.

Disturbances to the system, such as hypoxia, may act at both the breathing system and the feedforward controller. Physiological evidence that the respiratory system does behave as a controlled system is abundant, so the objective of this chapter is to examine in detail the elements of the control system and the responses of the system to various disturbances.

It is usual to discuss phenomena common to all animals in rough phylogenetic order. In the case of respiratory control, however, so little is known about the invertebrates that the reverse approach is taken here. That is, the control systems of vertebrates will be discussed first and then one invertebrate class examined for important similarities and differences. In both cases we are concerned with the same general elements; i.e., the total metabolic demand for O_2 delivery and CO_2 excretion are the primary feedforward elements, and the values of P_{O_2}, P_{CO_2}, and pH are the primary determinants of the feedback (control) elements.

FISHES: CONTINUOUS WATER-BREATHING

Receptors

Fish not only have to deal with a highly variable external gas environment but also require homeostasis in the face of internal changes, as from exercise. In addition, respiratory responses to various noxious external stimuli may also be required. As Johansen (1970) surmised, then, there are several classes of receptor which may conveniently be categorized as internal, external, and defense receptors (Jones & Milsom, 1982; Shelton et al., 1986; Smatresk, 1986, 1988a,b).

Defense Receptors

Gas exchangers are typically delicate tissues with a high surface area and a thin blood-medium barrier. It is therefore important that an animal be able to sense potential dangers and to have receptors with appropriate location and characteristics. These receptors may warn of mechanical damage or of damaging or simply unfavorable conditions in the environment. Receptors that respond to high-level stimuli representing damaging conditions are called *nociceptors*.[1] Not a great deal is known about defense receptors in fish, but Poole and Satchell (1979) have described nociceptors in gills of dogfish that are responsive to mechanical stimulation and to various chemical stimuli, and Smatresk (1988a) has characterized receptors in the nares of the gar (*Lepisosteus osseus*) which reflexly inhibit gill ventilation in response to chemical irritants and high salinity. There is also a startle response in many fish: vibration or light changes can induce a brief apnea and bradycardia (Cameron, unpublished observations). Finally, the mechanical stimulus of foreign objects in the inhalant water can induce a "cough" reflex, a brief but forceful reversal of the ventilatory current, sometimes accompanied by extreme adduction of the jaws and operculi (Satchell & Maddalena, 1972).

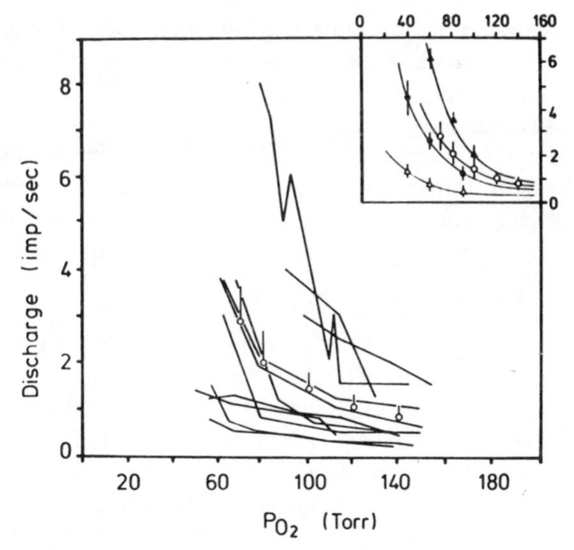

Figure 8.4: Frequency discharge characteristics of O_2 chemoreceptors from tuna gills compared with those from the cat (inset, upper right). The open circles indicate the mean for all fibers. (From Milsom & Brill, 1986, by permission, ©Elsevier North Holland, Inc.)

External Chemoreceptors

Because the response of gill ventilation to external hypoxia is quite rapid (see below), the existence of receptors sensitive to external oxygen in fish has been suspected for a long time (Daxboeck & Holeton, 1978; Smith & Jones, 1978; Smith & Davie, 1984). To date, however, the actual cells have not been identified. Dunel-Erb et al. (1982) have described a type of neuroepithelial cell in the primary lamellae that appears very similar to mammalian chemoreceptors, however, and there is now clear evidence of oxygen chemoreception from the gill nerves in both tuna (Fig. 8.4)(Milsom & Brill, 1986) and channel catfish (Burleson & Smatresk, 1988). In the tuna study only the first gill arch was examined, but in channel catfish the oxygen chemosensitivity seemed generally distributed over all the gills, since all branchial nerves had to be sectioned to abolish the reflex responses. Some earlier workers had focussed on the pseudobranch as a receptor site (Laurent, 1967; Laurent & Rouzeau, 1972), but it now appears that the reason pseudobranch denervation has little effect on respiratory reflexes (Randall & Jones, 1973; Bamford, 1974) is the apparent redundancy of receptors in other gills.

In the elasmobranch fishes Butler et al. (1977) have presented evidence for oxygen-sensitive receptors widely distributed within the branchial cavities and innervated by cranial nerves V, VII, IX, and X. This pattern seems consistent with the general distribution of gill receptors found in teleosts.

It should be noted that the designation of these oxygen chemoreceptors in the gills as "external" does not necessarily mean that they directly face the water. It would be sufficient for the receptors to be located within a short diffusion distance of the water. Indeed, Smatresk's (1986, 1988a) work on gill receptors in gar shows a response time of a few seconds, consistent with a somewhat (functionally) removed location within the gills but still too short to point to a location downstream in the circulation from the gills, due to circulation lags. Milsom and Brill (1986) found that the response to a cessation of perfusion was more rapid than to a decrease in the PO_2 of the bathing medium in most, but not all, of the fibers, perhaps indicating mixed internal and external orientation. A high rate of oxygen consumption in the receptors themselves is also implied, a characteristic common to oxygen chemoreceptors in other phyla (Fidone & Gonzalez, 1986).

Internal Chemoreceptors

Using internal NaCN injections as a probe, Smatresk (1986) and Smatresk et al. (1986) have presented evidence for internal oxygen chemoreceptors in gar (facultative air-breathers) located between the ventral aorta and the gill vasculature. Similar receptors have been identified in the water-breathing channel catfish by Burleson and Smatresk (1988) using small CN^- pulses. In this position such receptors would provide effective monitoring of the venous oxygen tension in water-breathing fish and of the mixed venous systemic return and the return from the air-breathing organ in the gar. Barrett and Taylor (1984) have also presented evidence for venous (internal) O_2 chemoreceptors in elasmobranchs. A ready acceptance of this evidence is based somewhat on a "backward" perspective; i.e., since a separate population of internal receptors is found in other vertebrate groups, one is inclined to look for similar structure and function in the fishes. So far, however, there is no information on their exact location, cell identification, or afferent neural "wiring."

Mechanoreceptors

Various classes of mechanoreceptor provide feedback information to fish for following the progress of muscular movements, including the rhythmic movements of ventilation and slower postural movements (Sutterlin & Saunders, 1969; Ballantijn & Roberts, 1976; Ballantijn, 1982). Pasztor and Kleerkoper (1962) described the importance of the smooth musculature of the gill filaments in positioning and maintaining the proper shape for the water sieve, and Sutterlin and Saunders (1969) speculated that the proprioceptors in the filaments might provide phasic feedback during the respiratory cycle. More recently de Graaf and Ballantijn (1987) and de Graaf et al. (1987) demonstrated that the proprioceptors of the filaments of carp have high thresholds, respond more to mechanical disturbance, and are probably best categorized as nociceptors. Another class of proprioceptor relays information about the position of the gill arches, and these nerves do display phasic discharge patterns linked to the respiratory cycle. Milsom and Brill (1986) also described many kinds of bursting fibers in the first gill arch of tuna and thought they might be associated with transmittal of rhythmic and tonic muscle information. Presumably these various mechanoreceptors provide feedback information to the motor and pattern generator areas of the brain through the cranial nerves. Clamping of the jaw in a fixed position produces

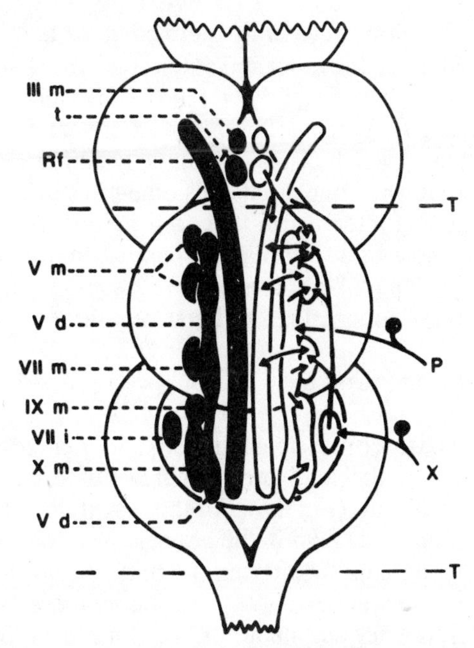

Figure 8.5: Centers of the brain involved in generation of the respiratory drive in fish, projected onto the dorsal surface. (Left) Main sites (in black) located below paired optic lobes and in medulla below cerebellum and facial lobe. (Right) Main interconnections (arrows). Dashed lines (T), area defined as essential for normal rhythmic breathing by transection experiments. III m, oculomotor nucleus; t, tegmental respiratory neurons; Rf, reticular formation; V m, trigeminal motor nucleus; V d, descending trigeminal nucleus; VII m, facial motor nucleus; IX m, glossopharyngeal motor nucleus; VII i, intermediate facial nucleus; X m, vagal motor nucleus; P, muscle proprioceptive input; X, vagal sensory input. (From Shelton et al., 1986, by permission, ©Amer. Physiol. Soc.)

modification of the motor discharges in the respiratory muscles within 0.2 second, suggesting a rapid reflex modulation of the respiratory rhythm (Ballantijn & Roberts, 1976).

Afferent Neural Connections

Although the nervous systems of fishes have been studied for a long time (Nicol, 1952; Campbell, 1970), information on afferent pathways in the respiratory control system could hardly be worked out without the sensors having first been identified. Work on the location of receptors within the gills dictates that the vagus (X) nerve carries the principal afferent input (Smith & Jones, 1978; Smith & Davie, 1984; Smatresk et al., 1986), but cranial nerves V, VII, and IX also carry input (Smatresk, 1988b), particularly from sensory elements in the buccal and branchial chambers (Butler et al., 1977; Daxboeck & Holeton, 1978; Burleson & Smatresk, 1988).

Central Integration

A large number of studies have been concerned with mapping the distribution of the respiratory elements of the fish brain (see review by Shelton et al., 1986). Respiratory information is apparently processed within the medulla; both afferent and efferent (motor) pathways are confined to the cranial nerves and do not extend caudally into the spinal cord. Removal of the fore- and midbrains does not interfere with the respiratory rhythm. Within the medulla, however, the situation is evidently quite complex and is presently understood incompletely. Some of the input and output paths are known from intracellular microelectrode recording (Fig. 8.5), but the basic respiratory pattern appears to be generated in concert by rather large populations of interneurons and motor neurons, both classes of which display rhythmicity. Different groups of cells can be found that exhibit rhythmic discharges timed to a brief interval of the respiratory cycle, or to longer intervals, or that have continuous phasic output. Besides the reflex arc of sensor-brain-ventilatory muscles, there must also be myriad connections allowing input of other sensory information from other regions of the brain, but about these connections we know very little.

Effector Limbs

The three principal functions under the control of the respiratory integrating center are the ventilation minute volume, cardiac output, and the perfusion pattern of the gills (Fig. 8.6). For the first of these functions, the motor pattern generator function resides in the brain, as just discussed, and the resultant muscle action is modulated by altering the pattern and intensity of discharges from the motor neurons activating the ventilatory muscles (Ballantijn & Alink, 1977). Both the frequency and depth of breathing are modified in response to various stimuli, with attendant adjustments of the gill filaments, mouth parts, and operculi.

The "pattern generator" for the heart is the rhythmic discharge of the sinoatrial (SA) node (see Chapter 4). The major afferent pathway between the brain and the heart is the vagus nerve, which exerts an inhibitory effect on the heart when active. Normally the vagus supplies a tonic inhibitory signal; when vagal input is blocked by atropine or by sectioning, the heart accelerates; and when acetylcholine is applied experimentally, the heart slows (Randall, 1970a). Car-

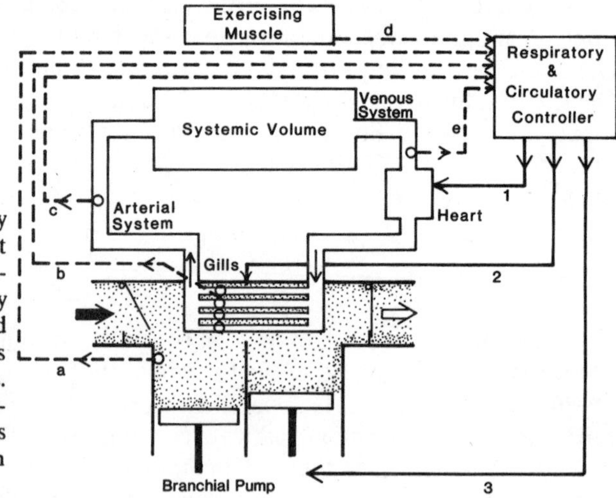

Figure 8.6: The respiratory control system in a teleost fish. Efferent (sensory) pathways (a–e) are indicated by the heavy dashed lines, and afferent (motor) pathways (1–3) by the heavy solid lines. Only scanty evidence is available for all efferent pathways except b. (Modified from Shelton, 1970.)

diac output is determined by several factors in addition to the neural input via the vagus, with the next strongest influence exerted by preload, i.e., the venous filling pressure (Farrell et al., 1982). The vagus evidently also contains cardioacceleratory fibers: Gannon and Burnstock (1969) found adrenergic fibers using fluorescence staining methods, but epinephrine has a much smaller effect than either vagal activity or changes in preload (Farrell et al., 1982).

Reflex Responses

With the overall objective of respiratory homeostasis in mind, we may now examine the various reflex responses of the respiratory system in fish to see how the several elements of the control system operate. A primary problem for fish is the limited availability of oxygen in the aquatic environment and the sometimes extreme variability of ambient gas conditions. A common, perhaps primary response of fish to hypoxia is to increase swimming activity in order to move out of the hypoxic area (Høglund, 1961; Smatresk, unpublished data). Nearly all fishes respond to decreases in ambient oxygen by increasing ventilation (Shelton et al., 1986). Early reports indicated that responses of elasmobranchs were much smaller than those of teleosts, but more recent work on well-rested and undisturbed dogfish has demonstrated responses of similar magnitude (Metcalfe & Butler, 1984a). There is a great deal of species variability in tolerance to low oxygen and in the pattern of ventilatory changes as a function of oxygen. Some animals maintain a nearly constant oxygen consumption by proportional increases in ventilation, some seem to overcompensate, and others increase ventilation only enough to partially compensate for the reduced oxygen supply (Fig. 8.7). There is often little correspondence between the ventilation volume and frequency, illustrating the changes in stroke volume and the complexity of the central response. The hypoxic reflex is demonstrably related to the changes in discharge pattern of the O_2 chemoreceptors in the gills (Milsom & Brill, 1986; Smatresk et al., 1986; Burleson & Smatresk, 1988) and is usually accompanied by a bradycardia. Hypoxic bradycardia is a bit of a puzzle: in order to maintain oxygen delivery to tissues, cardiac output should be increased. The bradycardia, however, is usually accompanied by an increase in stroke volume which may result in constant or increased blood flow (Holeton & Randall, 1967; Taylor et al., 1977; Taylor & Barrett, 1985). Coupled with changes in blood flow distribution within the gills (Metcalfe & Butler, 1984b), transit time and gas exchange may actually be increased during hypoxia. Although the bradycardia is mediated by vagal inhibition, the central connections for this reflex are only beginning to be worked out (Barrett & Taylor, 1985).

The response of ventilation to hyperoxia is more or less opposite the hypoxic response: Ventilation decreases, resulting in little change in overall oxygen delivery (Randall & Jones, 1973; Dejours et al., 1977). There is a bradycardia in hyperoxia as well, but it is not clear if cardiac output actually declines. Interestingly, when ventilation (and cardiac output?) are reduced in hyperoxia, the P_{CO_2} of the blood rises. In air-breathing vertebrates this rise in arterial P_{CO_2} would be sufficient to restimulate ventilation, but it does not do so in the fish (Dejours et al., 1977). This brings up a general controversy as to whether there is a true ventilatory response to hypercapnia in fishes. The question was confused by older studies in which mammalian CO_2 concentrations in the range of 5 to 10% were employed. These concentrations serve as strong irritants or even anaesthetics in fish and so confound any hypercapnic reflex with behavioral and other non-

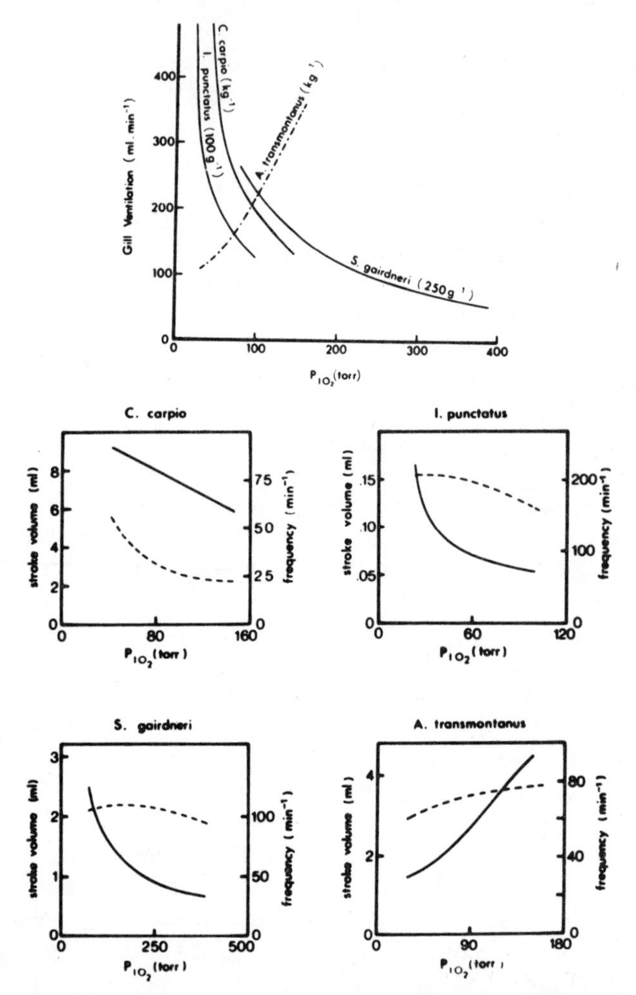

Figure 8.7: Responses of ventilation in various species of fish to hypoxia. The upper panel shows total minute volume, and the lower panels stroke volume and frequency responses. The catfish (*Ictalurus punctatus*) data are from Gerald & Cech, 1970; carp (*Cyprinus carpio*) data from Lomholt & Johansen, 1979; sturgeon (*Acipenser transmontanus*) data from Burggren & Randall, 1978; and rainbow trout (*Salmo gairdneri*) data from Randall & Jones, 1973. (Figure from Shelton et al., 1986, by permission.)

physiological reactions. The most careful studies of this question to date conclude that there is a small response to CO_2 but that it is probably an indirect effect of the reduction in oxygen content of the blood caused by elevated CO_2; i.e., an indirect oxygen response (Shelton et al., 1986).

There is also a coordinated set of reflex responses to exercise (swimming) in fish consisting of increased ventilation, increased cardiac output, and increased perfusion of secondary lamellae in the gills. Part of the reflex is mediated by adrenergic receptors in the heart and in the gill vasculature (Steen & Kruysse, 1964; Payan & Girard, 1977; Farrell et al., 1982), part by oxygen responses of the internal chemoreceptors, and part by direct input from skeletal muscle mechanoreceptors (Fig. 8.6; Shelton, 1970; Cunningham et al., 1986). Since infrared photographic studies (Davis, 1972) as well as model calculations (Cameron & Polhemus, 1974)(see Chapter 7) show that only about 20% of the secondary lamellar area is required to meet resting demands, there is room for a large increase in functional gill surface during exercise, temperature increase, or other demand-stimulating situations.

AIR-BREATHING FISH

A common environmental stress encountered by fish is reduction or complete absence of oxygen in the water, so it is perhaps not surprising that nearly every group of fish, including all primitive fish groups, have evolved mechanisms for air-breathing. The most common structural modifications are those involving the evolution of a lung or conversion of a swimbladder to a lung, but many other organs have been modified variously to serve the same purpose. The South American electric eel, *Electrophorus*, for instance, has a highly vascularized and modified buccopharyngeal chamber (Carter, 1957), and the gouramis a labyrinth organ branching off the opercular chamber (Burggren, 1979) Many catfish have modified stomach air-breathing organs (ABOs), and some fish breath through skin and fins (see Chapter 11). The addition of a second gas exchange organ operating in a different medium imposes a new level of complexity upon the respiratory control system.

Chemoreceptors

In order to monitor the efficacy of gas exchange in the lung, it would seem advantageous to possess a set of sensors that would provide information on the gas concentrations in the aerial medium. Since atmospheric air is for all practical purposes constant in its composition, however, sensors positioned to monitor the gas actually transferred to the blood would provide even better information. According to studies by Smatresk et al. (1986), the internal O_2 chemoreceptors of gar are ideally placed to provide this information. Increases in oxygen transfer across

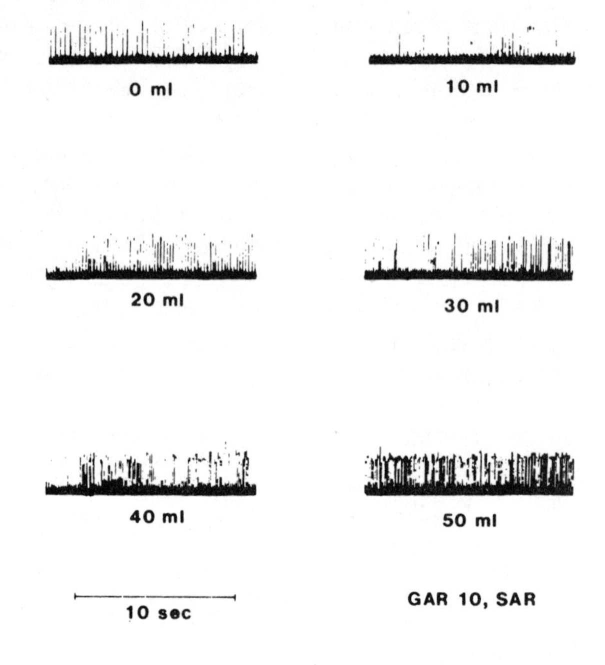

Figure 8.8: Discharge patterns of a single slowly adapting receptor (SAR) from gar lung at different degrees of lung inflation. (From Smatresk & Azizi, 1987, by permission, ©Amer. Physiol. Soc.)

the ABO (the swimbladder/lung in the case of gar) show up directly as increases in the mixed venous oxygen tension. So far as anyone knows there are no additional oxygen receptors actually in the ABO's of fish. Externally oriented chemoreceptors appear to be located on the first three gill arches in the lungfish *Protopterus*, and they provide information relevant to air versus water-breathing, as in aquatic hypoxia (Lahiri et al., 1970).

The question of CO_2 chemoreception arises again with the air-breathing fishes. Were the ABO the only gas exchange organ, control of P_{CO_2} in the lung air might be physiologically important. Nearly all of the air-breathing fish, however, continue to excrete the majority of their CO_2 across either the gills or a combination of gills and skin (Johansen, 1972; Romer, 1972; Smatresk & Cameron, 1982a). The primary function of the ABO, then, is to supplement O_2 supply, and CO_2 seldom reaches significant concentrations. There is actually some CO_2 sensitivity shown by certain stretch receptors in the ABO (see below), but it is arguable if it constitutes a CO_2 receptor class.

Mechanoreceptors of the ABO

In fish with swimbladder/lungs three classes of mechanoreceptors provide information about the degree of distension of the lung, the rate of change of distension, and to a lesser extent the P_{CO_2} in the lung. These receptors are known as slowly adapting receptors (SARs), which are essentially tonic receptors; rapidly

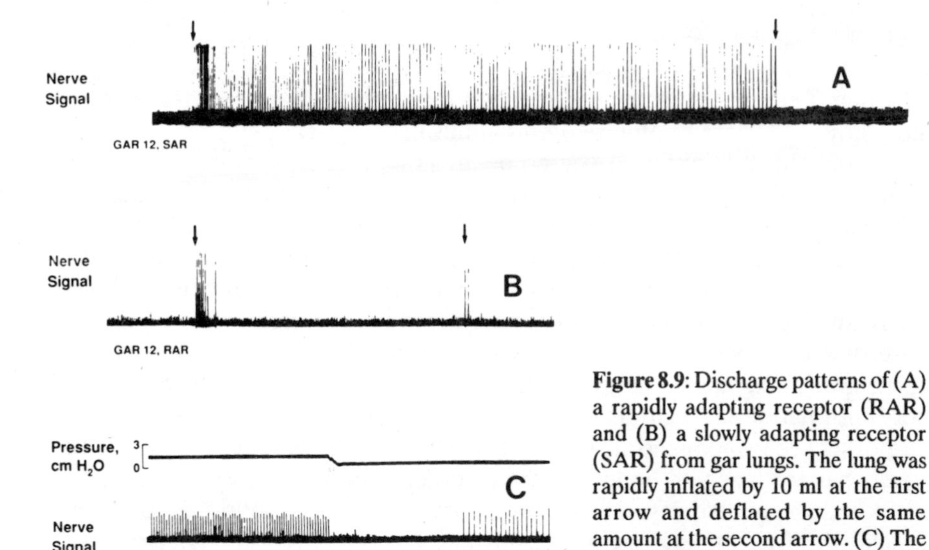

Figure 8.9: Discharge patterns of (A) a rapidly adapting receptor (RAR) and (B) a slowly adapting receptor (SAR) from gar lungs. The lung was rapidly inflated by 10 ml at the first arrow and deflated by the same amount at the second arrow. (C) The response of a slowly adapting receptor to step deflation. (From Smatresk & Azizi, 1987, by permission, ©Amer. Physiol. Soc.)

adapting receptors (RARs), phasic receptors which give rate-of-change information; and CO_2-sensitive SARs. The existence of these receptors was suggested by Johansen (1970), and they were first described in lungfish (*Protopterus* and *Lepidosiren*) by DeLaney et al. (1983). SARs were described for a holostean, the bowfin *Amia calva*, by Milsom & Jones (1985) and all three classes of receptor have recently been described in another holostean,[2] the spotted gar *Lepisosteus oculatus*, by Smatresk and Azizi (1987).

The tonic discharge pattern of SARs is typical of that found in all vertebrate air-breathers (Fig. 8.8). As the degree of lung inflation increases, and therefore the degree of stretch increases, the frequency of SAR discharge increases. In the gar and bowfin the response to an increase in lung inflation is not purely tonic: There is an initial overshoot in frequency lasting some seconds, which represents a phasic signal (Fig. 8.9A). There is also a range of volume sensitivity in the SAR population such that at zero lung pressure some units fire and some do not. As the lung is inflated, more and more of the units are activated, with each maintaining a steady long-term discharge frequency at various constant volumes.

RARs, on the other hand, provide a purely phasic, or rate-of-change signal (Fig. 8.9B). In the gar the activity burst produced in the RARs by increasing lung volume lasts only 1 to 2 seconds.

Finally, some of the SARs exhibit CO_2 sensitivity. High levels of ambient CO_2 inhibit the discharge, but since the concentrations necessary for inhibition are so high in relation to normal physiological values of lung P_{CO_2}, the significance of these CO_2-sensitive SARs in fish lungs is not clear.

Central Integration

Since air breathing in fish is supplemental to gill ventilation, there is apparently no central pattern generator for air breathing in the air-breathing fish. Rather, it is an "on-demand" system triggered by integration of sensory input from four sources. These sources are the externally oriented oxygen chemoreceptors associated with gill reflexes in both air- and water-breathing fish, the internally oriented oxygen chemoreceptors described for air-breathers but presumed to exist in water-breathers as well, the mechanoreceptor information from the lung itself, and modulating input from various other sensory pathways including the defense receptors described above from the nares. Each of these sources of information is no doubt complex, and the possible combinations of peripheral sensory information and central response probably account for the diversity of reflex responses seen among the air-breathing fish (Smatresk, 1988b; and below). The whole pattern may also be influenced by other behavioral considerations; e.g., air breaths may be delayed when predators are present above the water, and synchrony of air breaths is conspicuous in schooling air-breathers (Kramer & Graham, 1976).

Effector Limbs

Besides the efferent pathways involved in controlling gill ventilation (previous section), the air-breathers must possess a second complete set of reflexes to ventilate whatever ABO they may possess. In some cases this may amount to no more than filling the branchial chamber with air instead of water, but in a fish such as the gar a complex sequence of events involving the swimming musculature, branchial muscles, air duct sphincter, and lung smooth muscle must occur (Rahn et al., 1971; Smatresk & Cameron, 1982b). The diversity of ABO structure and ventilatory mechanisms is great, and knowledge about any one of them is so scarce that it is not possible to make any general statements.

Reflex Responses

The varied responses of air-breathing fish to a variety of stimuli have been reviewed recently by Smatresk (1988b). From Table 8.1 it is clear that general rules are difficult to state. Closer examination of the species involved, however, may allow some generalizations about the response to P_{O_2} of the inspired water. In the most accomplished air-breathers such as *Protopterus*, there is usually little response to ambient aquatic hypoxia. In fish with only a limited ability to breathe air, the response is not very different from wholly aquatic fishes. In the gar, a facultative air-breather, the initial response to decreased ambient oxygen is to increase ventilation; but as P_{O_2} falls further, gill ventilation is inhibited and air breathing is stimulated. There is thus some central modulation of the response to hypoxia to help optimize the use of the gas exchange organs.

Table 8.1: Summary of responses to various stimuli that affect air-breathing (either frequency or tidal volume), gill ventilation (frequency or stroke volume), and heart rate in air-breathing fishes.

Stimulus	ABO	Gill	fH
Aquatic hypoxia	+, nc	+, –, nc	+, nc
Aquatic hyperoxia	–, nc	–, +	–
Aerial hypoxia	+, nc, –	+	
Aquatic hypercapnia	+, nc	+, –, nc	
Aerial hypercapnia	+, nc	+	
Activity	+	+	+
Increased temperature	+	+, –	+
Lung deflation	+, nc		–
Lung inflation	–, nc		+
Emersion	+	–	+, –
Irritants		–	

ABO, air breathing; Gill, gill; fH, heart frequency; +, stimulation; –, inhibition; nc, no change. Data from various sources compiled by Smatresk (1988a).

Within a particular species the reflex responses may also vary depending upon other sensory input. For example, in well-aerated water the response of the gar to lung deflation is absent, whereas at high temperature or in poorly aerated water lung deflation stimulates an air breath (Smatresk & Cameron, 1983). While this may actually be a chemoreflex rather than a stretch response (Smatresk & Azizi, unpublished data), the point about modulation is still valid.

The situation with regard to CO_2 control is no clearer than for the fully aquatic fish. As shown in Table 8.1, studies of hypercapnic responses show increases, decreases, or no change in both gill and lung ventilation. Even in those cases where there does appear to be a hypercapnic stimulation of breathing, the same question of whether the effect is direct or indirect via a reduction in oxygen content (Bohr or Root effect) has not been settled. The question does have evolutionary interest, however, since not only are there clear CO_2 responses in the higher vertebrates, they usually predominate over oxygen reflexes (see next sections).

A final category of reflexes concerns the distribution of blood flow between the systemic and gill vasculature. In a sense these two circuits compete for the available cardiac output; a decrease in the vascular resistance of the lung circulation, for example, will result in an increase in lung blood flow at the expense of the dorsal aortic (systemic) flow. Various authors have presented evidence that lung blood flow increases immediately following an air breath and declines throughout the breath interval, which may be an hour or more under some conditions (Farrell, 1978). Although the changes in blood flow may not provide exact matching of ventilation and perfusion, the purpose is probably matching of perfusion to gas exchange occurring in the ABO, which is nearly the same thing. The sensory input triggering this vasomotor reflex is not known; it could be information from the chemoreceptors, from lung mechanoreceptors, or even a local response of the lung vasculature itself. Whatever the source, the result would be to maximize the exchange capacity early in the breath interval when the lung PO_2 is highest.

As a final note, in anoxic water most air-breathing fish rely entirely on their ABOs for oxygen supply and continue to excrete most of their metabolic CO_2 through the gills, the skin, or a combination of both. Breathing in the ABO is an intermittent phenomenon, and gas tensions in the blood oscillate in phase with the air breath cycle. Under these conditions, very few features of their respiratory systems distinguish them from their (supposed) descendants, the ectothermic amphibians and reptiles.

INTERMITTENT AIR-BREATHERS: AMPHIBIANS AND REPTILES

If the air-breathing fish were like the water-breathers except with the addition of an air-breathing organ, the reptiles and amphibians can be considered like the air-breathing fish except that most lack gills. The on-demand, intermittent breathing associated with the ABO of fish has been fine-tuned, to be sure, but remains as the principal gas exchange modality in these two groups (Shelton et al., 1986). Complicating the picture overall is the great range of habits and habitats of both amphibians and reptiles. Although the amphibians are on the whole the more aquatic of the two, turtles and many lizards are partly or wholly

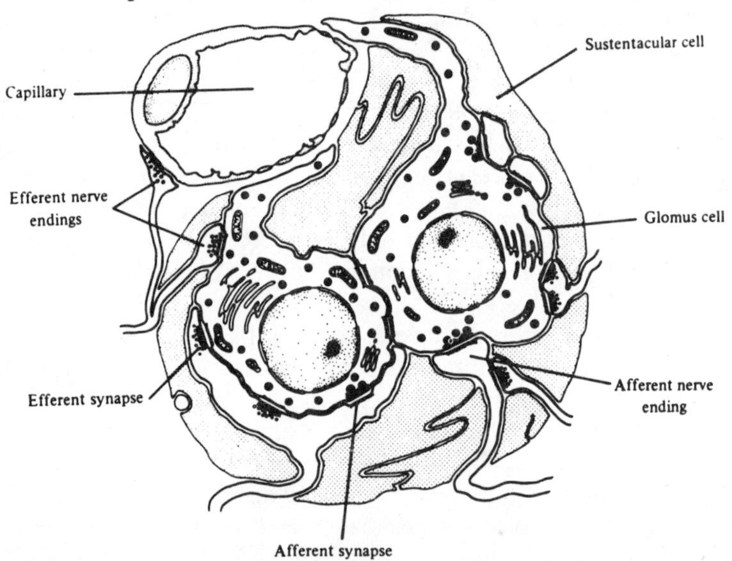

Figure 8.10: Structure of a typical vertebrate chemoreceptive unit. The unit is composed of glomus cells, sustentacular cells, and both afferent and efferent nerve endings. The calyx type nerve ending makes reciprocal synaptic contact with the glomus cell. Sympathetic efferent nerves are shown innervating both a glomus cell and a fenestrated capillary. (From Jones & Milsom, 1982, by permission, ©The Company of Biologists, Ltd.)

aquatic, and many are good divers. Turtles are among the most tolerant of anoxia of any animals known (Ultsch & Jackson, 1982), and even some snakes are aquatic, carrying out much of their gas exchange cutaneously (Heatwole & Seymour, 1975). There is therefore a continuous spectrum within these two groups from unimodal aquatic breathers (e.g., neotenous salamanders), to bimodal but principally aquatic forms (ranid frogs), to bimodal but principally terrestrial forms (the toad, *Bufo*; various turtles), and finally unimodal air-breathers with only a minor cutaneous respiration component (e.g., tortoises). Nevertheless there is a comforting similarity in the basic elements of the control systems, both within the two groups and when compared with other vertebrate classes.

Peripheral Sensory Input

Defense Receptors

The necessity for sensing adverse environmental conditions is no less acute in the amphibians and reptiles than in the fish, although the stimuli may differ. When frogs submerge they normally do not take water into the mouth and of course need to keep water out of the lung. Receptors around the nares and mouth respond to water stimuli, provoking an apneic reflex not only in frogs but in newts and other amphibians (West & Jones, 1976; Jones and Milsom, 1982). The response involves mechanoreceptors innervated by the fifth (V)cranial nerve. A second line of defense is provided by water-sensitive taste receptors on the tongue (Zotterman, 1949).

In diving and swimming reptiles there must also be a means for preventing water from entering the lung. Instead of reflexes for mouth closure, however, the mouth remains open in many diving reptiles, and the receptors that were the second line of defense in amphibians become the primary one. That is, receptors in the posterior nares, larynx and glottis all form part of a reflex arc that produces apnea upon submergence and may also stimulate lung contraction and a reduction in blood flow to the lung (Jones & Milsom, 1982).

Pulmonary Stretch Receptors

In both the amphibians and the reptiles pulmonary stretch receptors have received considerable study. In both frogs and turtles the stretch receptors are generally distributed in subepithelial connective tissues in the walls and dividing septa of the lung. The response characteristics of the receptors fall into the same three categories as described above for air-breathing fish; i.e., there is a class of tonic, slowly adapting stretch receptors that provide volume (distension) information; a second class of rapidly adapting receptors that provide rate-of-change information; and a third, intermediate class that show burst activity upon volume change but also have a tonic discharge. The situation is complicated further by a variable chemosensitivity to CO_2. Milsom and Jones (1977; Jones & Milsom, 1979) have described a polymodal distribution of CO_2 sensitivity in both frog and

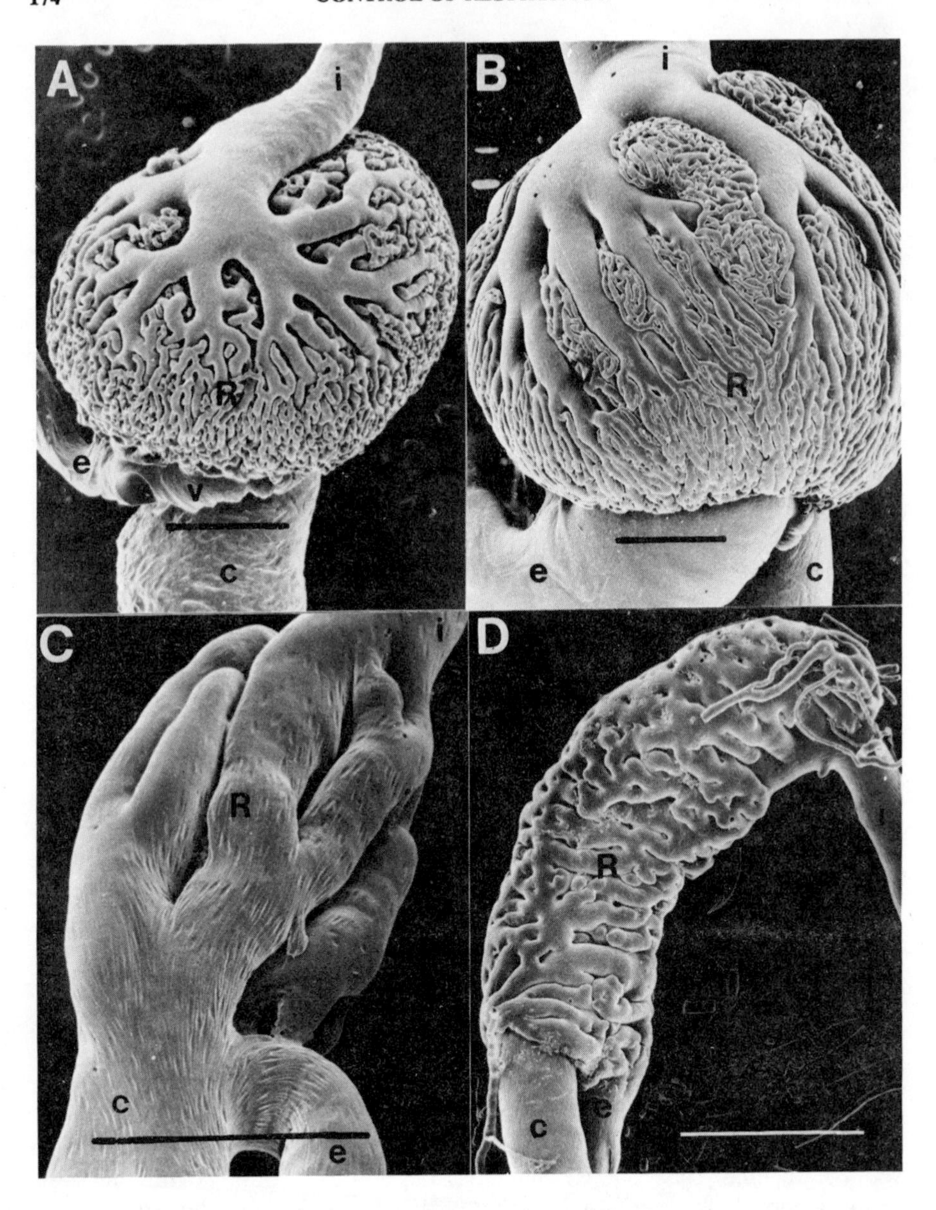

Figure 8.11: Scanning electron micrographs of vascular casts of carotid labyrinths of various anuran and urodele amphibians. The internal carotid (i) originates in the labyrinth rete (R) and flows to the upper jaw, brain, and eyes. The external carotid (e) supplies the lower jaw region and arises recurrently from the labyrinth, sometimes forming a vascular ring (v) around the common carotid artery (c). (A) *Rana catesbiana*, bar = 100 μm. (B) *Bufo marinus*, bar = 1 mm (C) *Amphiuma tridactylum*, bar = 100 μm. (D) *Ambystoma maculatum*, bar = 1 mm. (Toews et al., 1982, by permission; photographs courtesy of D. P. Toews.)

turtle, and there appears to be some interaction between lung stretch and CO_2 concentration (Jones & Milsom, 1979, 1982).

Peripheral Chemoreceptors

In all of the vertebrates, including amphibians and reptiles, particular associations of cell types are associated with chemoreceptor activity. These associations consist of glomus cells, sustentacular cells, and nerve terminals (Fig. 8.10). The glomus cells are relatively large epithelial cells with extensions coming into close contact with arterial capillaries. The glomus cells contain a variety of neurotransmitter substances and intracellular machinery typical of endocrine synthetic cells (Jones & Milsom, 1982). The nerve terminals make synaptic contact with the glomus cells, and the sustentacular cells surround both, presumably providing nourishment and support, as their name suggests.

The distribution of these chemoreceptive units is variable but suggests a common derivation from the third aortic arch. Such cells have been described in fish gills (Dunel-Erb et al., 1982; see above) and occur in amphibians in the carotid labyrinth (Fig. 8.11), a structure so derived. The elaborate nature of the carotid labyrinth of amphibians is not understood. Although it certainly has chemoreceptor function (Ishii et al., 1966, 1985a), the chemoreceptor function is still present in amphibians that have virtually no labyrinth (Toews et al., 1982). Various regions of the carotid circulation in reptiles, similarly derived embryonically, also contain the typical chemoreceptor cell associations, but in diffuse patches rather than in a discrete body (Jones & Milsom, 1982; Ishii et al., 1985b).

Despite differences in structure, the function of the carotid-type chemoreceptor unit appears to be the same wherever it is found. Denervation of the carotid chemoreceptor reduces normal breathing, and either abolishes or depresses the breathing responses to hypoxia and hyperoxia (Jones & Milsom, 1982; Ishii et al., 1985b).

In addition to the chemoreceptors of the carotid region, there are other islands of chemoreceptor tissue in the central cardiovascular system. In frogs there appears to be a chemoreceptor response from the pulmocutaneous arch (Lillo, 1980; Ishii et al., 1985a), and in reptiles there is a distinct area near the heart and supplied by the left aorta (Jones & Milsom, 1982; Ishii et al., 1985b). The functional significance of these secondary areas is not known at present, but it may be that the carotid units are more concerned with control of ventilation, and the aortic units are more concerned with blood flow distribution, responding more to systemic PO_2 and PCO_2.

Central Chemosensitivity

As was the case in the study of fish, prior to the identification and functional characterization of peripheral chemoreceptors there was a tendency to relegate the chemoreceptive function to the central nervous system. Although we now have ample evidence of peripheral chemoreceptor function, there is still the question of whether there is also significant central chemoreception and what the physiological significance of it might be. Hitzig et al. (1985) have shown that alteration of the acid-base status of the brain ventricles of the turtle by a perfusion technique does alter the chemical drive for ventilation. This study provides the only information for lower vertebrates, and it remains to be seen how such central responses may interact with peripheral input to produce the final system output.

Central Integration – No Pattern Generator?

In amphibians and reptiles the dominant breathing mode is intermittent, often characterized by a fairly predictable group or burst of breaths punctuated by an apneic period of variable length. This pattern is in strong contrast to the regular, rhythmic breathing pattern found in the fish, and once again in birds and mammals. For the aquatic amphibians and reptiles this breathing pattern is more easily understood than in animals such as lizards which are wholly terrestrial. The diving species usually complete a series of breaths when surfacing and of course are apneic when submerged. The details of breathing patterns in both amphibians and reptiles have been reviewed by Shelton et al. (1986).

There is not a great deal of information on the central organization of breathing centers in either amphibians or reptiles. Since an entirely different muscle set in the trunk is employed for aspiration breathing in animals such as turtles and alligators compared with the buccal breathing of the frog, motor and command fibers more caudal in the CNS must be involved. Work on the turtle has demonstrated bulbar respiratory neurons with discharge patterns similar to the mammalian respiratory neurons (Takeda et al., 1986), but how they are controlled during periodic breathing is not understood. There is clearly room for much future work.

Effector Limbs

For lung and buccal breathing the effector limbs are comprised of the motor pathways and muscle groups that lead to sequential expansion and contraction of the lung. Among the amphibians and reptiles there is great diversity of muscle groups utilized for this purpose, however. In alligators, for example, the liver is moved back and forth, acting as a piston pump for lung filling, and in many turtles there are axillary pockets of the air spaces that participate in ventilation. In frogs, the floor of the buccal cavity is used alternately to move air in and out of the mouth alone or to ventilate the lung. Many of the muscle groups used for respiration also function in feeding, locomotion, sound production, etc. There is clearly much plasticity in how the respiratory control system is applied to full effect in different species.

An interesting sidelight, too, is the question of if there is active control of gas exchange in the skin. Frogs, in particular, conduct the larger part of their CO_2 excretion through the skin, so there is at least the potential for control of this gas exchange via vasomotor reflexes involving the cutaneous circulation. By selectively exposing either lungs or skin of the bullfrog *Rana catesbiana* to hypercapnia, Jackson and Braun (1979), however, concluded that the skin was an uncontrolled or poorly controlled avenue for CO_2 excretion, and that the lung was the principal effector organ of P_{CO_2} control. More recent work, however, has shown considerable flexibility in cutaneous exchange: Many animals actively ventilate the skin, especially during aquatic hypoxia (Feder & Pinder, 1988); the absolute rate of blood flow as well as its distribution can be controlled to influence cutaneous gas exchange (Burggren, 1988a); and regulation of flow through the capillaries of the skin has an important regulating effect on gas exchange (Malvin, 1988).

Reflex Responses

Hypoxia

Since the primary challenge of environmental hypoxia is a reduction in the amount of oxygen supplied per unit volume ventilated, the appropriate response is an increase in ventilation minute volume. In both amphibians (Boutilier & Toews, 1977) and reptiles (Glass et al., 1983) there is a hyperbolic relation between inspired P_{O_2} and ventilation (Fig. 8.12) that serves to maintain Pa_{O_2} within fairly narrow limits. This reflex response may be mediated via the chemoreceptive areas in the carotid labyrinth (amphibians)(Ishii et al., 1985a) and the carotid regions (reptiles; Ishii et al., 1985b) involving the vagal afferents, central integra-

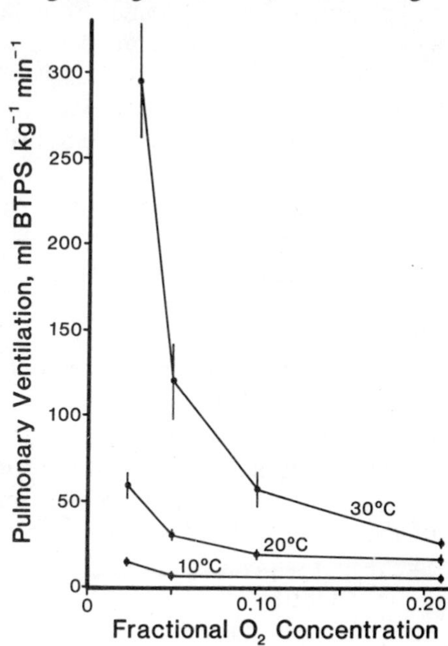

Figure 8.12: Relationship between ventilation and inspired O_2 in turtles at three body temperatures. (From Glass et al., 1983, by permission, ©Springer-Verlag.)

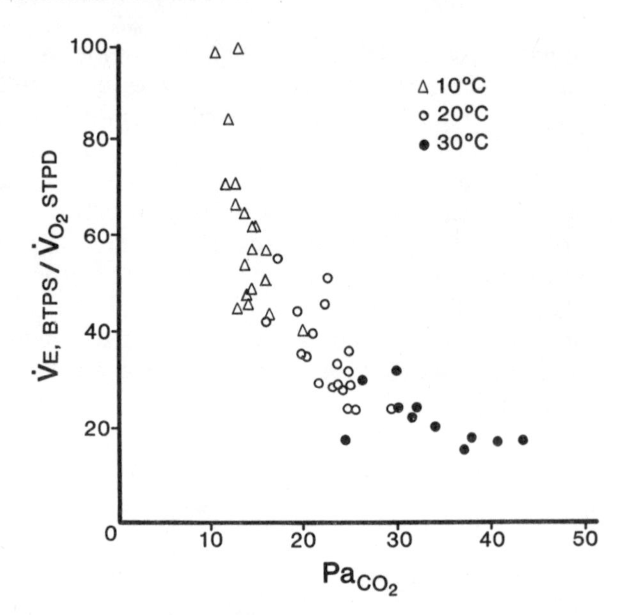

Figure 8.13: The ratio of ventilation to oxygen consumption declines in turtles as temperature increases. (From Jackson et al., 1974, by permission, ©Elsevier North Holland, Inc.)

tion of the reflex, and motor signals to the ventilatory musculature. Jones and Chu (1988), however, have found that the responses to hypoxia in *Xenopus*, the African clawed frog, are no different after carotid denervation, raising the possibility of either redundant sensors or a central oxygen chemoreceptor. During diving hypoxia and anoxia these reflexes are obviously overridden either by input from the defense receptors of the nares and glottis or by direct central (behavioral) input, i.e., commands from the feedforward controller. The response to hypoxia is quite temperature-sensitive, with only small responses at 10°C and very large ones at 30°C (Glass et al., 1983). Wood (1984) argued that part of the temperature sensitivity is explained by sensitivity of the oxygen chemoreceptors to oxygen content, as opposed to partial pressure. An alternative explanation is that at low temperatures in aerated water a considerable portion of the gas exchange is supplied through the skin, even in turtles (Ultsch & Jackson, 1982). The cutaneous supply combined with a much lower metabolic rate may serve to decrease the lung ventilation response.

Hypercapnia

CO_2 is a powerful stimulus to ventilation in both amphibians (Jackson & Braun, 1979) and reptiles (Jackson et al., 1974; Milsom & Jones, 1980). In the turtle *Chrysemys scripta* inhalation of 10% CO_2 produces a 7- to 20-fold increase in ventilation minute volume (Jackson et al., 1974; Milsom & Jones, 1980). The increase is accomplished in normal animals mainly by a decrease in the interbreath interval, with only minor changes in tidal volume; in vagotomized animals some response persists, but tidal volume is the principal variable and frequency is little changed (Milsom & Jones, 1980). It will be interesting to see if future research supports the pattern that these results suggest, i.e., that CNS chemosensitivity exerts more control over tidal volume, and the peripheral receptors (via the vagus) have the primary influence on frequency.

Temperature

Since metabolic O_2 consumption and CO_2 production increase with increasing temperature with a Q_{10} of about 2, it is to be expected that ventilation and cardiac output might increase accordingly. The requirement for maintaining an appropriate acid-base balance, however, dictates that ventilation must increase relatively less at higher temperatures in order to produce an inverse temperature-P_{CO_2} relation (see Chapter 6). Jackson et al. (1974) found a regular decline in the ratio of ventilation to O_2 consumption (Fig. 8.13), a relationship subsequently confirmed by Glass et al. (1985). The mechanisms by which this matching is controlled are unknown.

CONTINUOUS AIR-BREATHERS: BIRDS AND MAMMALS

Receptors

Carotid Bodies

Situated near the bifurcation of the carotid artery into its internal and external branches in mammals and close to the origin of the common carotid in birds are small ovoid organs known as the carotid bodies (Jones & Milsom, 1982; Scheid & Piiper, 1986). Their blood supplies branch off the carotid artery, and they are innervated by a branch of the vagus (X) nerve (Scheid & Piiper, 1986) in birds, and by the glossopharyngeal (IX) nerve in mammals (Fidone & Gonzalez, 1986). The interior of the carotids consists of closely packed associations of glomus, sustentacular and nerve cells, as described above (Fig. 8.10). The carotid bodies appear to have as their primary function the transmittal of information about arterial P_{O_2} and are important in the responses of both birds and mammals to hypoxia and hyperoxia. It is less clear the extent to which the carotid bodies are important in the responses to hypercapnia; recordings from single afferent fibers show CO_2 sensitivity, but the hypercapnic reflexes are not so strongly affected by deafferentation of the carotid bodies.

Aortic Bodies

The so-called aortic bodies are rather diffuse clusters of tissue similar in type to the carotid bodies but scattered among various locations near the heart along the aorta and major arteries. Although they display chemosensory responses, they seem to be more involved in pressure regulation, and their role in the chemoreflex responses is not nearly so clear as that of the carotid bodies (Cunningham et al., 1986). There is some experimental evidence that they respond to oxygen content as well as partial pressure (Fitzgerald & Lahiri, 1986). Evidence for the existence of aortic bodies in birds is sketchy (Jones & Milsom, 1982).

Pulmonary Receptors

The distribution and histology of receptors in birds and mammals is not very different from what has been found in amphibians and reptiles, but there are some phylogenetic puzzles presented by their functional properties. The bulk of the receptor population in amphibian and reptile lungs consists of stretch receptors, some of which are CO_2-sensitive.

In birds almost the entire population of lung receptors is sensitive to CO_2 exclusively (IPCs), and nearly insensitive to stretch stimulation. Similar IPCs are found in lizards (Douse & Mitchell, 1987), snakes (Furilla & Bartlett, 1987), alligators (Milsom et al., 1987) and a tortoise (Ishii et al., 1986). Lunged fish, amphibians, and turtles, on the other hand, have only pulmonary stretch receptors (PSRs)(Milsom, personal communication). Mammals appear to have only PSRs which are almost as CO_2-sensitive as those in turtles and amphibians (Jones & Milsom, 1982; Fitzgerald & Lahiri, 1986; Scheid & Piiper, 1986). Mammals are currently thought to have evolved from the stem reptiles near the chelonia, before the rest of the reptiles and birds, so these chemoreceptor distributions are consistent except for the presence of IPCs in the tortoise. In view of the structural homology and histological similarity (when known) of these various receptors, it is tempting to agree with Jones and Milsom's (1982) hypothesis that these varying stimulus modalities simply represent group-specific adaptations of the same receptor mechanism. That is, both stretch and intrapulmonary P_{CO_2} are indices of the efficacy of ventilation. With its air sac arrangement and unidirectional ventilation pattern, the bird lung changes volume less, and so the CO_2 sensitivity gives a better index of lung function (see Chapter 13). In mammals, lung distensibility is greater, and stretch receptors predominate.

Other Receptors

In addition to the known, discrete receptors described above, there is mounting evidence for chemosensitivity in other locations that is physiologically significant in respiratory control. Tissue similar to that of the carotid body (i.e., glomus tissue) and not innervated by the cranial nerves IX or X has been described in mammals, and this tissue has been shown to provide at least some of the normal reflex response to hypoxia (Martin-Body et al., 1985). Whether it is simply another example of redundancy in sensory systems, or these other peripheral receptors have a unique function is not yet known.

Central Chemoreception

The brains of birds and mammals do not respond much to oxygen, except perhaps indirectly as a result of metabolic depression or secondary metabolic acidosis. They do respond to P_{CO_2} and pH, although it is still not clear which is the primary stimulus. Chemosensory areas along the floor of the medulla have been identified, and it is possible to isolate reflex responses by various experimental manipulations. The central response is usually slower than the peripheral reflexes, and the current view is that the central chemoreceptors provide a level of "chemical drive" appropriate for chronic regulation of P_{CO_2} and pH (Cunningham et al., 1986).

Central Integration

The subject of brain involvement in the regulation of respiration in birds and especially mammals is extremely complex and the literature voluminous (e.g., Fishman et al., 1986). From the comparative standpoint, however, the basic scheme is similar to what has been discussed in connection with other groups. One important difference is that in birds and mammals we see the return to a rhythmic pattern generator for continuous breathing, much as in the unimodal aquatic fish. The generation of this pattern involves a number of loci in the brain acting as subcenters, including areas in the cortex, pons and medulla. Vagal and glossopharyngeal afferents impinge in a complex manner on several areas, and the nature of interactions among the many kinds of stimuli arriving at the brain is a very active area of current research. The brain generates the patterns for an oscillatory sequence of inspiration and expiration, the frequency and depth of which must be adjusted to satisfy the integrated demand computed from the afferent inputs.

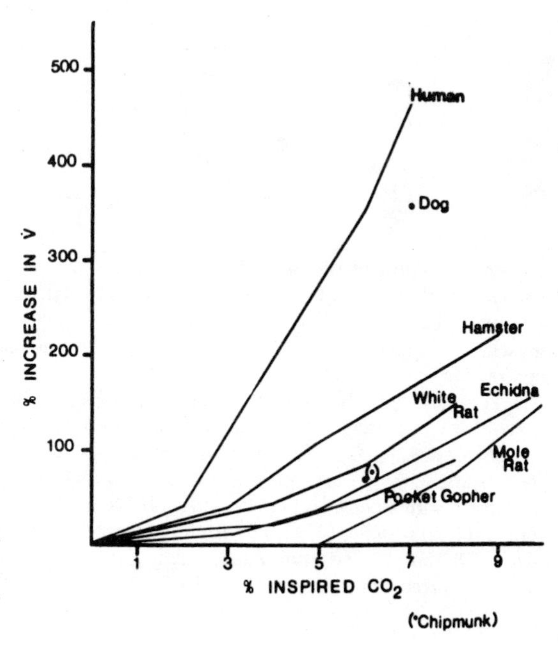

Figure 8.14: CO_2 sensitivity in various burrowing versus non-burrowing mammals. (From Tenney & Boggs, 1986, by permission, ©Amer. Physiol. Soc.)

Reflex Responses

Hypoxia

Both birds and mammals respond to hypoxia by increasing ventilation, though there may be considerable species differences (Scheid & Piiper, 1986). The fact that inhalation of pure oxygen depresses ventilation also indicates that a basal level of hypoxic drive exists under normoxic conditions. The primary stimulus for the hypoxic response clearly comes from the oxygen chemoreceptors of the carotid bodies, since deafferentation of the carotid bodies abolishes most of the response to hypoxia and hyperoxia (Bouverot et al., 1979; Martin-Body et al., 1985).

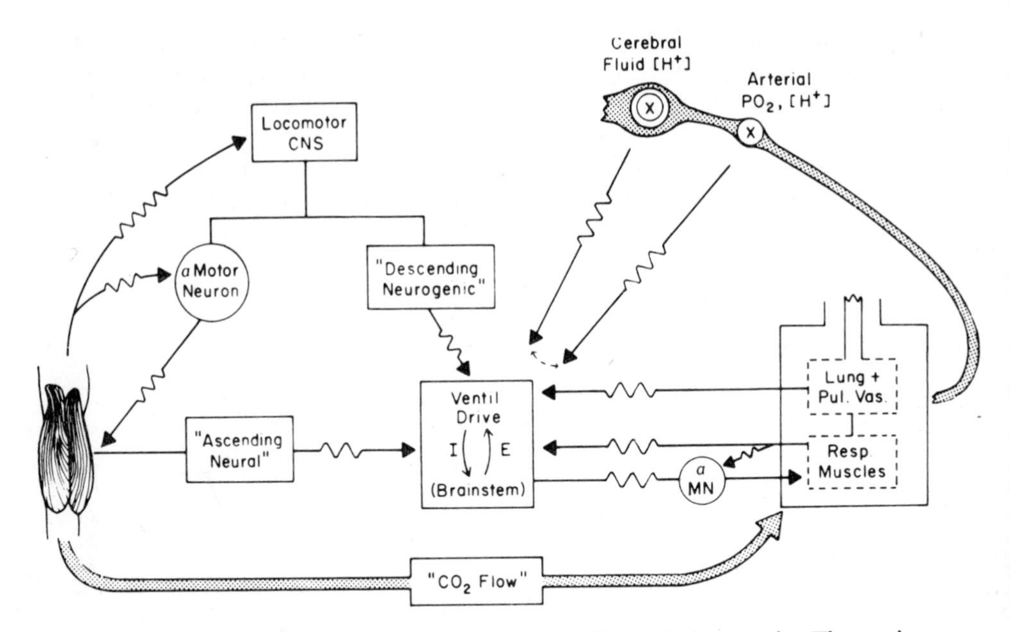

Figure 8.15: Functional pathways for the control of ventilation during exercise. Three primary stimuli are shown (feedforward command elements), each with input to respiratory center neurons: (1) A descending neurogenic drive linking central commands to locomotor skeletal muscle with input to respiratory center neurons; (2) an ascending neural drive with afferents from contracting skeletal muscle to respiratory center neurons; and (3) CO_2 flow (i.e., Q times venous CO_2 content), sensed in the lung and mediated via the vagus nerves. Four types of error detection or feedback effects are shown: (1) from working skeletal muscle to higher CNS for fine control of locomotion; (2) from intercostal muscle spindles to phrenic nerves and to the CNS; (3) from lung and airway receptors to the CNS. These feedbacks from the chest wall and lung are concerned with mechanical events and probably their optimization via control of breathing pattern and lung volume. (4) From peripheral (carotid body) chemoreceptors affected by arterial blood PO_2 and pH, and from central (medullary) chemoreceptors affected by extracellular (and probably intracellular) cerebral fluid pH. A potential interaction between the two types of chemoreceptor is also shown. (From Dempsey et al., 1985, by permission, ©FASEB.)

Hypercapnia

Not only is there a significant increase in ventilation during hypercapnia in birds and mammals but the response is extremely rapid, less than 1 second in the bird (Scheid & Piiper, 1986). Attribution of the response to a single receptor system is not so easy as with oxygen, however, since questions remain about the relative roles of carotid chemoreceptors, lung receptors, and the CNS itself. Certainly there is synergism between lung and carotid receptors but not a simple additive relation (Scheid & Piiper, 1986). In the mammals there appears to be a considerable reduction in the CO_2 sensitivity in burrowing species (Fig. 8.14)(Tenney & Boggs, 1986), and the same is true for certain burrowing birds (Boggs & Kilgore, 1980). In any case the result of the reflex response is to increase the pulmonary washout of CO_2, thus lowering alveolar and arterial P_{CO_2}s to maintain respiratory and acid-base homeostasis. In chronic hypercapnia, particularly when the inspired CO_2 is high enough to prevent increased ventilation from maintaining normal values for pH and P_{CO_2}, pH compensation is brought about by activation of other organ responses, notably in the kidney (Pitts, 1973). Even in fairly short-term hypercapnia, moderate pH compensation alters the relation between ventilation and pH in the duck in such a way as to suggest that ventilation is a single function of pH rather than P_{CO_2} (Dodd & Milsom, 1987).

Exercise

During moderate to heavy exercise oxygen consumption and CO_2 production may increase 8- to 12-fold, with corresponding requirements for increases in ventilation, cardiac output, etc. Although it is easy to demonstrate that ventilation increases along with the increased metabolic rate, the control pathways involved in the exercise response have been controversial for many years. The potential pathways for control inputs and outputs are summarized in Fig. 8.15. Part of the necessary reflex responses are provided by sensory input from arterial oxygen receptors, part from intrapulmonary CO_2-sensitive receptors, part from the central chemoreceptor areas, and a substantial portion from the feedforward command centers. The earliest ideas focussed on feedback information from the peripheral chemoreceptors, but hyperventilation during exercise is too rapid to be accounted for on the basis of changes in peripheral blood, and the responses largely persist after denervation of the peripheral chemoreceptors (Dempsey et al., 1985; Eldridge et al., 1985). Clearly some sort of feedforward command originates within the CNS, but how much of it is generated from locomotory centers, and how much might actually be a feedback from the working muscles is controversial (Cunningham et al., 1986). There is certainly a role for the peripheral chemoreceptors, but they probably act as a fine-tuning mechanism to prevent mismatching of ventilation and CO_2 production; the ventilation response is usually larger than the increase in CO_2 production, so that P_{CO_2} falls slightly during exercise.

Some evidence has been presented showing that "CO_2 flow" to the lung is an important parameter of the exercise response (Dempsey et al., 1985), although the sensors and neural pathways for such a mechanism are not clear. The overall reflex responses to exercise are well coupled to demand, and the ratio of ventilation to oxygen consumption changes very little over a wide range of oxygen consumption values (Dejours, 1975).

Actually there are a number of other reflexes that have not been considered here and that are important particularly during exercise. At rest many of the capillaries in muscle are closed, for example, and they open during active contraction, producing increased muscle blood flow, a decreased diffusion path from blood to mitochondria, and decreased vascular resistance in the muscle bed. If a large mass of muscle is involved in the exercise, there will be further reflex responses to maintain blood flow and pressure in the face of declining peripheral resistance. Postural movements associated with the exercise will require further pressure regulation; increased heat production may require changes in skin blood flow or provoke panting, etc. Only the basic receptor-CNS-effector reflexes affecting the most obvious components of respiration have been considered here; for more detail any of the various reviews referred to will provide a more complete summary of the large literature.

INVERTEBRATE CONTROL SYSTEMS

For the vast majority of invertebrates, virtually nothing can be said of the control systems involved in respiration. One feels intuitively certain that such control systems exist, at least in the more complex invertebrates, but due to experimental difficulties, small animal size, and a general paucity of scientific effort not much has been learned, particularly of receptors and neural information pathways. Only in the crustaceans and a few molluscs have any significant studies been conducted, and they are as yet fragmentary. For comparative purposes the

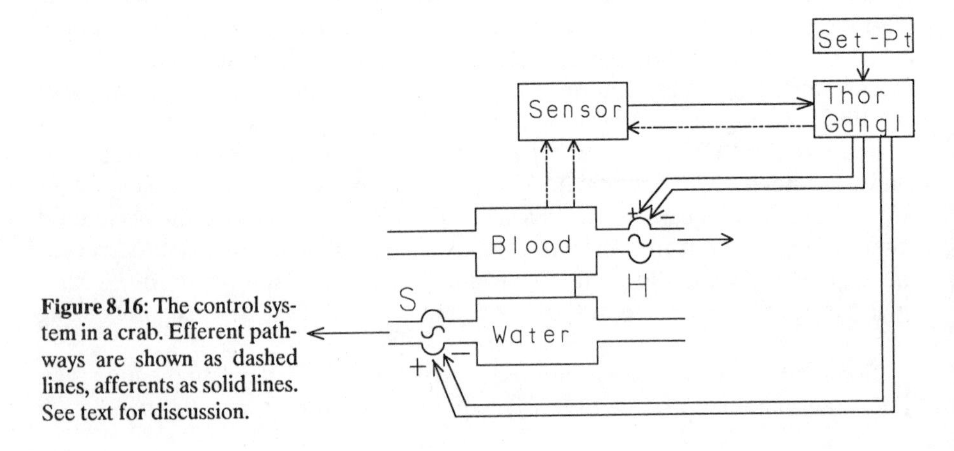

Figure 8.16: The control system in a crab. Efferent pathways are shown as dashed lines, afferents as solid lines. See text for discussion.

respiratory control system of the crab will be discussed, although it may not adequately represent the vast diversity of the invertebrate phyla.

Receptors

No chemoreceptor has yet been described in an invertebrate. There are, however, some studies showing oxygen sensitivity of nerves coming from the gills in the horseshoe crab (a chelicerate, *Limulus polyphemus*; Crabtree & Page, 1974) but only unpublished reports of oxygen receptor-like activity in the gills of crustaceans (McMahon & Wilkens, 1983). In neither case have the actual oxygen receptor cells been identified. There may be receptors at sites other than the gills, since it is not necessary for the receptors to be in the gills themselves. McMahon & Wilkens (1975) suggested on the basis of oxygen pulse experiments in hypoxic lobsters that oxygen sensitivity was a property of the CNS; i.e., the lag time for response was consistent with the longer circulation times for blood to flow from gills to the central sites. In this case the sensor function might be an intrinsic property of other central elements. This argument is somewhat reminiscent of the history of study of other groups of animals, however. In the absence of information on peripheral receptors, chemoreceptor function was often ascribed to the CNS by default, but in most cases the CNS component of chemosensory input has been found to be minor.

The control system in a crab is represented somewhat hypothetically in Fig. 8.16. Although a set-point element is shown, we do not know what or where it is; similarly, the sensor element is shown with uncertain input from either the blood or water side of the gills, or from the CNS itself.

Central Integration and Effector Limbs

The ventilatory system of the crab consists of two separate branchial chambers, each ventilated by a modified mouth-part appendage and each having separate inhalant and exhalant water openings (see Chapter 9). The muscles that drive

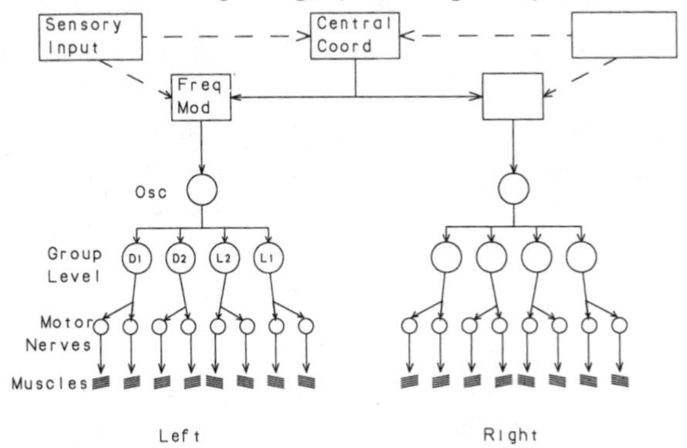

Figure 8.17: Neural connections of the ventilatory motor system in a crab. (Modified from McMahon & Wilkens, 1983.)

ventilation are innervated by various tracts from the thoracic ganglion. Wilkens and McMahon have summarized what is known about the control of the ventilatory apparatus, and a modified version is shown in Fig. 8.17. Each side receives motor nerves from one of a pair of symmetrical centers. Nonspiking oscillators on each side provide the basic frequency signal, which is modified by interneurons that probably receive a variety of sensory inputs. The oscillator entrains a series of group level neurons, which in turn activate motoneurons. The motoneurons are grouped into two depressor groups and two levator groups for control of the beating movements of the scaphognathite, the mechanical structure propelling water through the gill chamber. There appears to be loose and variable coordination between the ventilatory machinery on the two sides (Batterton & Cameron, 1978; McMahon & Wilkens, 1983), presumably effected by central neural connections. We have not, however, discovered the neural links between the receptors, the set-point comparator, and the motor tracts.

Respiratory Reflexes of the Crab

Most crabs respond to hypoxia by increasing ventilation, at least down to some critical value for PO_2 (Taylor et al., 1977; Batterton & Cameron, 1978; Burnett & Johansen, 1981; McMahon & Wilkens, 1983). Responses of crabs to hypoxia are discussed in more detail in Chapter 9, but from the standpoint of reflex responses little more can be said since we do not know the receptor, the afferent pathway, or the sites of central integration.

Environmental hypercapnia does not appear to be a respiratory stimulus in the crabs, in common with aquatic fish (Batterton & Cameron, 1978). As discussed in connection with other water-breathers, oxygen limitation appears to be the primary driving force for respiratory control systems, and CO_2 excretion represents only a minor problem. Interestingly, several terrestrial crab species do display a ventilatory reflex in response to elevated inspired CO_2 (Cameron & Mecklenburg, 1973; Cameron, 1975b; Smatresk & Cameron, 1981); but as in the case of oxygen responses, nothing is known of the receptors or of the reflex pathway involved.

Exercise, temperature and other internal and external disturbances provoke responses that have been more or less well characterized in crustaceans (Truchot, 1983; Cameron, 1986), but until more has been learned of sensory systems and neural pathways, discussions of control systems in crustaceans must be brief. It will be interesting indeed to learn in the future what similarities and differences exist between vertebrate and invertebrate chemoreceptors and their associated reflex connections.

Footnotes

[1]In mammalian lungs, nociceptors located between the pulmonary capillaries and the alveolar wall are called juxtapulmonary capillary, or "J" receptors. The term is sometimes applied generally to nociceptors responsive to edema or tissue damage.

II
Case Studies of Animals

Chapter 9

Case Study: The Blue Crab

(*Callinectes sapidus* Rathbun: Brachyura: Crustacea)

Four species account for more than 90% of the research done on the respiratory physiology of crustaceans: the crayfish (*Astacus* species), lobster (*Homarus americanus*), blue crab (*Callinectes sapidus*), and green shore crab (*Carcinus maenas*). The reasons are probably the obvious ones: availability, size, tractability in the laboratory, and economic importance. Of these crustaceans, the blue crab is probably the athlete of the lot since it is by nature a swimming crab and more aggressively predatory than many species. Scientific work on the animal has been conducted for at least 150 years, and the animal is familiar to both shore-goers and commercial fishermen.

FUNCTIONAL MORPHOLOGY

Gills

Although the eight pairs of gills in the blue crab are not visible from the outside, they are external organs in the sense of not being enclosed within the body cavity proper. The covering of the gill chamber, the *branchiostegite*, is formed as a lateral and ventral extension of the dorsal side of the thorax. In the blue crab, the outer edge of the branchiostegite is very tightly apposed to the body, almost appearing to be fused to it. Toward the anterior edge, at a point just above the merus of the cheliped, there is an opening, sometimes called the Milne-Edwards opening, through which water enters the gill chamber. This opening is guarded by setae on both margins and can be nearly occluded by retraction of the cheliped. No doubt the location of the opening and the arrangement of setae serve to prevent larger particles from entering the gill chamber and impeding water flow.

The gills themselves are wedge-shaped, lying in the gill chamber like pieces of a pie (Fig. 9.1). There are eight pairs of gills, conventionally numbered from anterior to posterior, the posterior ones being much larger than the anterior. Gills can arise from three portions of the body and are given different names accordingly. The first gill pair arises from the membrane attaching the coxa of the second maxilliped to the body wall and is therefore called an *arthrobranch*. This first gill pair is quite small and has lamellae only on the posterior side. The second pair arises directly from the coxa of the second maxilliped and is therefore called a *podobranch*. Gills 3 and 4 are arthrobranchs attached to the third maxilliped, gills

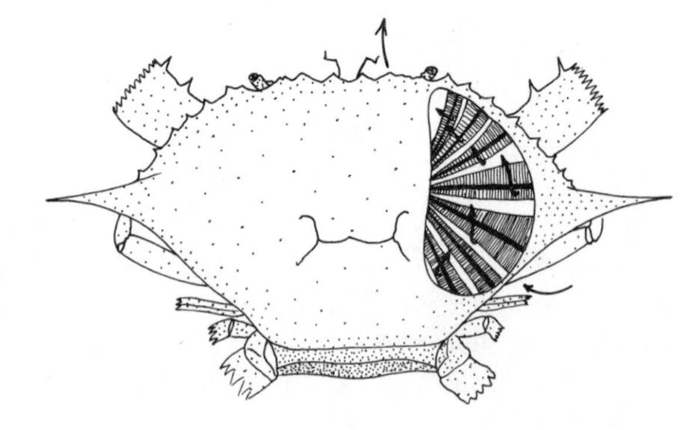

Figure 9.1: Dorsal dissection of the blue crab's branchial chamber, showing the arrangement of gills within the chamber. The anteriormost gill is quite small, so only seven pairs are shown. The arrows indicate the direction of water flow.

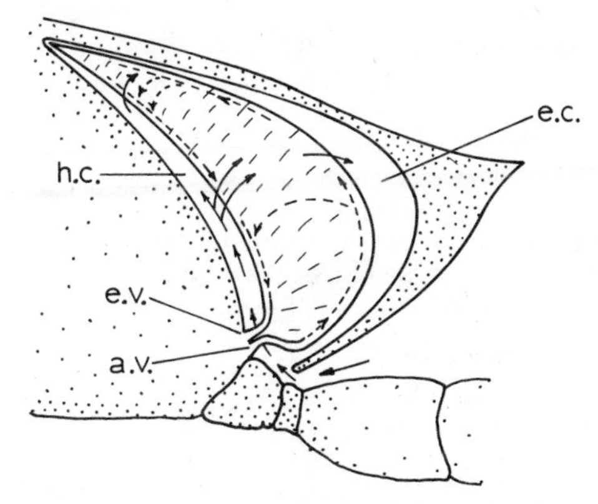

Figure 9.2: Section through the gill chamber, showing the arrangement of a gill and the directions of blood and water flow (arrows). e.c., epibranchial chamber; h.c., hypobranchial chamber; e.v., efferent vessel; a.v., afferent vessel. (After Hughes et al., 1969.)

5 and 6 also arthrobranchs of the first periopod, and gills 7 and 8 arise singly from the second and third periopods. These last two pairs arise from the body wall above the attachment of the periopods and are therefore called *pleurobranchs* (Pyle & Cronin, 1950; McLaughlin, 1982).

The bases of gills 3 and 4, which bracket the Milne-Edwards opening, are thinned and slightly stalk-like, providing a path for water to flow into the space beneath the gills, the so-called *hypobranchial* chamber (Fig. 9.2). Water is distributed through this space and then flows up through the gills, passing between the gill lamellae. The water flow is then directed anteriorly, where it eventually exits through the exhalant water opening in the region of the mouth parts. The scaphognathite lies in the excurrent water channel and by its beating action provides a continuous flow through the gill chamber (Young, 1975; McMahon & Wilkens, 1983).

Extending into the gill chambers from the maxillipeds are three long feather-like structures called *flabellae*. The one derived from the first maxilliped lies atop the gills, whereas those from the second and third maxillipeds lie below the gills. All three are adorned on their margins by long, fine setae and can be moved back and forth over the surfaces of the gills by the maxilliped musculature. They presumably have a cleaning function, though it has never been directly demonstrated.

The structure of a typical gill is shown in detail in Figs. 9.3 and 9.4 (see also Fig. 3.7). The vessel carrying blood to the gill, the afferent vessel, lies along the dorsal midline of the gill, and the somewhat larger return (efferent) vessel lies along the ventral midline. In the center, between these two vessels, there is an area of supporting tissue, and the lamellae extend to either side. Each lamella receives its blood flow from the afferent vessel via a small opening; the blood flows outward and downward between the two surfaces of the lamella and finally drains back into the efferent vessel. The surface of the lamella consists of an outer chitin layer, usually about 3 μm thick, and an underlying layer of respiratory epithelial cells, usually 2 to 4 μm thick. At intervals there are *pillar cells*, which extend across the blood space, giving support to the lamellae and maintaining the dimensions of the blood space. On the outer margins of the lamellae there

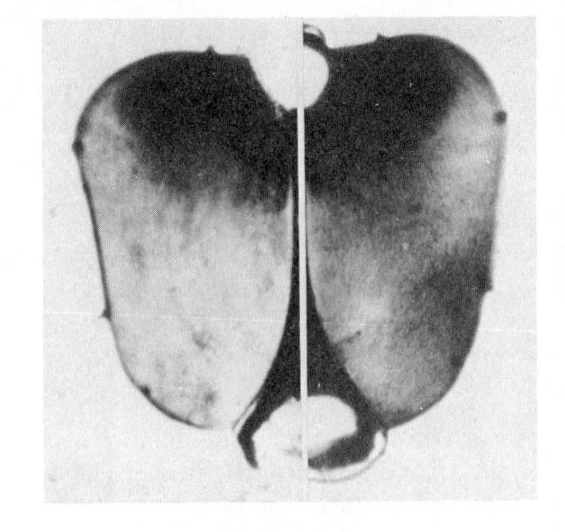

Figure 9.3: Photomontage of lamellae cut from two different gills. The lamellae were stained with osmium to selectively darken the area of thick epithelium. The lamella on the right shows a darker area of thick epithelium, whereas the one on the left, from an anterior gill, is uniform in density. (From Cameron & Iwama, 1987, ©The Company of Biologists, Ltd.)

are papillae of various shapes (Fig. 3.7) that serve to maintain the spacing between adjacent lamellae.

The Scaphognathite; Mechanics of Ventilation

The propulsive force for flow of water through the gills is provided by paired scaphognathites. The scaphognathites (from Gr. *scaphos*, shovel, and *gnathos*, tooth) are actually large, spatulate exopods of the maxillules (mouth appendages). They extend to the side and slightly posteriorly, lying in the water channel connecting the gill chambers to the excurrent water openings alongside the mouth. Their borders are fringed with fine setae that help form a seal against the wall of the water channel. The beating action is rapid and complex (Young, 1975), but the result is to produce lower pressures in the branchial chambers than ambient, drawing water in through the Milne-Edwards openings and through the gill sieve.

Gill Surface Area and Volume

Although tedious, measurement of the gill surface area of crabs is relatively easy. The lamellae are reasonably flat and hold their shape well because of the external chitin layer. Some estimates for the blue crab and for some other fully aquatic crabs are given in Table 9.1. There is an inverse relation between body size and gill surface that would account for some of the difference between the two estimates for the blue crab. These gill surface areas are wholly within the range of values for various fish (Hughes, 1966).

Estimation of the functional surface area is complicated by the presence of two distinct types of epithelium in the gill lamellae. The normal respiratory epithelium is a rather undifferentiated squamous layer averaging about 2 to 3 μm

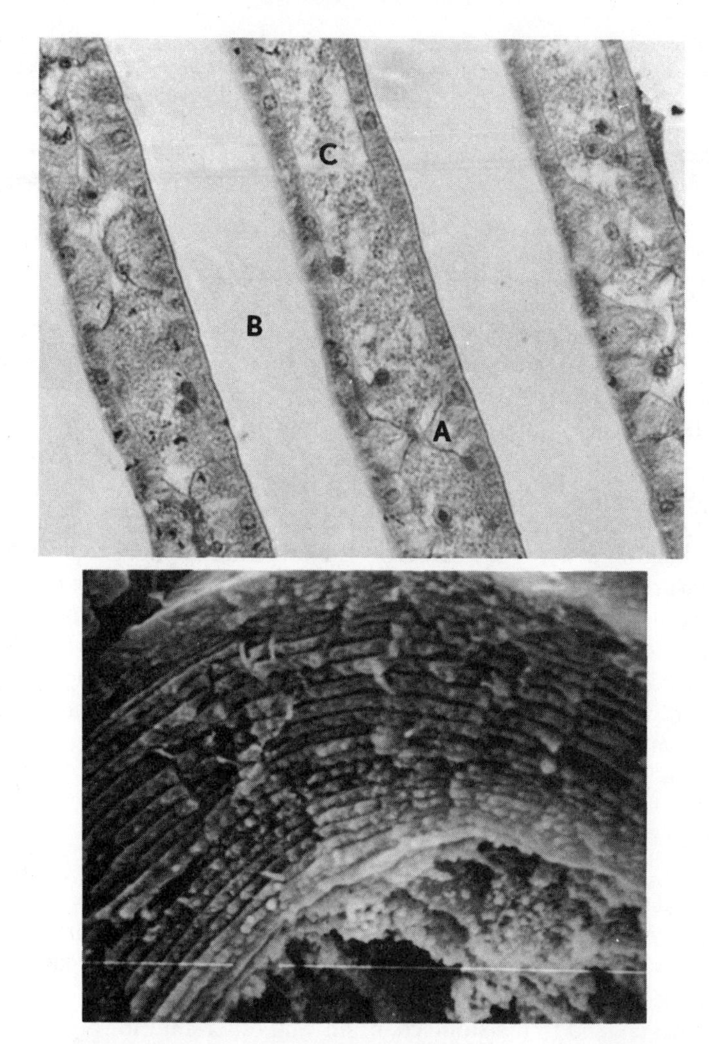

Figure 9.4: (Upper) Cross section through a gill, showing the water space between lamellae (B), the blood space filled with fixed protein (C), and a partial section of a supporting pillar cell (A). The lamellar thickness is approx. 50 μ)m thick. (Photo courtesy of J. B. Aldridge). (Lower) Scanning electron micrograph of the torn edge of a single lamellae from a blue crab's gill, showing the laminar arrangement of chitin in the cuticle. The chitin is about 1 μm thick.

in thickness. There is a second type of cell, however, that bears the hallmarks of chloride cells (Copeland and Fitzjarrell, 1968) and is thought to be involved primarily in osmoregulation (Henry & Cameron, 1982a). These cells are found in discrete patches toward the dorsal ends of the lamellae and occur in the blue crab in lamellae on the anterior half of gill 5 and both halves of gills 6 through 8. In seawater-adapted crabs these osmoregulatory patches represent about 10% of the total gill surface area, but in freshwater-adapted crabs the area increases

Table 9.1: Gill surface area in the blue crab and two other brachyurans.

Species	Weight	Area, $mm^2\ gm^{-1}$	Reference
Callinectes sapidus	143	1367	Gray (1953)
Callinectes sapidus	260	710	Aldridge & Cameron (1979)
Carcinus maenas	41	777	Hughes et al. (1969)
Menippe mercenaria	163	887	Gray (1953)

to about 45% (Aldridge & Cameron, 1982). Since the 10 to 12 μm thickness of these patches would significantly reduce their capacity for diffusive gas exchange, the effective respiratory surface is accordingly reduced.

The presence of chitin also reduces the diffusive permeability of the gill surface. Krogh (1919a), working on a single piece of chitin from a beetle larva, reported oxygen and CO_2 permeabilities about one-tenth those of other tissues. Some 70 years later these values have been confirmed independently by two groups (Mangum et al., 1985; Cameron & Kormanik, unpublished data).

By measuring a large sample of the total gill complement, it is possible to make further estimates of the water volume contained in the gills at any one time as well as the blood volume contained in them. The water volume in the gills is approximately 3.5 ml for a 200 gm crab, and the blood volume is about 2.9 ml. At normal ventilation and circulation minute volumes, the residence or contact times for water and blood are 1.9 and 6.7 seconds, respectively (Cameron, 1986)(Table 9.2).

The Vascular System

As discussed in an earlier chapter, the blue crab has what is known as an "open" circulatory system, primarily differentiated from vertebrate systems by the lack of capillaries and by different cellular structure of the vessels (Johnson, 1980). Beginning at the heart, however, there is a considerable network of vessels, the largest of which lead to the stomach, brain, and other anterior organs. The system of vessels is not as extensive in the blue crab as in some other crabs; flow within the skeletal musculature appears to be in channels formed by muscle bundles and is boosted appreciably by muscular contraction. Return (venous) flow is channeled into numerous sinuses, the most accessible of which are the infrabranchial sinuses at the bases of the walking legs. The blood flows from there through the gills, returning to the heart in paired return channels that enter the pericardial space at its anterolateral corners. The pericardium is rigid, so that when the heart contracts there is a significant *vis a tergo* effect, promoting return flow.

At present there are no known valves in the circulatory systems of crabs, but physiological evidence has been presented for valve action in the gills of the green shore crab, *Carcinus maenas* (Taylor & Taylor 1986). The volume of the blue crab's circulatory system is large compared to that of vertebrates: 30% of body weight (Gleeson & Zubkoff, 1977; Cameron & Batterton, 1978b).

BLOOD OXYGEN PIGMENT: HEMOCYANIN

Carriage of oxygen in the blood of the blue crab is enhanced by the copper-protein, hemocyanin. Without any oxygen carrier, the maximum capacity of the blood equilibrated to a standard atmosphere at 25°C and full-strength seawater would only be 0.21 mM L^{-1} (= 0.47 vol%, i.e., ml O_2 per 100 ml blood). Reported blood oxygen capacity for the blue crab is 0.71 mM L^{-1} (1.6 vol%)(Mangum & Weiland, 1975), more than three times the capacity for physical dissolution.

Like all blood pigments, oxygen is bound reversibly within the physiological range of oxygen partial pressures, and due to cooperativity among binding sites the polymeric hemocyanin has a sigmoid curve (Fig. 9.5)(see Chapter 5). In the normal physiological pH range blue crab hemocyanin has a normal Bohr shift, i.e., the affinity is decreased by increasing P_{CO_2} and decreasing pH. Temperature

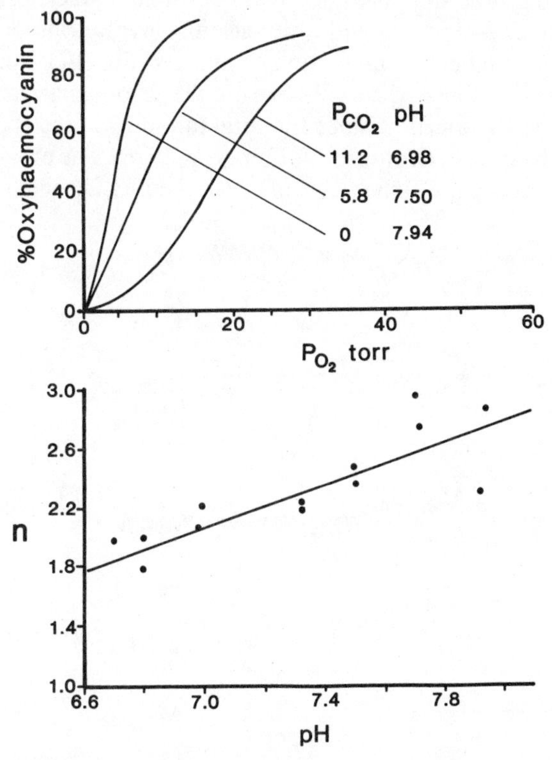

Figure 9.5: (Upper) Oxygen dissociation curves from blue crab at three CO_2 tensions and 28°C. (Lower) The influence of pH on cooperativity. (From Young, 1972a, by permission, ©Elsevier North Holland, Inc.)

causes a marked right shift of the curve, and increasing salinity causes a left shift, mostly as a result of specific effects of Ca^{2+}, Mg^{2+} and Na^{+} (Mason et al., 1983). Very few measurements of the in vivo partial pressures of oxygen have been made, particularly from the prebranchial ("venous") side, but the values fall in the 5 to 15 torr range, indicating a significant contribution of the hemocyanin to normal respiratory oxygen delivery (Mangum & Weiland, 1975; Cameron, unpublished data).

GAS EXCHANGE AT REST

Ventilation

The volume of water pumped through the gills at rest (Table 9.2) is the result of quite complex patterns of beating of the two partially independent scaphognathites. The beat is first of all mechanically complex and produces correspondingly complex pressure and flow waves (Fig. 9.6). The patterns are complicated further by the periodic stoppages and reversals that characterize ventilation in the crustaceans. These apneas and reversals may involve one or both scaphognathites, leading to several combinations of effects (Fig. 9.7). Fortunately, however, neither the resistance of the gill sieve nor the scaphognathite stroke volume appear to change significantly, since the total minute volume is predicted fairly well by beating frequency (Batterton & Cameron, 1978). The pressures measured across the gills are usually in the range of –1 to –2 cm H_2O, which correlates fair-

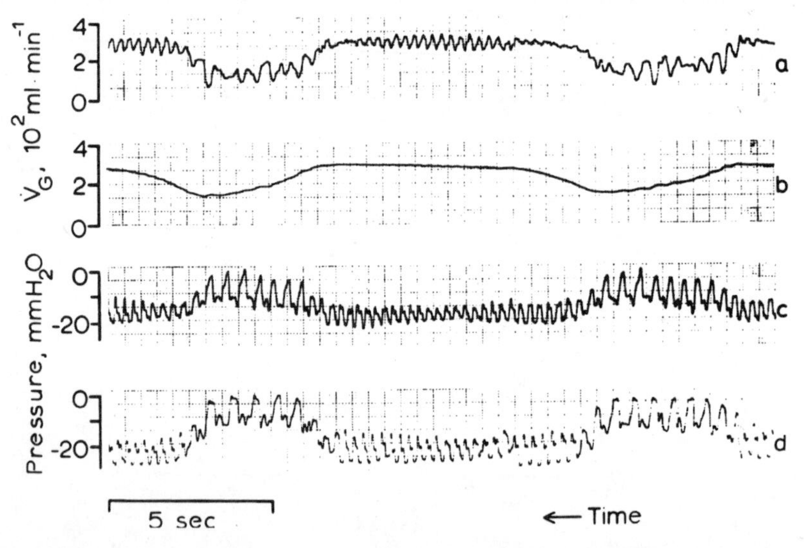

Figure 9.6: Pressure and flow in the gill chambers of the blue crab. Simultaneous changes in the pressure and frequency in both branchial chambers (lower, c and d) produce increases in instantaneous (upper, a) and mean flow (b). (From Batterton & Cameron, 1978, ©Alan R. Liss, Inc.)

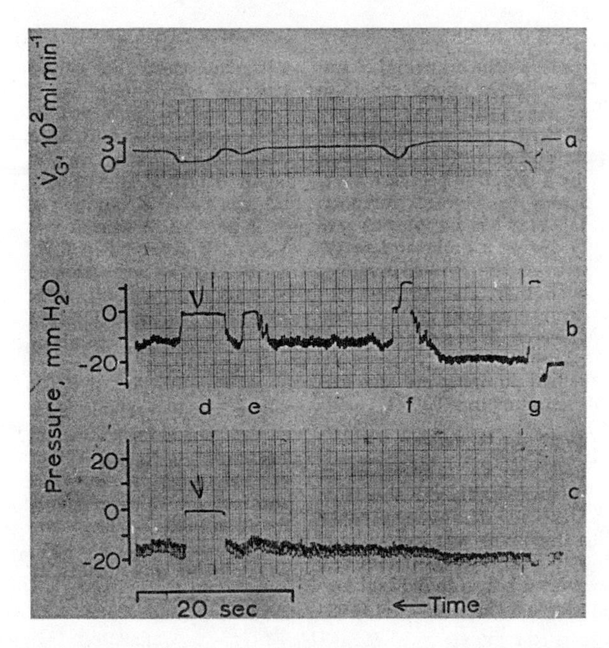

Figure 9.7: Simultaneous recordings of ventilation (upper trace) and pressure in the branchial chambers (b and c) showing representative patterns of bilateral and unilateral scaphognathite activity. d, bilateral apnea, causing pressure on both sides and flow to fall to zero; e, unilateral apnea, causing pressure to fall to zero in the right branchial chamber; f, unilateral reversal indicated by strong positive pressure in right chamber; g, bilateral reversal indicated by strong positive pressure in both branchial chambers. (From Batterton & Cameron, 1978, ©Alan R. Liss, Inc.)

ly well with the measured flow and the resistance calculated for the gill sieve (see above; Aldridge & Cameron, 1979).

Typically the extraction of oxygen from the water flowing over the gills is about 50%, but there is usually an inverse relation between extraction efficiency and flow, as there is in the fish (Chapter 10).

Perfusion

Technical difficulties in the study of crabs have prevented direct measurements of most circulatory parameters. The primary problem is the coagulation of blood, a process that is biochemically different from the clotting of vertebrate blood. The coagulation time varies from species to species and from one individual to another, but it can be as short as a few seconds in the blue crab, particularly at higher temperatures. The blood forms a solid gel that prevents implantation of catheters or sensors and sometimes even confounds measurements in drawn blood samples. A second problem is that surgery is nearly impossible due to the coagulation and to excessive tissue autolysis, which seems to occur after injury.

We therefore have little direct information on cardiac output, blood pressure, and blood flow. Total cardiac output has been calculated from total oxygen con-

Table 9.2: Respiratory system parameters of the resting blue crab at 25°C.

Symbol	Parameter	Value	
W	Weight	200 g	
A	Gill surface area	710 $mm^2\ g^{-1}$	Aldridge & Cameron, 1979
GBV	Gill blood volume	2.9 ml	Aldridge & Cameron, 1979
GWV	Gill water volume	3.5 ml	Aldridge & Cameron, 1979
t_b	Blood contact time	6.7 sec	Cameron, 1986
t_w	Water contact time	1.9 sec	Cameron, 1986
$\dot{V}_G$	Ventilation	111 ml min^{-1}	Batterton & Cameron, 1978
$\dot{M}O_2$	Oxygen consumption	9.24 μM min^{-1}	Batterton & Cameron, 1978
$\dot{M}CO_2$	CO_2 excretion	8.30 μM min^{-1}	Assumed
RQ	Respiratory quotient	0.9	Assumed
P_I	Inspired oxygen tension	145 torr	Batterton & Cameron, 1978
P_E	Expired oxygen tension	68 torr	Batterton & Cameron, 1978
$\%E_w$	% Extraction of O_2 from water	53 %	Batterton & Cameron, 1978
αBO_2	Blood oxygen capacity	0.71 mM L^{-1}	Mangum & Weiland, 1975
PaO_2	Arterial oxygen tension	75 torr	Booth *et al.*, 1982
PvO_2	Venous oxygen tension	14 torr	Booth et al., 1982
$\%E_b$	% Extraction of O_2 from blood	50%	Mangum & Wieland, 1975
CaO_2	Arterial oxygen content	0.62 mM L^{-1}	Calculated
CvO_2	Venous oxygen content	0.31 mM L^{-1}	Calculated
ΔCO_2	A–V O_2 difference	0.31 mM L^{-1}	Calculated
$\dot{Q}$	Cardiac output	26 ml min^{-1}	Calculated, Fick
$\dot{V}_G/\dot{Q}$	Vent.–perfusion ratio	4.3	Calculated
$PaCO_2$	Arterial CO_2 tension	1.5 torr	Cameron & Batterton, 1978a
$PvCO_2$	Venous CO_2 tension	3.8 torr	Cameron & Batterton, 1978a
pH_v	Venous pH	7.96	Cameron, 1978b
C_T	Total CO_2	8.81 mM L^{-1}	Cameron, 1978b
EPR	Exchange potential ratio	0.50	Calculated

sumption and blood oxygen values (Table 9.2) using the Fick equation (see Chapter 3), but the attainment of truly mixed venous samples is difficult to verify. Thermal dilution techniques have also been applied, but with limited success (Burnett et al., 1981). Blood pressure measurements can be made on an acute basis, but they are apt to be from disturbed animals. Best estimates of peak systolic pressures are in the range of 10 to 20 mm Hg (12 to 25 cm H_2O), with diastolic and peripheral systemic pressures barely above ambient. Recordings from the walking legs of *Cancer magister* show no detectable pulse pressure but large excursions due to leg movement (unpublished data).

Efficiency Indices

The various indices used to describe efficiency of gas exchangers are defined in Chapters 3 and 7. Perhaps the most relevant for the blue crab at rest are the ventilation-perfusion ratio of 4.3 and the capacity-rate ratio (water to blood) of 1.26.

Theory predicts that maximum efficiency is attained at capacity-rate ratios near 1, so at least for oxygen the blue crab's respiratory system appears to be reasonably optimized. Taking the nonlinear dissociation curve into effect, the exchange potential ratio (EPR; see Chapter 7) is the ratio of the amount of oxygen actually exchanged compared to the maximum possible exchange. For the data given in Table 9.2 this value is 0.50 at rest, less than optimum, but still quite efficient in overall operation.

For CO_2 the values are quite different: Capacity-rate ratios are not straightforward to calculate owing to the complex equilibria involved, but the capacities of blood and water are nearly the same. Consequently, the capacity-rate ratio is about 4, and the EPR is also low. These values are typical for water-breathers; the theory is that since obtaining oxygen is a much more serious problem in water than getting rid of CO_2, the gas exchange systems appear to be optimized for oxygen exchange (Cameron, 1986; Randall & Cameron, 1973). In effect the gills are strongly hyperventilated with respect to CO_2.

Acid-Base Status

As in all aquatic animals, the P_{CO_2} is low in the blue crab, held to low values by the low metabolic rate and high relative ventilation. Average acid-base data are given in Table 9.3 for resting, relatively undisturbed blue crabs at 27°C. Intracellular pH is maintained about 0.5 pH units below that of the blood, although there are some variations from one tissue to another (Wood & Cameron, 1985).

The pH for any given temperature (see below) tends to vary somewhat from one study to another. At lower external salinity the blood pH tends to rise (Mangum et al., 1976; Henry & Cameron, 1982a), but handling conditions and even slight CO_2 build-up in holding tanks can change the resting acid-base status. For-

Table 9.3: Normal acid-base values for resting blue crabs at 27°C.

Parameter	Value	Source
pH_e, blood	7.754	1
pH_i, skeletal muscle	7.206	1
Pa_{CO_2}, torr	3.05	2
C_T, mM L^{-1}	7.61	2
$[HCO_3^-]$, mEq L^{-1}	7.50	Calculated
$[CO_2]$, dissolved, mM L^{-1}	0.11	Calculated
$[CO_3^{2-}]$, mM L^{-1}	<0.05	Calculated
pK′	5.865	2
Total extracellular CO_2 pool, mM kg^{-1}	1.86	3
Total intracellular CO_2 pool, mM kg^{-1}	1.27	3
Total body fluid pool, mM kg^{-1}	3.13	Calculated
Total exoskeletal CO_2 Pool, mM kg^{-1}	1050	3

1: Wood & Cameron, 1985
2: Cameron & Batterton, 1978a
3: Cameron & Wood, 1985

tunately the handling necessary for blood sampling does not affect acid-base status nearly so much as it does in other groups of animals, especially fish (unpublished data).

Carbonic Anhydrase Distribution and Function

Unlike vertebrates, crabs do not have their respiratory pigment packaged in blood cells, nor do they have any carbonic anhydrase activity in the blood (Maren, 1967; Burnett et al., 1981; Henry & Cameron, 1982b). There is, however, considerable carbonic anhydrase activity in the gill tissue, which has led to a lively controversy over its function. Henry and Cameron (1983) have argued that it functions primarily in osmoregulation, possibly maintaining a supply of counterions for Na^+ and Cl^- exchange. The evidence for this theory is that the activity is concentrated in those gills with high proportions of the thick (chloride cell) epithelium, and that activity increases during low salinity acclimation. Inhibition of carbonic anhydrase with sulfonamide drugs produces a primarily nonrespiratory alkalosis. Burnett et al. (1981), on the other hand, have argued that its primary function is to accelerate the dehydration of HCO_3^- from the blood to CO_2 for diffusive excretion. In order for this to be so, either the carbonic an-

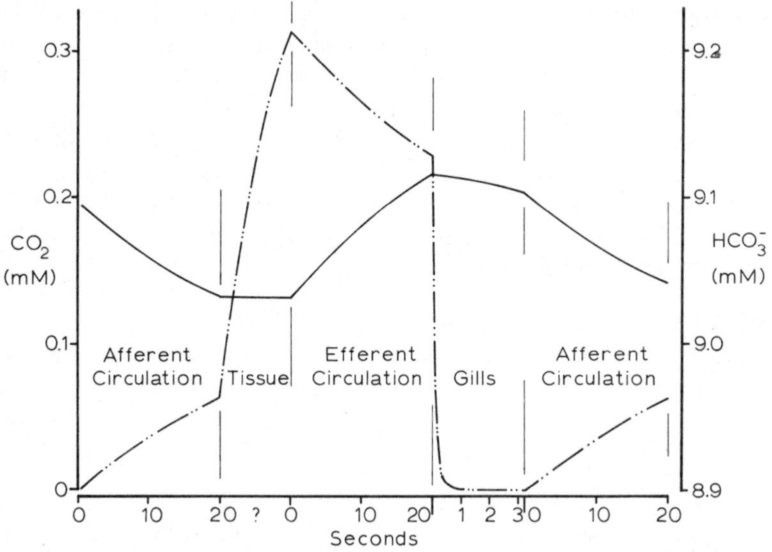

Figure 9.8: Results of a computer simulation model of CO_2 exchange in the blue crab without carbonic anhydrase activity in the blood. The solid line shows HCO_3^- concentration and the dashed line the P_{CO_2} as blood moves around the circulation. At the far left, as blood moves from the gills toward the tissues, there is a slow equilibration that is not yet complete when the blood reaches tissues. In the tissues, CO_2 diffuses rapidly into the blood, increasing the P_{CO_2}, but having little effect on HCO_3^-. During the return circuit toward the gills there is partial equilibration, but the blood is still significantly out of equilibrium when it enters the gills. In the gills, the dissolved CO_2 is rapidly lost by diffusion, but only a little HCO_3^- has time to react to form more dissolved CO_2. The cycle then repeats. See text and Chapter 7 for further discussion. (From Cameron, 1979, ©Elsevier North Holland, Inc.)

hydrase would have to be bound on the outside of the basolateral membranes of the gill epithelium (Henry, 1987, 1988) or the HCO_3^- permeability of the epithelial cells would have to be very high. So far experimental evidence for an important role of carbonic anhydrase in normal CO_2 excretion is, in this author's opinion, unconvincing.

Possible Disequilibrium of CO_2 and HCO_3^-

Some years ago I published a model of CO_2 excretion in the blue crab proposing that due to the lack of circulating carbonic anhydrase activity there might be a significant disequilibrium in the blood (Cameron, 1979, 1986). That is, the uncatalyzed reaction rates of CO_2 hydration and dehydration are slow; so if the time elapsed between when CO_2 diffuses from tissue into blood and when the blood enters the gills is fairly short, equilibrium will not have been reached. If it were so, the proportion of CO_2 staying as dissolved gas in the circulation would be much higher than predicted from equilibrium equations, thus facilitating diffusive loss (Fig. 9.8). Model studies of the circulation and the CO_2 equilibrium supported this theory but there has not yet been a good experimental test. No inferences can be drawn from measurements made in the usual manner, since any disequilibrium existing when the samples are drawn will disappear during the time required to make the measurement.

TEMPERATURE EFFECTS

Metabolism

Like most poikilothermic animals, the total metabolic rate is highly dependent upon temperature. Data of Scholander et al. (1953) show the oxygen consumption of the blue crab changing from 10 ml kg^{-1} h^{-1} at 5°C to approximately 300 ml kg^{-1} h^{-1} at 30°C. The temperature slope is much steeper at low temperatures, following the pattern first reported for fish by Ege and Krogh (1914). Owing to the technical difficulties of measurement, almost no data have been published on the CO_2 excretion rates or RQ values at different temperatures.

Such high temperature sensitivity implies that the gas exchange system must be able to accommodate at least 30-fold variations in rates of oxygen delivery and CO_2 clearance, requiring proportionate adjustments in ventilation, perfusion, and diffusive exchange at gills and tissues.

Gas Exchange

In one study of temperature effects on the blue crab, an increase from 10° to 27°C increased oxygen consumption by a factor of 3.1 and ventilation by 3.0 (Cameron & Batterton, 1978a). The extraction of oxygen from the water rose from 24% to 48% but was offset by the decrease in oxygen solubility at higher temperature.

As a result the ratio of water ventilated to oxygen consumed (the convection requirement) did not change significantly. Assuming that the RQ did not change with temperature, the ratio of ventilation to CO_2 production also did not change.

Less information is available on changes in gas transport by the blood, but cardiac output must increase roughly in proportion to the rate of metabolism. Both higher temperature by itself and the attendant lower pH (see below) cause right shifts in the oxygen dissociation curve, i.e., decreased affinity (Young, 1972a). This shift would enhance the oxygen diffusion gradient from blood to tissue, at the expense of reducing the gradient favoring oxygen loading at the gills.

Acid-Base Status

As temperature rises from 10° to 27°C, pH in the blood of the blue crab falls, as is the general pattern in poikilotherms (Fig. 6.7)(Cameron & Batterton, 1978a)(see Chapter 6). Comparing the in vivo data with both open- and closed-system in vitro data, it is clear that the pH adjustment with temperature is achieved with only small changes in P_{CO_2} and larger changes in the $[HCO_3^-]$ (Fig. 6.11). This behavior requires active acid-base exchanges with the environment (Wood & Cameron, 1985; Cameron, 1986) which are probably Cl^-/HCO_3^- and Na^+/H^+ exchanges (Cameron, 1976; 1986; Heisler, 1986b). They are almost certainly branchial exchanges, since urine production is low in seawater, and the crab's antennal gland (a functional analog of the kidney) does not appear to play a role in acid-base regulation (Cameron & Batterton, 1978b). Acute changes in temperature lead to transient acid-base disturbance, no doubt reflecting the time required for ionic exchanges (Wood & Cameron, 1985). Notably absent is any significant ventilatory control of P_{CO_2} as shown by the in vivo P_{CO_2} data and the constant ratio of ventilation to oxygen consumption. Here again the gas exchange system of the crab appears to be cued to the uptake of oxygen, whereas ionic exchanges are invoked for acid-base control.

The pH-temperature slope for the blue crab was –0.012 pH/°C, well below Reeves' (1972) predicted –0.017 for "alphastat" regulation (see Chapter 6). In a more recent study of the blue crab a slope of –0.015 was found for the range 10° to 30 °C, but the slopes of intracellular pH data for various tissues ranged from –0.012 to –0.016 (Fig. 6.8)(Wood & Cameron, 1985). Since we do not know the relevant $\Delta H°$ values for either blood or the intracellular proteins, alpha imidazole cannot be calculated (Cameron, 1984).

AMBIENT GAS EFFECTS

Hypoxia

The relation between metabolic rate (i.e., oxygen consumption rate) and ambient oxygen concentration is highly variable among invertebrates (Mangum & van

Winkle, 1973; Mangum & Towle, 1977). Even within the brachyuran crabs, many patterns have been described. Some crabs are good regulators, maintaining a nearly constant rate of oxygen consumption down to fairly low oxygen concentrations, others are almost complete conformers, and still others intermediate in pattern. The blue crab seems to be a partial regulator, with oxygen consumption maintained reasonably well down to about 40% saturation but falling steeply beyond that until the lower lethal limit of about 15% saturation is reached (Batterton & Cameron, 1978).

As the available oxygen declines in the water, the supply is maintained by increases in ventilation volume. In the blue crab this increase is accomplished entirely by increased scaphognathite beat frequency, but in the lobster there is also an increase in the stroke volume (McMahon & Wilkens, 1975). Increased extraction efficiency would also be helpful, but it does not occur in the blue crab.

From other studies on crustaceans it is clear that the degree to which hypoxic regulation is achieved depends upon several variables. At higher temperatures there is usually less regulatory ability. When animals' metabolism is already elevated by exercise or disturbance, regulatory ability declines (Mangum & van Winkle, 1973). Finally, the thickness of the chitin over the gills may have an important influence: The Dungeness crab (*Cancer magister*) is a good regulator and has only about 0.5 μm of chitin covering the gill epithelium; the king crab (*Paralithodes kamtschatica*) is a complete conformer, and has a very thick (10 to 12 μm) chitin covering (Cameron & Mecklenburg, unpublished data).

Adjustments must also occur in the blood as ambient oxygen falls, but no specific data are available for the blue crab. In other crustaceans venous oxygen tension and saturation fall, offsetting the decline in arterial oxygen content, so that at least over some range of ambient oxygen the delivery to tissues can be maintained by the circulatory system (McMahon & Wilkens, 1983). How cardiac output or blood flow distribution may change, however, is not known; there is little change in heart rate (McMahon & Wilkens, 1975, 1983), but cardiac output is often adjusted more by stroke volume changes than by frequency.

Hyperoxia

Hyperoxia is not a very common environmental circumstance, though it sometimes occurs during the daytime due to high photosynthetic rates. Hyperoxia is studied more as a tool for understanding the respiratory control system. The most common response to hyperoxia is a large decrease in ventilation, with little change in the overall metabolic rate (Dejours & Beekenkamp, 1977). As oxygen tension in the blood rises and ventilation falls, the P_{CO_2} of the blood becomes elevated, sometimes considerably, causing a respiratory acidosis. Over a period of hours or days the acidosis is compensated by ionic exchange mechanisms but without any restimulation of ventilation. The persistent acidosis is a further piece of evidence that the respiratory system of crustaceans is controlled mainly by oxygen supply, and not significantly by CO_2 excretion requirements.

Hypercapnia

The blue crab exhibits what could now be called a classic response to hypercapnia. There is first an increase in the P_{CO_2} of the blood, causing a decrease in pH (respiratory acidosis) with a small increase in [HCO_3^-]. Over a period of about 24 hours the acidosis is compensated by branchial ion exchanges, leading to partial restoration of normal blood pH and a much greater increase in the blood [HCO_3^-] (Cameron, 1978b)(Fig. 6.15). The ionic fluxes can be measured as apparent net H^+ excretion (Fig. 9.9); they are partially inhibited by amiloride, which blocks of sodium entry into cells, and by SCN^-, which blocks Cl^-/HCO_3^- exchange (Cameron & Kormanik, unpublished data).

Interestingly, pH compensation is never 100%, even though at higher ambient P_{CO_2} values higher HCO_3^- concentrations are possible. That is, pH compensation at an ambient CO_2 of 7.5 torr may stop at 65% with blood HCO_3^- at 15 mM L^{-1}, but when the ambient CO_2 tension is raised to 15 torr, 65% compensation is again achieved at a HCO_3^- concentration of 25 or 30 mM L^{-1} (Cameron & Iwama, 1987). Since no absolute upper limit of HCO_3^- compensation is being reached, as had been suggested earlier (Claiborne & Heisler, 1986), the data suggest that some compromise set-point is reached at each level of hypercapnia (see Chapter 6).

Another interesting point is that during hypercapnia ventilation is not stimulated (Batterton & Cameron, 1978), at least in water. Blue crabs are able to breathe air for extended periods, however, but when doing so they also show no ventilatory response to hypercapnia, unlike other air-breathing crabs (Cameron, 1975b) and vertebrates.

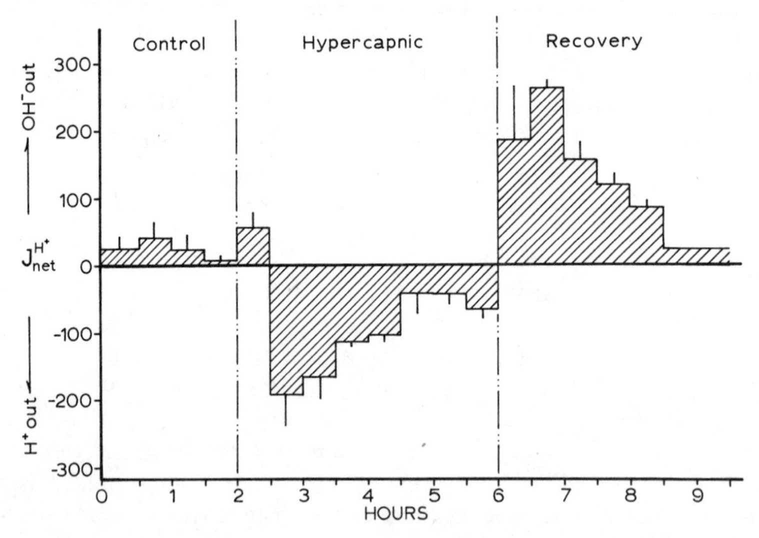

Figure 9.9: Net H^+ fluxes across the gills during hypercapnic compensation. (Cameron & Kormanik, unpublished data.)

In the air-breathing crabs some buffering of hypercapnic acidosis by the carbonate pool of the shell has been observed (DeFur et al., 1980; Henry et al., 1981). In the blue crab and other aquatic crabs, however, there does not appear to be any significant contribution of shell buffering; probably the ionic exchanges with ambient seawater are sufficient for compensation (Cameron, 1985a).

EFFECTS OF EXERCISE

Metabolism

In the most thorough study of exercise in the blue crab to date, Booth et al. (1982) found that oxygen consumption increased 2.6-fold within a few minutes of the start of moderate exercise at 20°C in a forced laboratory situation. The blue crab is notable among crustaceans for its ability to sustain activity: In the laboratory an hour of steady exercise could usually be elicited. Despite this ability to exercise and to increase aerobic metabolism, there was a substantial anaerobic component indicated by a lactate buildup of from 0.7 mM/L at rest to 9.8 mM L^{-1} after 25 minutes. One suspects that they could do at least this well under natural

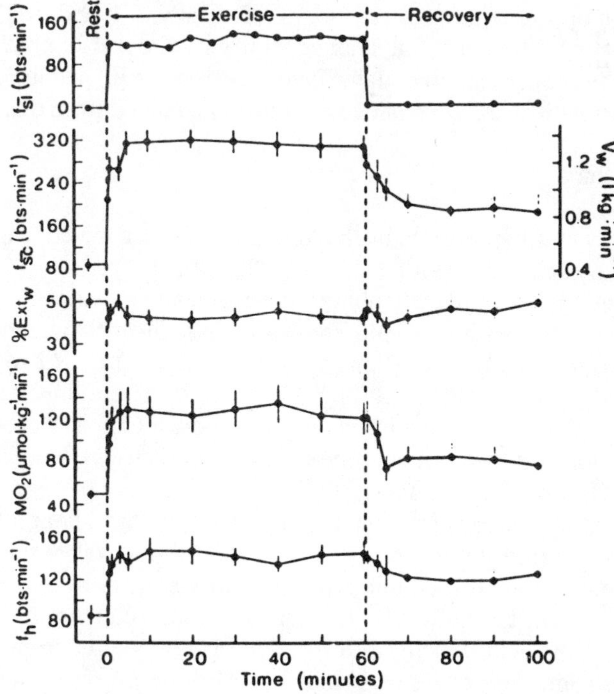

Figure 9.10: Ventilation, oxygen consumption, and heart rate increase rapidly during a period of swimming exercise but recover to normal fairly quickly. f_{sl}, frequency of swimming leg beating; f_{sc}, scaphognathite beat frequency; fh, heart rate. (From Booth et al., 1982, by permission, ©Springer-Verlag.)

conditions, since in many estuarine systems the blue crab undertakes quite long migration at various seasons. In contrast, many crabs such as the Ocypodid land crabs can exercise vigorously for periods ranging from a few minutes (Smatresk et al., 1979; Smatresk & Cameron, 1981) to only a few seconds in the case of escape running.

Gas Exchange

Changes in several parameters of the gas exchange system from Booth et al.'s (1982) study are shown in Fig. 9.10. The 2.6-fold increase in oxygen consumption is met by ventilation more than tripling after a few minutes of exercise and by only a slight decline in extraction efficiency. The A–V oxygen difference increases from 0.31 to 0.36 mM L^{-1} (0.7 to 0.8 vol%), but the venous oxygen reserve remains at about 20% of saturation. From Fick calculations, cardiac output rose approximately 2.3-fold, achieved by almost equal increases in rate and stroke volume.

Normally one might expect the increased CO_2 and reduced pH during exercise (see below) to produce a strong Bohr shift in the hemocyanin, which in turn might compromise oxygen delivery. Truchot (1980), however, has discovered that lactate has an opposing effect on hemocyanin affinity, acting during exercise to maintain an almost normal oxygen dissociation curve. Booth et al. (1982) found, in fact, that the changes about equally oppose each other in the blue crab, so that hemocyanin function in gas exchange is almost unchanged from rest to exercise.

Acid-Base Status

One hour of swimming exercise by the blue crab causes a mixed respiratory and non-respiratory acidosis (Booth et al., 1984)(Fig. 9.11). Arterial P_{CO_2} rises from about 3.5 to nearly 6 torr, pH falls almost half a unit, and the total CO_2 concentration is depressed by nearly one-half. These acid-base disturbances indicate a significant increase in CO_2 production, but if RQ measurements had been made simultaneously, they might have shown very large apparent RQ values. The reason is the effect of acidification on the CO_2 pools: CO_2 is excreted from de novo production and from shifts in the equilibrium between dissolved CO_2 and bicarbonate (Burggren & Cameron, 1979).

The respiratory portion of the acidosis disappears within 1 hour after exercise, but the non-respiratory portion, due to lactate buildup, takes many hours to eliminate. During the exercise and recovery periods the portion of the acidosis attributable to lactate is quite a bit less than the actual concentration of lactate in the blood. This acid deficit has also been observed after exercise in other aquatic crabs (Smatresk et al., 1979) and fish (Heisler, 1986b), and results from ionic exchanges which rapidly transfer excess H^+ ions to the external seawater. Lactate is not excreted to any significant extent (Bridges & Brand, 1979) but is instead remetabolized slowly. Presumably as the reconversion of lactate

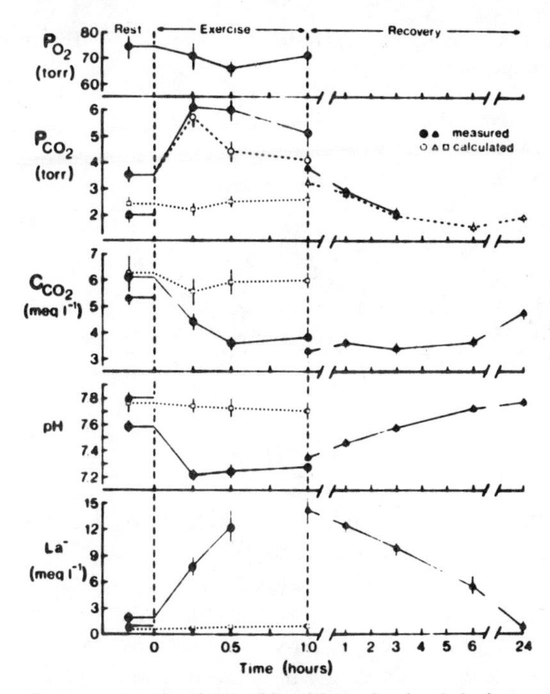

Figure 9.11: Although oxygen tension in the blood is maintained during exercise (uppermost panel), a fairly severe mixed respiratory and nonrespiratory acidosis develops, as indicated by the increased P_{CO_2}, decreased pH and CO_2 content, and increased lactate concentration. The acidosis requires several hours to correct after swimming exercise. (From Booth et al., 1984, by permission, ©Elsevier North Holland, Inc.)

proceeds the H^+ are taken back up from the seawater, but this process is a slow one lasting many hours and has not been measured successfully.

SPECIAL FEATURES OF THE BLUE CRAB'S RESPIRATORY SYSTEM

Equilibrium with the Calcareous Exoskeleton

One conspicuous difference between the blue crab and ourselves is the presence of a calcareous exoskeleton. The exoskeleton is complex, comprised of several organic and cellular layers, but the principal mineral salt it contains is $CaCO_3$. The formation and maintenance of $CaCO_3$ at physiological ion activities and pH values would be difficult, but the mineralized layers are contained within a physiologically separate fluid compartment and maintained at a pH about 0.5 unit alkaline to the blood (Wood & Cameron, 1985). During the formation of the shell following each moult, several ion transport processes are activated (Fig. 9.12) at impressive rates. When a net H^+ excretion occurs after exercise or during hypercapnia, for example, transport rates seldom exceed 1.5 mEq kg^{-1} h^{-1}, but by

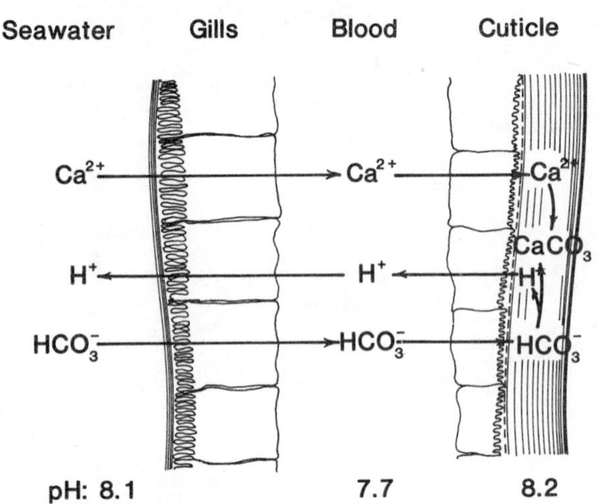

Figure 9.12: Several ion transport pathways are required for new shell formation after moult in the blue crab, both at the gills and across the endothelial boundary between the blood and the carapace fluid compartment. Movement of calcium from seawater to blood appears to be passive, whereas transport from blood to the cuticle is active. Bicarbonate is taken up directly from seawater.

12 hours after moult rates as high as 22 mEquiv kg^{-1} h^{-1} have been recorded (Cameron, 1985c; Cameron & Wood, 1985).

During the intermoult period the exoskeleton is protected by a combination of physiological processes and morphological barriers. The underlying epithelial cell layer maintains the protective alkaline pH, and exchange with other physiological fluids is probably reduced by the completion of a basement membrane between the epithelial sheet and the under surface of the mineralized layer. Still, some calcium dissolution is seen during severe acid-base disturbance, although much less than is seen in terrestrial crabs, which lack the large external sink represented by seawater (Cameron, 1985a).

Moulting

Moult (ecdysis) out of an old shell entails severe stress on the respiratory system. It is at the same time a period of high metabolic demand and a period of impaired functioning of the gas exchange system. The process of escaping from the old carapace takes anywhere from 10 minutes to an hour (personal observation), and during part of this time no ventilation occurs. Mangum et al. (1985) found that blood oxygen tensions fell as low as 7 torr during ecdysis, PCO_2 rose to 20 torr, and pH fell perhaps 0.2 unit. The acid-base data show a large nonrespiratory alkalosis component of unknown origin which almost offsets the respiratory acidosis. Lactate concentrations vary, showing some rise during and immediately after moult, but the increase is only moderate, suggesting a continuation of mainly aerobic metabolism. Finally, ecdysis is accomplished partly by pronounced uptake of water and swelling, which has the effect of diluting the blood more than 4-fold, decreasing hemocyanin concentrations to almost useless values.

Chapter 10

Case Study: The Rainbow Trout

(*Salmo gairdneri* Richardson)

The popularity of the rainbow trout as a research subject is probably due to a combination of circumstances. The fish is immensely popular as a game fish and is widely distributed throughout the northern states and Canada. It has also proved adaptable to fish farming operations, providing substantial income to various regions of the United States, Canada and Europe. It is available in convenient sizes from fish farms near many major universities and has large and easily accessible blood vessels. Finally, much of the research on fish during the productive period after 1960 was done in Canada.

Some cautionary notes are in order, however. The rainbow trout is not necessarily a "typical" fish. It is evolutionarily primitive, has requirements for fairly high dissolved oxygen, and prefers low temperatures and running water, yet it is not a continuous swimmer but a "sit-and-wait" predator. The hatchery stocks from which most research subjects are drawn are often quite different from wild stocks: Hatchery stocks often suffer from chronic nutritional deficiencies (G. R. Bell, personal communication), may carry diseases, have different body form (personal observation), and in fact may be slightly to considerably different in every physiological parameter from the wild-caught fish. Hematocrit, for example, is typically 20 to 25% in hatchery stocks, whereas a wild population from a lake in central British Columbia average 40 to 50% (Cameron, 1971b).

The work on rainbow trout, however, is so extensive that it is difficult to discuss the respiratory physiology of fish without constant reference to it. With these reservations, then, the rainbow trout is described in the following chapter as an example of the respiratory physiology of an aquatic vertebrate.

FUNCTIONAL MORPHOLOGY

Gills: The Gas Exchanger

The Water Sieve

The structure of gills of fish in general and trout in particular has received thorough review treatment (Laurent & Dunel, 1980; Hughes, 1984; Laurent, 1984), so only a general description will be given here. The trout has four pairs of gill arches (Fig. 10.1). The rainbow trout is a species that has retained the rem-

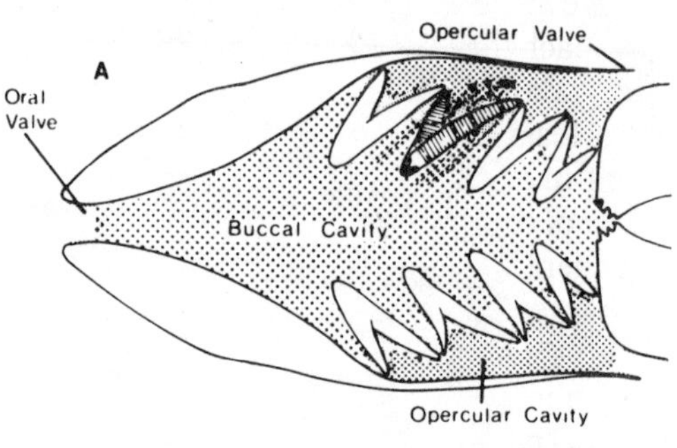

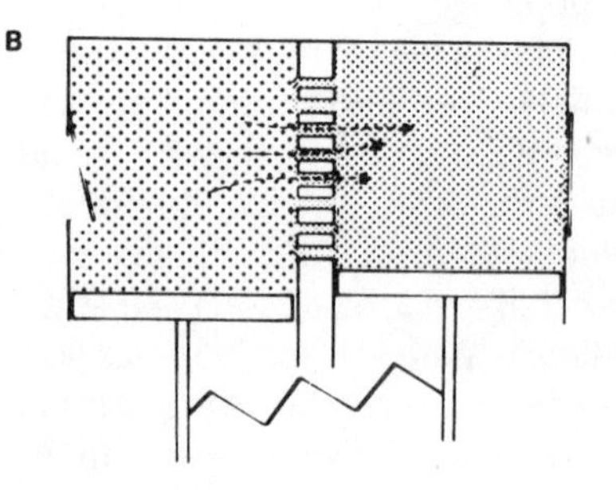

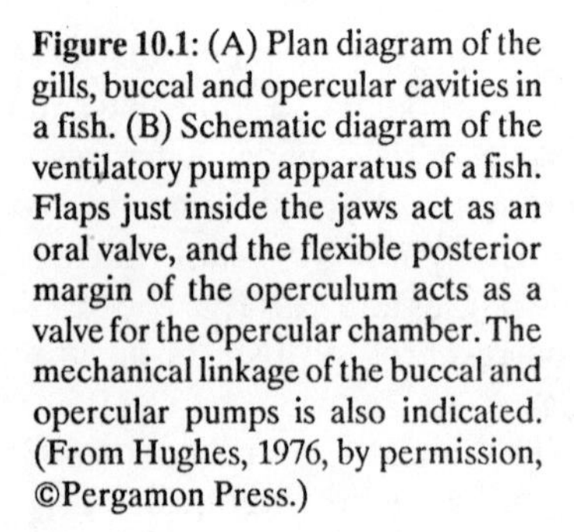

Figure 10.1: (A) Plan diagram of the gills, buccal and opercular cavities in a fish. (B) Schematic diagram of the ventilatory pump apparatus of a fish. Flaps just inside the jaws act as an oral valve, and the flexible posterior margin of the operculum acts as a valve for the opercular chamber. The mechanical linkage of the buccal and opercular pumps is also indicated. (From Hughes, 1976, by permission, ©Pergamon Press.)

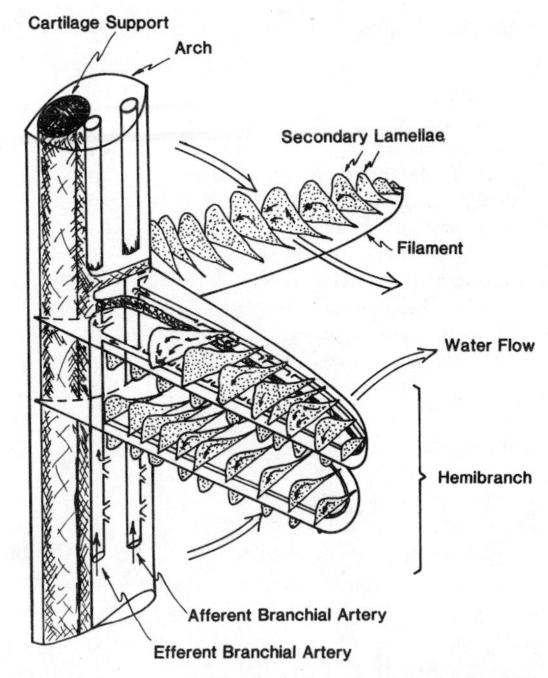

Figure 10.2: The arrangement of arches, filaments, and secondary lamellae in a trout gill. Each filament is supplied from the afferent branchial artery, and the filamental arteries supply each secondary lamella through short branches. The countercurrent arrangement of blood and water flow is indicated. (Modified from Randall, 1970b.)

nant anterior arch, the pseudobranch, which when present is attached to the wall of the buccal chamber. Gill filaments are arrayed in two rows on each gill arch, each row termed a hemibranch (Fig. 10.2). In terms of the degree of subdivision of the exchange surface, the filaments correspond to the lamellae of the crab gill, but each filament is further supplied with dense rows of perpendicularly oriented secondary lamellae, forming a finer sieve of higher order than the crab's gills.

Measurement of the total gill surface area in fish is somewhat more difficult than in crabs; not only are there tens of thousands of lamellae, but the shape and area of secondary lamellae vary depending upon the gill arch and the position on the hemibranch (Hughes & Morgan, 1973; Hughes, 1984). By systematically sampling representative lamellae in several regions of each gill arch, accurate estimates of a number of parameters of the gills may be obtained. Some morphometric data for rainbow trout gills from various sources are given in Table 10.1. These data have been used not only to compare respiratory exchange area among species but to analyze pressure, flow, and gas profiles in the water contained in the pores of the sieve formed by the secondary lamellae.

Water flow is probably more complex than indicated by the simple models employed. The surfaces of the filaments and lamellae are pitted and ridged in elaborate ways, for example (Fig. 3.7), and the water flow must be deflected at either end of the lamellar channels. The driving force is provided by the combined action of the buccal and opercular pumps (Fig. 10.1), and even under quiescent conditions the flow is probably pulsatile in the lamellar channels (Hughes & Shelton, 1962; Holeton & Jones, 1975).

Table 10.1: Morphometric data for trout gills, excluding the pseudobranchs.

Parameter	Value	Source
Weight	175	Assumed
No. of arches	4 pairs	
No. of hemibranchs	16	
Total No. of filaments	1500	Hughes & Morgan (1973)
Sec. lamellae/mm filament	21	Hughes (1966)
d, distance between 2° lamellae	0.023 mm	Hughes (1966)*
l, length of 2° lamellae	0.70 mm	Hughes (1966)*
½b, ave. height of 2° lamella	0.20 mm	Hughes (1966)*
Total pores in gill sieve	14.8×10^4	Hughes (1966)*
Ave. surface area of 2° lamella	0.2 mm^2	Hughes (1966)*
Total gill surface area	220 mm^2 g^{-1}	Hughes & Morgan (1973)

*Data for *Salmo trutta.*

The Gill Vasculature

The primary pathway for blood flow in the fish gill is relatively simple: there is an afferent artery in each gill that derives from the ventral aorta (Fig. 10.2). This afferent branchial artery gives rise to filamental arteries that pass out along the trailing edge of each filament. Small, short branches at the level of each secondary lamella allow blood to enter the downstream end of the secondary lamella, and the flow at this point appears to be controlled by an adrenergic sphincter (Boland & Olson, 1979; Laurent & Dunel, 1980). Flow through the secondary

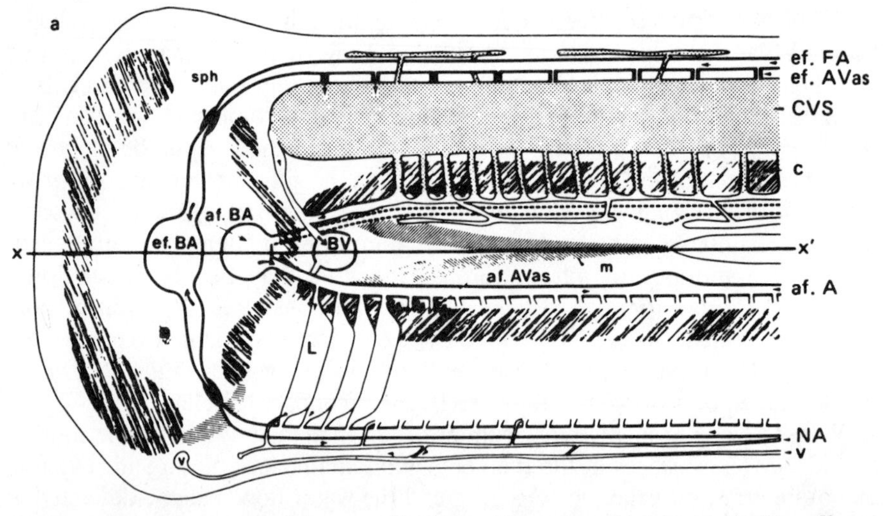

Figure 10.3: The vasculature of the trout gill. af.BA, afferent branchial artery; ef.AVas, efferent arteriovenous anastomosis; BV, branchial vein; c, cartilage; CVS, central venous sinus; ef.FA, efferent filamental artery; ef.BA, efferent branchial artery; m, abductor muscle; n, nerve; NA, nutrient artery; L, lamella; sph, sphincter. (From Dunel & Laurent, 1980, by permission, ©Cambridge Univ. Press.)

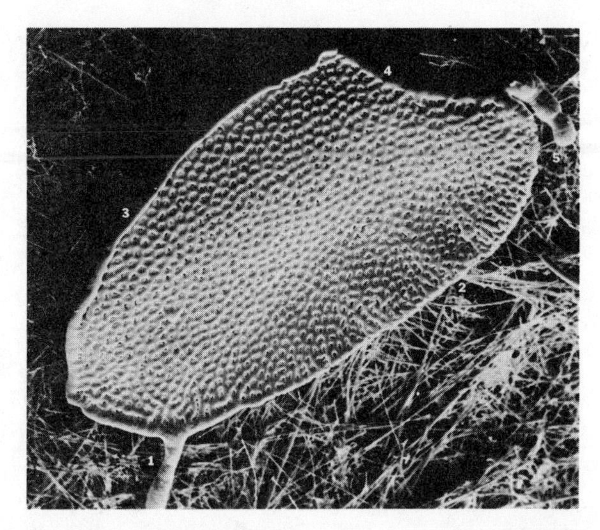

Figure 10.4: Scanning electron micrograph of an acrylic cast of the vascular spaces in the rainbow trout gill. Holes left by the pillar cells are visible in the cast of the lamella. Blood enters through a branch of the afferent filamental artery (1, lower left), flows through the lamella and through basal (2, lower right) and marginal channels (3, upper left) toward the leading edge (4, top), finally leaving through the efferent vessel (5, top right). (Photo courtesy of Dr. B. Gannon.)

lamella is counter to the direction of water flow, emptying into efferent filamental arteries at the upstream (water) edge of the filament. The efferent filamental arteries collect into efferent branchial arteries in each arch, and the eight efferent branchial arteries unite into a single mid-dorsal aorta situated just beneath the skull at the rear of the buccal chamber (Smith & Bell, 1975).

In addition to this respiratory circulation, however, there is a complex ancillary circulation within the filament (Fig. 10.3). A large central venous sinus (CVS) receives arterialized flow from the efferent filamental arteries, and the efferent blood from this CVS returns to the heart via the branchial veins. Work on this recurrent circulation has shown that it diverts about 7% of the cardiac output and has a much lower hematocrit than the rest of the blood (Ishimatsu et al., 1988). The low hematocrit apparently results from "plasma skimming" (Olson, 1984). There is also a nutritive circulation consisting of small capillaries that appear to supply the chloride cells preferentially (Boland & Olson, 1979; Dunel & Laurent, 1980; Laurent, 1984; Olson, 1984). Within each secondary lamella, the blood space, when cast in acrylic plastic, looks like a sheet with numerous holes punched through it. These holes correspond to the spaces normally filled by pillar cells (Fig. 10.4).

Examining a cross-section through an individual secondary lamella, it is apparent that the diffusion path for gases is rather complex (Fig. 10.5). The outermost layer is made up of one or more overlapping squamous epithelial cells, underlaid by a thin collagen layer. Since such thin sheets of tissue would not maintain their shape under the high blood pressures in the gills, they are supported

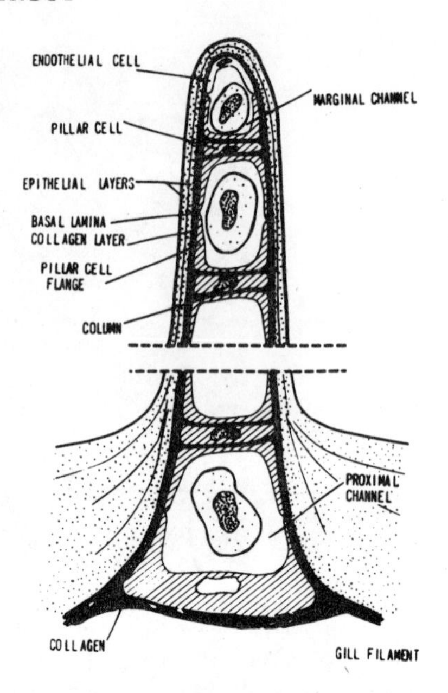

Figure 10.5: Transverse section through a secondary lamella of a teleost gill. The diffusion path consists of two epithelial layers, a basal lamina and collagen layer, and the flanges of the pillar cells. The basal portion of the lamella is embedded in the filament, presenting a much longer diffusion path. (From Hughes, 1980, by permission, ©Cambridge Univ. Press.)

by pillar cells, shaped as their name would suggest. These cells contain potentially contractile actin-like fibers (Bettex-Galland & Hughes, 1973), but they apparently do not play an active role in changing the dimensions of the lamellar blood space (Booth, 1979). The pillar cells also have flanged ends that meet beneath the collagen layer and add to the diffusion path. In most fish, including the rainbow trout, the secondary lamellae are partially embedded in the tissue of the filament, and there is a basal channel for blood flow that is slightly larger than the intralamellar channels. Thus part of the blood flow is presented with a very long diffusion path to the water space, and flow through the basal channel may represent an anatomical shunt (Smith & Johnson, 1977).

The Circulatory System

In general, the circulatory system is typical of that for fish as outlined in Chapter 4. It is a single-circuit system with blood flow channeled first through the capillary-like bed of the gills, back into major vessels for distribution to the rest of the body, through secondary tissue capillary beds, and finally through a collecting network of veins to return to the heart (Smith & Bell, 1975).

The blood volume in salmonids tends to be larger than most teleosts: Estimates range from 3 to 6%, with an average of 5% of body weight (Holmes & Donaldson, 1969). Compared to marine perciform fish of the same weight, the blood vessels of trout are considerably larger, and so the trout is relatively much easier to work with when sampling procedures are required. There is a small coronary vessel in trout that supplies approximately 25% of the myocardial mass with arterialized blood. The remainder of the myocardial muscle mass receives only a venous supply from blood passing through the lumen (Cameron, 1975a; Santer & Greer-Walker, 1980).

BLOOD PIGMENT: HEMOGLOBIN IN ERYTHROCYTES

Unlike the blue crab just discussed, trout have hemoglobin in erythrocytes, like other vertebrates. The erythrocytes are of medium size, about 10 μm diameter, and the total hemoglobin concentration averages 74 gm L^{-1} at a hematocrit of 31.6% (Houston & DeWilde, 1968). Erythropoietic tissue appears to be found mainly in the head kidney, but cell production rates in adult trout must be very low. When made experimentally anemic, the slow rate of replacement of erythrocytes suggests that the erythrocytes normally have a very long life-span (Cameron, 1971b).

Some dissociation curves for rainbow trout are shown in Figs. 5.3 and 5.5. There is a considerable Bohr shift in the physiological range: Changing P_{CO_2} from 3 to 7 torr at 15°C changes the P_{50} from 20 to 30 torr. Various authors have reported widely differing Root shift values for the rainbow trout. Randall (1970b) showed dissociation curves with a large Root shift, whereas Cameron (1971a) found less than 5% change in oxygen content with a 3 torr change in P_{CO_2}, consistent with Eddy (1971), who reported reductions in oxygen capacity of up to 14% at P_{CO_2} of 7 torr.

GAS EXCHANGE AT REST

Ventilation

An analysis of the mechanical movements of the gill ventilation apparatus and the resultant pressure and flow patterns is given by the work of Hughes and Shelton (1962). During phase 1 of the ventilatory cycle, the mouth is open and water flows into the buccal cavity (Fig. 10.1). The opercular valves are closed, and the operculi are abducted, generating a negative pressure relative to the buccal cavity and a net water flow across the gills. During phase 2 the mouth is closed, generating a positive pressure that closes the buccal flap valve and then increases into phase 3. At some point in phase 2 the positive buccal pressure begins to predominate, and the opercular abduction stops. Phase 3 is marked by opening of the opercular valves and adduction of the operculi. During this phase the buccal volume is reduced by action of the jaws and the floor of the buccal cavity, sustaining the positive pressure and flow gradient across the gills. Finally there is a brief phase 4 during which the buccal valves open, the mouth opens, and opercular adduction is completed. There is typically a brief period of reversed pressure during phase 4. The ventilatory cycle has often been portrayed as a two-part pump (Alexander, 1967; Hughes, 1976), but this is an oversimplification, both

Table 10.2: Respiratory parameters for rainbow trout (weight = 200 gm, temperature = 10 ± 1°C).

Symbol	Parameter	Value	Source
W	Weight	200	Assumed
V_f	Ventilation freq.	74 min^{-1}	Davis & Cameron (1970)
V_{sv}	Ventilation stroke vol.	0.47 ml	Davis & Cameron (1970)
$\dot{V}G$	Ventilation, minute vol.	35.2 ml min^{-1}	Davis & Cameron (1970)
$\dot{M}O_2$	Oxygen consumption	11.5 μM min^{-1}	Davis & Cameron (1970)
$\dot{M}CO_2$	CO_2 excretion	10.4 μM min^{-1}	(If RQ = 0.9)
PIO_2	Inspired O_2 tension	160 torr	Cameron & Davis (1970)
PEO_2	Expired O_2 tension	86 torr	Cameron & Davis (1970)
PaO_2	Arterial O_2 tension	133 torr	Cameron & Davis (1970)
PvO_2	Venous O_2 tension	32 torr	Cameron & Davis (1970)
CaO_2	Arterial O_2 content	3.42 mM L^{-1}	Calc., Cameron (1971a)
CvO_2	Venous O_2 content	2.30 mM L^{-1}	Calc., Cameron (1971a)
$PICO_2$	Inspired CO_2 tension	0.2 torr	Assumed
$PaCO_2$	Arterial CO_2 tension	2.1 torr	Cameron & Davis (1970)
$PvCO_2$	Venous CO_2 tension	3.3 torr	Calc., Wood et al. (1982)
$CaCO_2$	Arterial CO_2 content	4.9 mM L^{-1}	Cameron & Randall (1972)
$CvCO_2$	Venous CO_2 content	5.8 mM L^{-1}	Calculated
pH	pH	7.90	Calculated
$\dot{Q}$	Cardiac output	7.6 ml min^{-1}	Cameron & Davis (1970)
$\dot{V}G/\dot{Q}$	Vent.–perf. ratio	10.4	Cameron & Davis (1970)
CRR	Capacity–rate ratio	1.2	Cameron & Davis (1970)
$\%EbO_2$	Extraction, blood	33%	Calculated
$\%EwO_2$	Extraction, water	46%	Cameron & Davis (1970)
Hct	% Packed cell vol.	23%	Cameron & Davis (1970)
Hb	Hemoglobin conc.	72 g L^{-1}	Cameron & Davis (1970)
αBO_2	Total oxygen capacity	3.5 mM L^{-1}	Holeton & Randall (1967)

anatomically and functionally. Many sets of muscles are involved in accomplishing the entire cycle (Ballantijn, 1969) and the mechanical and pressure forces interact extensively.

The net driving force, however, is the integrated mean pressure. With earlier techniques there was little alternative but to make recorder tracings of buccal and opercular pressures and to compute the integrated (i.e., time-averaged) mean pressure by manual methods. The advent of inexpensive computer-based data acquisition systems should make continuous calculation of integrated mean pressure relatively simple.

At a given force, the resulting flow is determined by the resistance of the gill sieve, which is not constant. The resistance is affected by the muscular movement of the hemibranchs (Pasztor & Kleerkoper, 1962) and to a lesser extent by the mechanical positioning of both mouth and operculi (Cameron & Cech, 1970). Ventilation can be increased, then, by either an increase in driving pressure or a decrease in the gill sieve resistance. Driving pressure can in turn be affected by changes in the rate (frequency) of ventilation and the strength of each stroke (the stroke volume). At rest at 9°C, Davis & Cameron (1970) found an average ventilation frequency of 74, stroke volume of 0.5 ml, and a total minute volume of 37 ml for rainbow trout averaging 210 g. The average extraction efficiency was 46%. A compilation of ventilatory measurements made in rainbow trout is given in Table 10.2.

The rainbow trout does not show the peculiarities of the crab's ventilatory system. Apnea is rare except temporarily as a startle response, the ventilatory apparatus functions as a single unit rather than two quasi-independent systems, and reversals of the sort seen in crabs do not occur. There is a sort of "cough" reflex (see Chapter 8), in which the ventilatory current and pressure is momentarily reversed. The function of these brief reversals is not really known, but they are too short and infrequent to have much influence on gas exchange.

There has been some controversy over the energetic cost of breathing in fish, since the combined effects of high density and low oxygen content mean that fish must ventilate a mass of respiratory medium that is about 10^4 greater than that for any air-breather. The cost of breathing is not simple to measure, however, and estimates have varied from as little as less than 1% of the total oxygen consumption (Alexander, 1967) to several percent (Cameron & Cech, 1970; Jones & Schwarzfeld, 1974) to a range of 30 to 70% (Schumann & Piiper, 1966). The latter results were obtained by methods that have been questioned, however, and the best current estimate is probably in the range of 3 to 10%.

Perfusion

Experimental data have suggested (Davis, 1972; Booth, 1979) and theoretical studies (Cameron & Polhemus, 1974) have supported the idea that not all secondary lamellae are perfused at rest. Gas exchange at rest appears to be adequately accounted for if only 20 to 30% of lamellae are perfused, so water flow through the rest of the gill sieve represents a functional dead space. The mixed expired water data (Table 10.2) cannot therefore be used to calculate a relevant exchange potential ratio (EPR), since this index is based on the conditions at the entry and exit to an exchanging unit. Several studies have shown that various pharmacological agents affect the resistance of the gill vasculature (Wood, 1974), and coupled with the anatomical studies of the gill vasculature it seems safe to conclude that a considerable amount of control can be exerted over both the total amount and the distribution of blood flow through the gills at rest.

In the rest of the body the blood flow is disproportionately distributed to various internal organs, such as kidney, spleen, and liver, and to the red muscle mass (Stevens, 1968; Cameron, 1975c; Neumann et al., 1983). Although red muscle constitutes only about 1 to 4% of body weight, it receives 8 to 14% of the total blood flow at rest. White muscle, on the other hand, comprises about one-half the body weight and receives roughly half the total flow. The difference in blood flow between red and white muscle corresponds to the 2.5-fold greater capillary density in red muscle (Bone, 1966; Cameron & Cech, 1970) and to the generally greater aerobic capability of red muscle (Gordon, 1968).

Dorsal aortic blood pressure in resting rainbow trout has a systolic peak of about 30 mm Hg and falls to about 25 mm Hg during diastole (Stevens & Randall, 1967a). Actual intracardiac pressures probably peak at about 60 mm Hg, but the narrow-bore cannulae that are usually employed for such measurements can cause significant damping of peak pressures, so the actual values may be higher (Jones, 1970). Trout are also very sensitive to disturbance in the laboratory, reacting to small vibrations with a bradycardia that may last for several minutes (Davis & Cameron, 1970).

Efficiency Indices

From the values in Table 10.2 several indices of the performance of the gas exchange system can be calculated. The extraction efficiency for obtaining oxygen from the water is nearly 50%, and when the effect of various anatomical and functional dead spaces are taken into account (Randall, 1970b), it is clear that the extraction of oxygen from the water passing over perfused secondary lamellae must be very high. The overall ventilation-perfusion ratio of about 10 implies a capacity-rate ratio very near one, the theoretically optimum ratio for a countercurrent exchanger. Although the superior features of the exchange potential ratio (EPR) were discussed earlier, this ratio really requires the use of gas tensions at the entrance and exit to perfused lamellar channels, so a valid number for this index cannot be calculated. The 33% extraction of oxygen from the blood implies a considerable venous reserve.

When similar indices are calculated for CO_2, is clear that the fish gill, like the crab gill, is substantially hyperventilated with respect to CO_2 excretion requirements. This mismatching for CO_2 is apparently a general characteristic of aquatic gas exchange.

Resting Acid-Base Status

The resting values for pH, P_{CO_2} and total CO_2 can vary quite a bit depending upon temperature (see below) and the composition of the water in which trout are living. Waters very low in inorganic ions can perturb the acid-base status, but within wide limits the values given in Table 10.2 are typical. At low temperatures

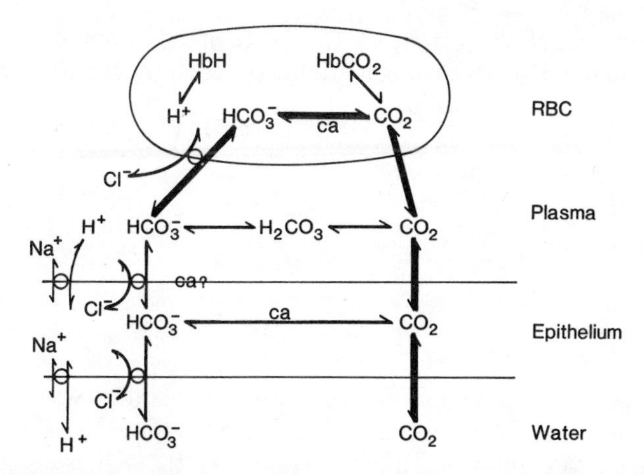

Figure 10.6: Major pools and pathways for CO_2 in fish. The heavy lines show the principal pathway for CO_2 excretion in the gill. NH_4^+ may substitute for H^+ in the Na^+ exchange pathways. See text for discussion. ca, carbonic anhydrase.

blood pH rises above 8, but the P_{CO_2} remains low, in the range of 1 to 3 torr, unless external P_{CO_2} is increased.

Carbonic Anhydrase and Normal CO_2 Excretion

Because of the equilibrium constants for the CO_2–HCO_3^- system, only about 4% of the total CO_2 is present as dissolved gas at any given time in the trout. In a single gill transit, however, 20 to 30% of the total CO_2 is excreted, making it clear that the primary reservoir for excreted CO_2 is the plasma HCO_3^- (Cameron, 1979). The generally accepted pathways for CO_2 diffusion and reaction are shown in Fig. 10.6. As blood enters the gill, the dissolved gaseous CO_2 begins diffusing out, unbalancing the equilibrium. The CO_2 lost can be replenished via two routes: first by (uncatalyzed) dehydration of plasma HCO_3^-, and second by a pathway through the erythrocyte. The erythrocytic pathway involves movement of HCO_3^- into the cell in exchange for Cl^-, the so-called *chloride shift* (or Hamburger shift, after the German physiologist, Hamburger). Inside the erythrocyte, the HCO_3^- is very rapidly dehydrated in the presence of carbonic anhydrase, producing the freely diffusible CO_2.

The kinetics of the movement of HCO_3^- into the erythrocyte, dehydration, and subsequent diffusion of CO_2 have been directly measured using a clever electrode and chamber apparatus invented by Lückner (1939). The device allows a thin film of blood to be rapidly exposed to changes in CO_2 and measures the time course of change in Cl^- activity, a direct measurement of the entire pathway (Fig. 10.6). Trout blood exhibits a typical chloride shift with a half-time less than 0.4 second, demonstrating clearly that the erythrocytic route is the principal conversion pathway for plasma HCO_3^- (Cameron, 1978a).

There is a substantial amount of carbonic anhydrase contained in the gill epithelium (Maren, 1967), some of which appears to be closely associated with the cell membranes (Rahim et al., 1988). If some of the carbonic anhydrase were actually on the outer surfaces, it could function in direct hydration or dehydration of HCO_3^- in either blood or water, but the best evidence to date is that it probably functions in osmoregulation and ion exchange pathways within the cells, much as it appears to do in the crab (Henry et al., 1988).

EFFECTS OF TEMPERATURE

Metabolism

The general pattern of metabolism with varying temperature was first described by Ege and Krogh (1914), working with a single goldfish. Their data showed a Q_{10} of about 2 at high temperatures, but constantly increasing slope at low temperatures, reaching Q_{10} values of nearly 10. More recent work has shown that these high Q_{10} values were probably the result of changing temperature too rapidly, unduly depressing metabolism at the lower temperatures (see Chapter 2). More complete work on rainbow trout and many other species shows Q_{10} values around 2 over wide ranges. Randall and Cameron (1973) reported a Q_{10} of about 2 between 10° and 20°C for rainbow trout, but many other studies of oxygen consumption have been made (see review by Fry, 1971). If a more eurythermal fish were considered, metabolism might change by 15- to 20-fold over the range of 0° to 40°C, and the gas exchange system would have to adjust its performance accordingly.

Gas Exchange

The increased oxygen demand at higher temperature must be met by changes in both the delivery of oxygen by ventilation and the distribution of oxygen by the circulation. The ventilatory supply could be met either by changes in the minute volume or increases in extraction efficiency, but the data show that the former is the more important mechanism (Fig. 10.7)(Randall & Cameron, 1973). Increased ventilation is accomplished partly by increased ventilatory frequency but more importantly by an increase in the volume pumped per cycle (ventilatory stroke volume). Whether there are compensatory changes in the gill sieve resistance is not known. Since ventilation and oxygen consumption change in concert, there is no significant change in the ratio of water ventilated to oxygen consumed.

There has not been very much work on temperature effects on cardiovascular function in the rainbow trout, but increased demand for oxygen transport appears to be met by a combination of adjustments. The arteriovenous oxygen difference increases (Heath & Hughes, 1971), as it does in another fish, the winter flounder (Cech et al., 1976), but not enough to meet the increased demand. In the winter flounder, increased cardiac output accounts for the rest of the increase. Heart rate increases account for all of the difference, with stroke volume not significantly changed. The increased cardiac output is not attended by an increase in blood pressure, so although the blood viscosity decreases at higher temperatures, there must still be some adjustment of vascular tone in order to regulate blood pressure (Cech et al., 1976).

Acid-Base Status

When trout blood is warmed in a closed system, it changes its pH by –0.019 unit per °C, whereas in an open (constant P_{CO_2}) system the temperature slope is only –0.004 to –0.005 (Randall & Cameron, 1973). In vivo the slopes vary from –0.010 to –0.017, depending upon previous acclimation temperature, but in general follow the pH-temperature relations described for other poikilotherms (see Chapter 6). Since the ratio of ventilation to oxygen consumption does not change (Fig. 10.7), assuming the RQ remains constant leads to the conclusion that the ratio of ventilation to CO_2 production also remains constant. One would therefore expect little change in the blood P_{CO_2}, and in fact none has been found (Randall & Cameron, 1973). Constant P_{CO_2} means that the decrease in pH as temperature rises must be achieved by a reduction in plasma $[HCO_3^-]$ through ionic exchanges (Cl^- for HCO_3^- and Na^+ for H^+ or NH_4^+)(see Chapter 6) in the gills and kidneys. Similar changes in plasma $[HCO_3^-]$ have been described now for several fish, and the net ionic fluxes at the gills have been measured (Cameron & Kormanik, 1982a; Cameron, 1984). The rainbow trout, and indeed all teleosts, carry out an open-system type of pH regulation in response to temperature change that

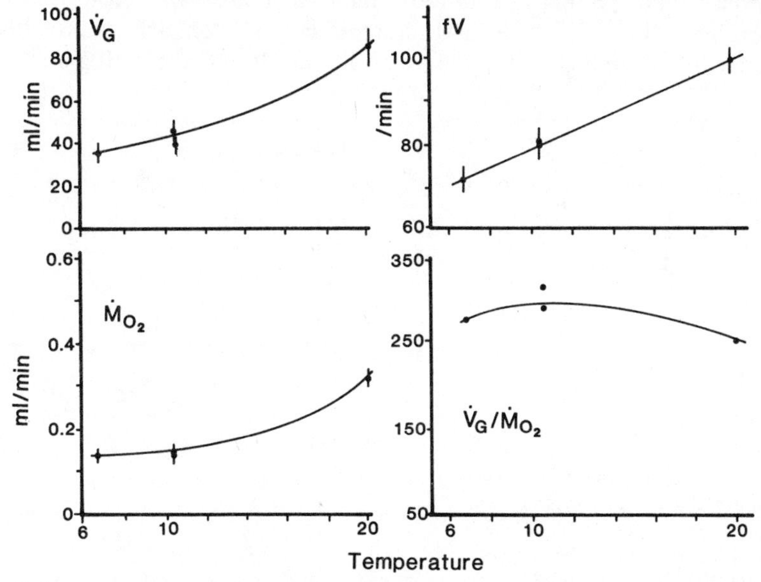

Figure 10.7: As temperature is increased, rainbow trout increase both oxygen consumption and ventilation proportionately, so that the ratio of the two remains essentially unchanged. (From Randall & Cameron, 1973, ©Amer. Physiol. Soc.)

is not based upon ventilatory control of PCO_2, but upon ionic regulation of $[HCO_3^-]$ and corresponding strong ions.

EFFECTS OF AMBIENT GASES

Hypoxia

Oxygen Consumption

When examining the effects of hypoxia upon metabolic rate (i.e., gas exchange) in fish, the same sort of division into conformers and regulators has been used as for invertebrates, but as with the invertebrates, fish seldom fall clearly into one category or the other. The rainbow trout doesn't either, as shown in Fig. 10.8, where oxygen consumption actually rises somewhat as oxygen tension is reduced from full saturation to about 80%, then falls steadily until the lower lethal limit of 20 to 30% saturation is reached. A rather different pattern was reported by Hughes and Saunders (1970), who found that oxygen consumption in the rainbow trout rose with an increasingly steep slope from fully saturated water down to one third saturation (50 torr). Holeton and Randall (1967) reported O_2 consumption increasing down to about half saturation and decreasing thereafter.

It is often difficult to distinguish between the direct effects of hypoxia on metabolism and indirect effects attributable to behavioral reactions. Several workers have shown that hypoxia stimulates swimming activity, presumably with the objective of finding better water conditions (Jones, 1952; Høglund, 1961).

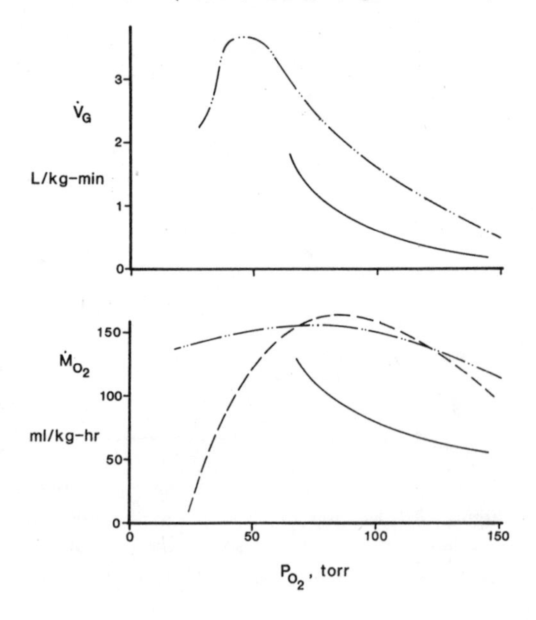

Figure 10.8: The response of ventilation (upper) and oxygen consumption (lower) to decreasing ambient oxygen. Solid line, data of Hughes & Saunders, 1970; dashed line, data of Marvin & Heath, 1968; dotted and dashed line, data of Holeton & Randall, 1967.

The portion of increased metabolism attributed to physiological processes such as ventilation must somehow be corrected for the energetic costs of swimming activity.

Ventilation

The increasing oxygen consumption was sustained and indeed was partly due to the increasing ventilation minute volume, which rose more than 8-fold in Hughes and Saunders' study. Increased ventilation during hypoxia is usually associated with decreased utilization (Hughes & Saunders, 1970; Davis & Cameron, 1970), but some earlier data must be discounted due to the opercular catheter technique that was used to compute them (Stevens & Randall, 1967b; Davis & Watters, 1970). The increased ventilation is achieved both by increases in frequency and stroke output.

Circulation

As environmental Po_2 is reduced, both arterial and venous oxygen tensions decline (Holeton & Randall, 1967). Because of the sigmoid nature of the dissociation curve, however, the decline in venous content is much greater than the decline in arterial content, so the A–V difference is increased, at least to moderate levels of hypoxia. At very low levels, this relation is reversed, perhaps contributing to a general failure of oxygen transport (Fig. 10.9).

Bradycardia has often been observed accompanying hypoxia in fish (Randall & Shelton, 1963), although its function is still not understood. Earlier suggestions

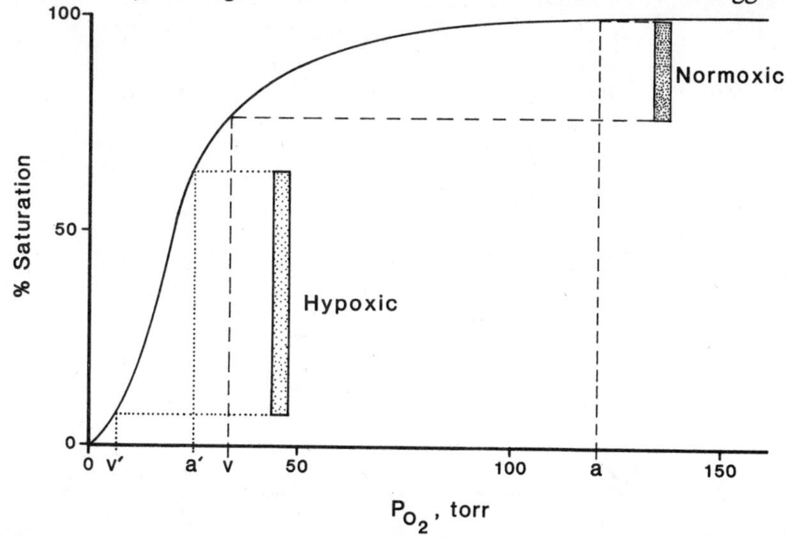

Figure 10.9: As ambient oxygen declines, depressing the blood oxygen tension, the sigmoid shape of the dissociation curve allows a greater A–V difference. At normal oxygen tensions there is a considerable venous reserve. The stippled bars show the A–V difference at normoxic arterial (a) and venous (v) oxygen tensions, and at severely hypoxic tensions (a′, v′). (Blood oxygen data from Holeton & Randall, 1967; dissociation curve from Cameron, 1971a.)

that blood flow declined have been partly disproved; the bradycardia is offset by increases in stroke volume (Holeton & Randall, 1967), so that there is only a slight decline in cardiac output in moderate hypoxia (when A–V difference increases) and then an increase in cardiac output at low oxygen tensions. One function of the bradycardia might be to increase synchrony between flow pulsations of blood and water in the gills, but the experimental evidence for that theory is weak (Hughes & Saunders, 1970). Other possibilities are either that the mechanical efficiency of the heart is improved at low rate and higher stroke volume or that the increased pressure pulsatility may promote lamellar recruitment (C. M. Wood, personal communication).

Acid-Base Status

There is also a lack of agreement upon what happens to acid-base status in rainbow trout during hypoxia. Holeton and Randall (1967) found no significant

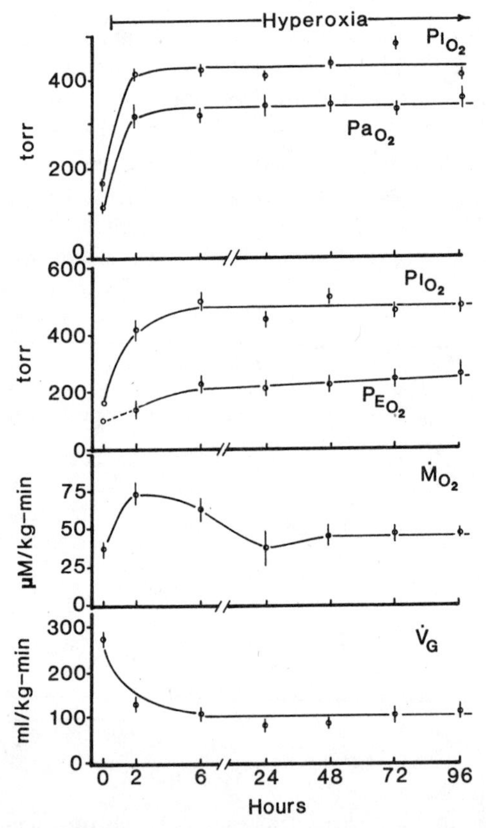

Figure 10.10: Respiratory changes during hyperoxia. (Upper) Inspired and arterial oxygen tensions during a 96-hour hyperoxic treatment compared to control (C) values. (Second from top) Inspired and expired oxygen tension. (Third from top) Oxygen consumption. (Bottom) Ventilation. (Re-drawn from Wood & Jackson, 1980, by permission.)

change in blood pH until very low P_{O_2} values were reached. At these low values very likely the contribution of lactic acid from anaerobic metabolism starts to become important. Thomas and Hughes (1982; and Thomas et al., 1986), on the other hand, reported a marked alkalosis at a P_{O_2} of 60 torr, with pH rising from 7.9 to 8.2, and P_{CO_2} falling by nearly half. This discrepancy is presently unresolved, but from a theoretical standpoint it is not surprising to see some reduction in the blood P_{CO_2}, even though the normoxic trout is already hyperventilated with respect to CO_2.

Hyperoxia

The principal effect of hyperoxia is a reduction in ventilation volume. Wood and Jackson (1980) found a 60% reduction in $\dot{V}_G$ after raising ambient oxygen to about 500 torr, a reduction that persisted for at least 4 days (Fig. 10.10). Oxygen consumption increased only temporarily (perhaps due to excitement of the fish), and pH was maintained almost unchanged. Arterial P_{CO_2}, however, increased from 2.5 to 4.3 torr; the acidosis that should have resulted was compensated by branchial ion exchanges acting to increase plasma $[HCO_3^-]$ (Wood et al., 1984).

Similar data have been published for other fish (Heisler, 1986b, for review), with only minor differences in the overall pattern of response. An interesting feature of all of the studies is that the increased P_{CO_2}, which must affect brain and cerebrospinal fluid pH, did not restimulate ventilation. Thus in the trout as well as the crab there is evidence that respiratory control is determined primarily by oxygen, with minor or insignificant sensitivity to CO_2.

Hypercapnia

The responses to normoxic hypercapnia are very similar to those described for the blue crab (see Chapter 9). That is, there is a rapid rise in blood P_{CO_2} and fall in blood pH, followed by a gradual compensation. The compensation is manifested by increasing $[HCO_3^-]$ at constant P_{CO_2}, with a partial return toward the original pH (Fig. 10.11). There is only a small contribution of the kidney to this compensation, most of which occurs via branchial Na^+/H^+ and Cl^-/HCO_3^- exchanges (Cameron, 1976; Cameron & Kormanik, 1982b; Heisler, 1986b).

There is a long-standing controversy as to whether hypercapnia stimulates ventilation in fish. Early literature reported increases, decreases, or no change in ventilation in response to hypercapnia, but each study could be criticized on one ground or another. In some, for example, CO_2 concentrations of 5 or even 10% were used – they are irritating and even toxic to fish. It is also difficult to sort out the behavioral effects, since fish sense changes in CO_2 and may become active, increasing ventilation as a response to the activity. There is a further problem of distinguishing between a direct CO_2 effect and an indirect one, since increased CO_2 will affect the blood oxygen tension and content. The best study of the problem to date concluded that there is either no direct response or that the

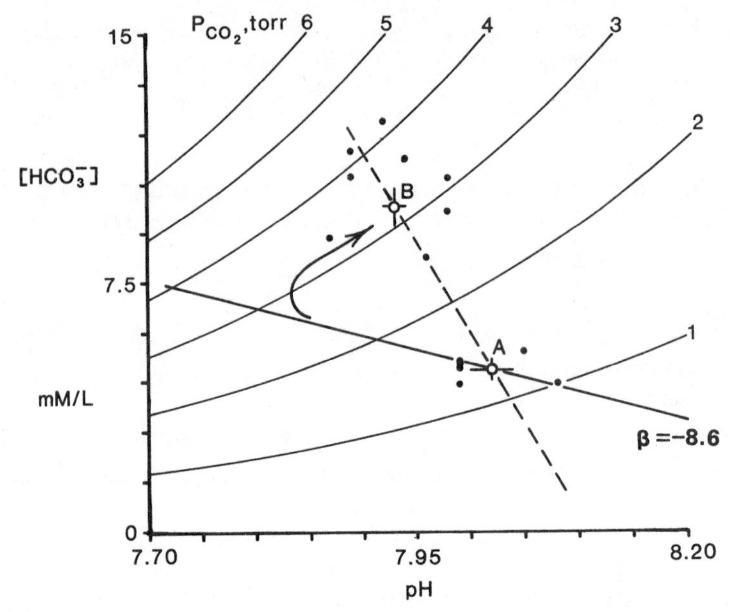

Figure 10.11: Responses to hypercapnia. (A) Control fish, ambient P_{CO_2} = 0.2 torr. (B) After 24 hours with ambient P_{CO_2} = 2.5 torr. (Data from Cameron & Randall, 1972.)

response can be attributed to reductions in tissue oxygenation due to Bohr and Root shifts (Smith & Jones, 1982).

EFFECTS OF EXERCISE

Metabolism

During swimming exercise the increased biochemical oxygen demand from muscle must be met by the gas transport system. Trout are swimmers of moderate ability, and adults can sustain swimming speeds of around 2 body lengths per second and juveniles about 5 lengths per second, far below the performance of some pelagic fish (Webb, 1971a,b), but nonetheless respectable. At maximum speeds, oxygen consumption is increased 7- to 8-fold (Kiceniuk & Jones, 1977). Only at speeds near fatigue level is any substantial increase in lactic acid evident, indicating that the majority of the swimming power is generated aerobically.

Gas Exchange

Ventilation

Increased ventilation during exercise was first reported by Stevens and Randall (1967b), but their values were calculated from oxygen consumption and from expired water samples obtained from an opercular catheter. Some of their values seem excessively high, probably due to heterogeneity in time and space of the water leaving the opercular margin (Davis & Watters, 1970). Kiceniuk and Jones employed a trailing latex "skirt" to provide mixing of the expired water, and achieved better results, albeit with considerable variability. They found that ventilation increased in direct proportion to the increased oxygen uptake, with little change in the percentage utilization of oxygen from the water.

There is an interesting switch to the "ram" mode of ventilation at higher swimming speeds, however. At intermediate speeds the trout still actively ventilates the gills but does not close the mouth completely during the cycle, allowing water to flow passively over the gills. As swimming speed increases, the ventilatory movements become gradually reduced, until at high speeds the trout simply holds the mouth open, regulating water flow by altering the gape of the jaws (personal observation). There may be an energetic advantage in what amounts to a transfer of the ventilatory work from muscles in the head region to the swimming musculature of the trunk; such an advantage is suggested by some oxygen consumption data that show a "notch" reduction at the point of transition to ram ventilation (Roberts, 1975).

Circulation

In Kiceniuk and Jones' (1977) study cardiac output increased about 3-fold, mainly through changes in stroke volume. The extra oxygen transport was provided by an approximate doubling of the extraction of oxygen by tissues, doubling

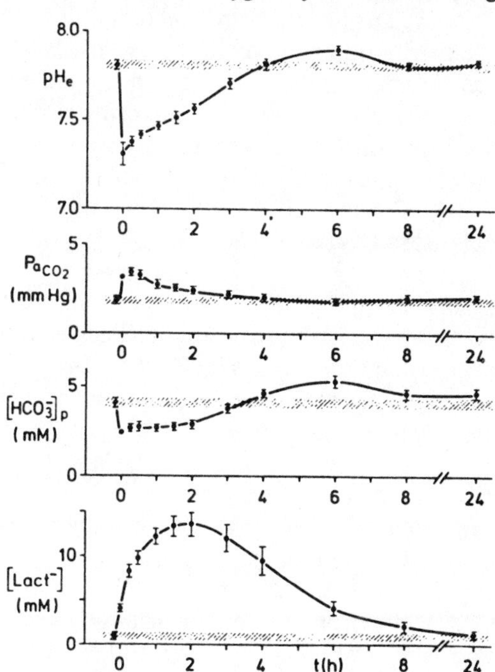

Figure 10.12: Acid-base responses of rainbow trout blood to swimming exercise. (From Holeton *et al.*, 1983, by permission, ©Elsevier North Holland, Inc.)

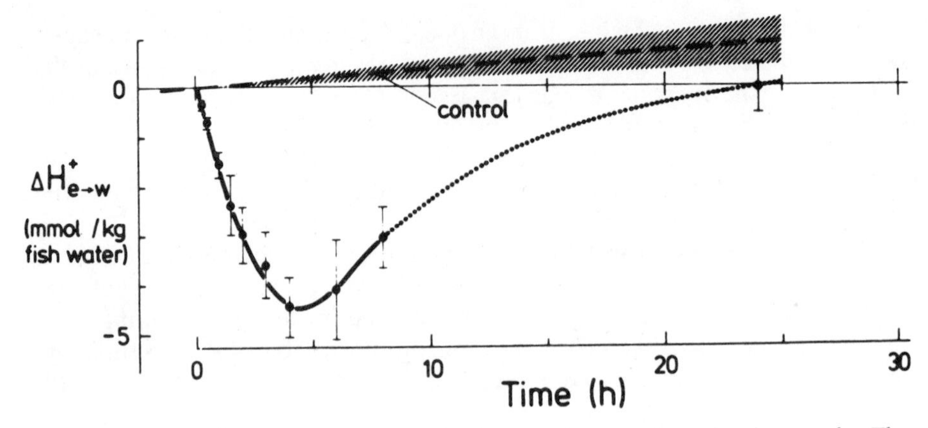

Figure 10.13: Net H^+ equivalents transferred to the water by trout after swimming exercise. The dashed line and hatched zone shows the small net base excretion that would have been expected from control (nonexercised) fish. After exercise there is a large transient H^+ excretion to the ambient water. (From Holeton *et al.*, 1983, by permission, ©Elsevier North Holland, Inc.)

the A–V oxygen difference. They noted that higher swimming speeds and oxygen consumption rates had been reported by Webb (1971a,b) and others, and also that greater increases in stroke volume appeared possible (Cameron & Davis, 1970). Since their work was performed at 15°C and in fully oxygenated water, one may speculate that the additional reserve would be used for activity at higher temperatures or in reduced ambient oxygen.

As well as increases in total blood flow, there appears to be some change in the distribution of blood flow. Stevens (1968) was unable to show any change by measurement of the blood volume contained in various tissues, but Neumann et al. (1983), using radioactive microspheres, showed 2.3- and 4.9-fold increases in the flow to white and red muscle, respectively, after strenuous exercise. Blood flow to some other tissues was actually reduced.

Acid-Base Status

Changes in the acid-base status of rainbow trout following exercise have been studied by a number of authors (Black et al., 1966; Holeton et al., 1983; Turner et al., 1983; Milligan & Wood, 1986). In all of these studies, the immediate result was an acidosis of mixed respiratory and nonrespiratory origin, the nonrespiratory component consisting entirely of lactic acid. The respiratory acidosis was usually of short duration, but the nonrespiratory portion had a much slower recovery in accordance with the dynamics of lactate movement from muscle to liver and remetabolism to glucose and/or glycogen (Fig. 10.12). As with the blue crab, the amount of lactate present in the plasma during recovery exceeded the apparent acidosis; the "missing" H^+ ions have been temporarily shuttled to the external environment by branchial ion exchanges and are replaced at a slower rate as recovery proceeds (Fig. 10.13)(Holeton et al, 1983; Heisler, 1986b).

Chapter 11

Transition from Water to Air:

Air-Breathing Crabs and Fish

Although life began in water, the abundance of terrestrial life today means that the transition from water to air must have come about through evolution of terrestrial forms from aquatic ancestors. An examination of phylogenetic relationships also indicates that the transition occurred not once, but perhaps many times during evolution. While they may not exactly mirror ancestral forms, various groups of extant animals contain species that are transitional in the sense of being partly adapted to aquatic life, and partly to terrestial life. Examination of their respiratory systems should illustrate the morphological and functional modifications that are necessary for the switch from one medium to the other.

AIR-BREATHING CRABS

General Ecology And Distribution

During the early part of this century some intrepid scientist-adventurers connected with the Carnegie Institute in Washington sailed a wooden ship from Chesapeake Bay to the Dry Tortugas, past Key West; and between hurricanes they assembled a field camp and laboratory. One of the early workers there was A. S. Pearse, who became interested in the variety of shore and land crabs abun-

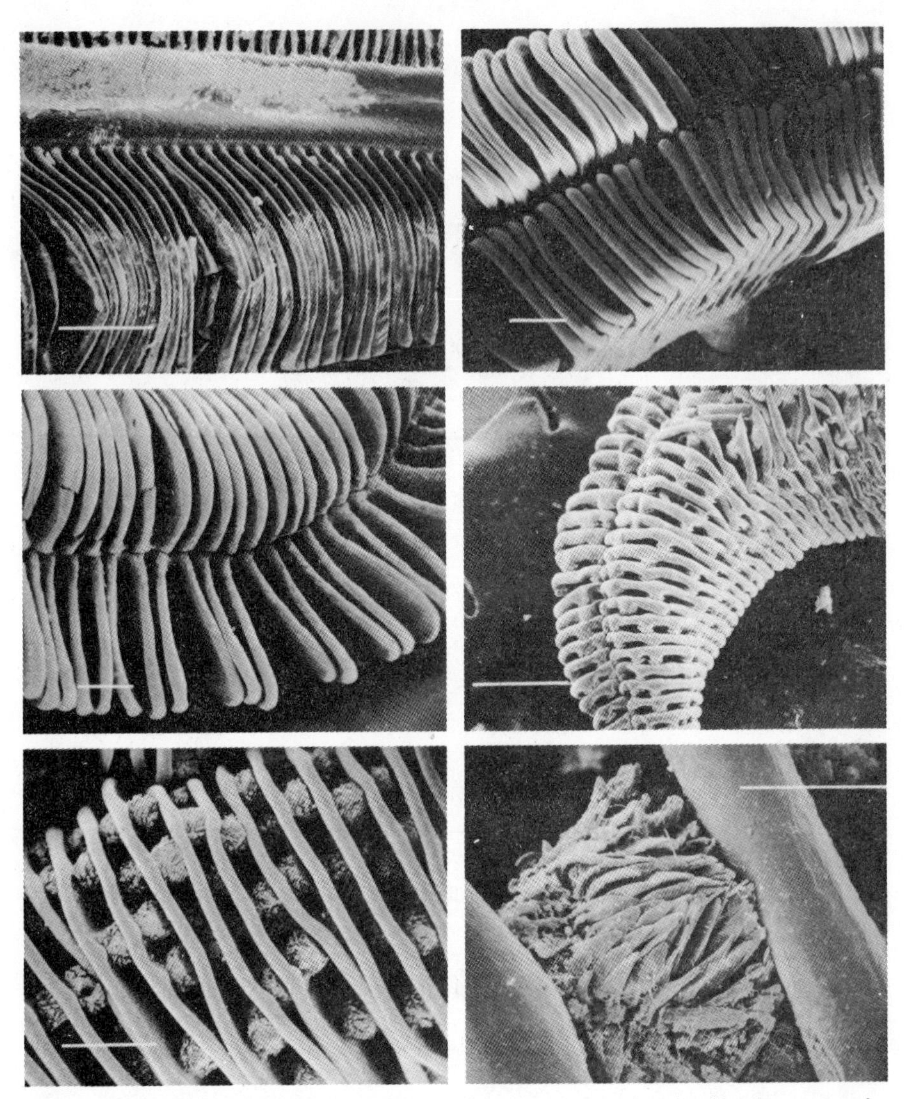

Figure 11.1: Scanning electron micrographs of gills of various crabs, showing the progressive reduction in surface area and modifications for preventing collapse in air. (Top left) *Callinectes sapidus*. (Top right) *Cardisoma carnifex*. (Middle left) *Birgus latro*. (Middle right) *Gecarcinus lateralis*. (Bottom left) *Gecarcoidea lalandii*. (Bottom right) Close–up of a tuft on a gill of *G. lalandii*. The white bar in each panel represents 0.5 mm, except in bottom right, 0.1 mm. (From Cameron, 1981a, ©Alan R. Liss, Inc.)

dant in the Tortugas. He observed that one could arrange these crabs in a series (the "Pearse series") ranging from strictly marine crabs, through amphibious crabs of the shore and splash zones, to those restricted to the dunes and drier areas of the islands (Pearse, 1929). This series has provided a perfect opportunity for the comparative physiologist to study the evolution of air-breathing in a non-vertebrate group. These species, of course, do not necessarily represent an evolutionary time series, but since each has arrived at a progressively greater degree of dependence upon air as the respiratory medium, the opportunity is present to examine which characteristics of the animals have changed.

Most marine crabs are able to survive in air for at least short periods of time, and when they succumb it may be from problems such as desiccation and not necessarily a failure of the respiratory system. The blue crab, for example, can survive for several days in air so long as the temperature is kept moderate and the humidity high (see Chapter 9). These crabs, however, do not normally emerge from the water, as do the many crabs of the beach and splash zone environment. The familiar green shore crab of the Atlantic (*Carcinus maenas*) and the Sally Lightfoot crab (*Pachygrapsus crassipes*) of the Pacific can be observed scuttling from land to sea in a seemingly uncaring fashion. Some grapsid crabs that I have observed on Pacific atolls appear to divide their time about equally between water and air. At the other end of the spectrum, crabs such as the purple land crab of the Caribbean (*Gecarcinus lateralis*) and the coconut crab of the Indo-Pacific (*Birgus latro*) can survive only in air, promptly drowning if forcibly submerged. These varying abilities for breathing air and water imply substantial modification of the respiratory systems.

Functional Morphology

An examination of the gills of a series of aquatic to terrestrial crabs quickly makes it obvious that there is a marked reduction in gill surface area as the crab becomes more terrestrial (Cameron, 1981a)(Fig. 11.1). This trend was first remarked by Gray (1953), even though he did not have fully terrestrial crabs to examine. The reduction in *Birgus latro* is more than 10-fold compared to the fully aquatic crabs. Besides a reduction in total surface, there is an obvious trend toward wider spacing of lamellae, thicker and stiffer lamellae, and a proliferation of supporting tubercles on the gills. The function of these modifications is to prevent the collapse of the gills in air. When the blue crab emerges, its thin and closely spaced gill lamellae are completely collapsed by surface tension of the water, rendering their functional surface area almost nil. Further examination of the small gill area that remains in crabs such as *Gecarcinus* and *Birgus latro* reveals that most of the epithelium is of the thick, osmoregulatory tissue type (unpublished data; see also Chapter 9), so the gill area still functional in respiration is minuscule.

Beginning with the partially terrestrial crabs such as the mangrove crab (*Cardisoma guanhumi*), the branchial chambers are progressively modified as air breathing organs. The volume of the chamber becomes larger, and a considerable air space is contained dorsal to the gills. The lining of the branchial chamber becomes thinned and well vascularized, and in some species is elaborated into ridges or tufts to enlarge the surface area (Diaz & Rodriguez, 1977).

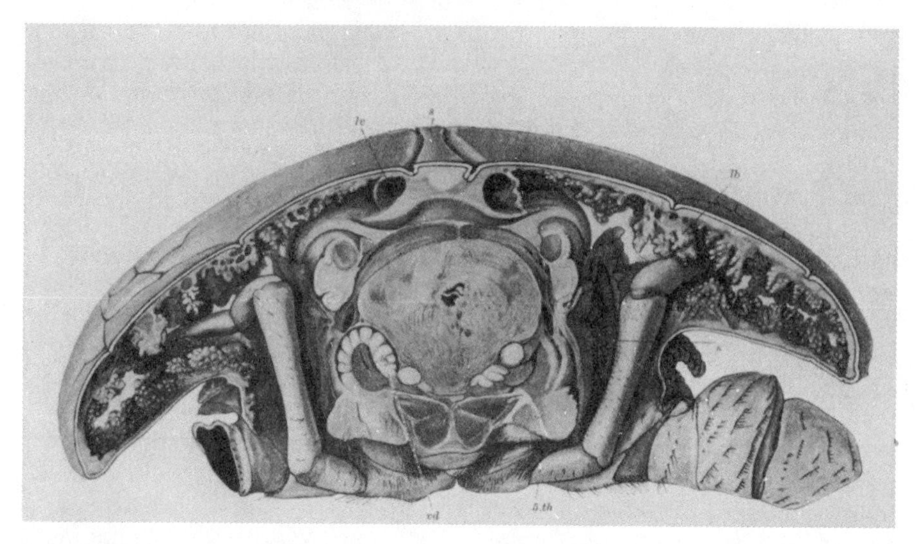

Figure 11.2: Cross section of *Birgus latro* at the level of the widest part of the lung chamber, showing the lung location and structure. lb, lung tufts; lv, lung vein; k, gills; 5.th, fifth thoracic leg. (From Harms, 1932.)

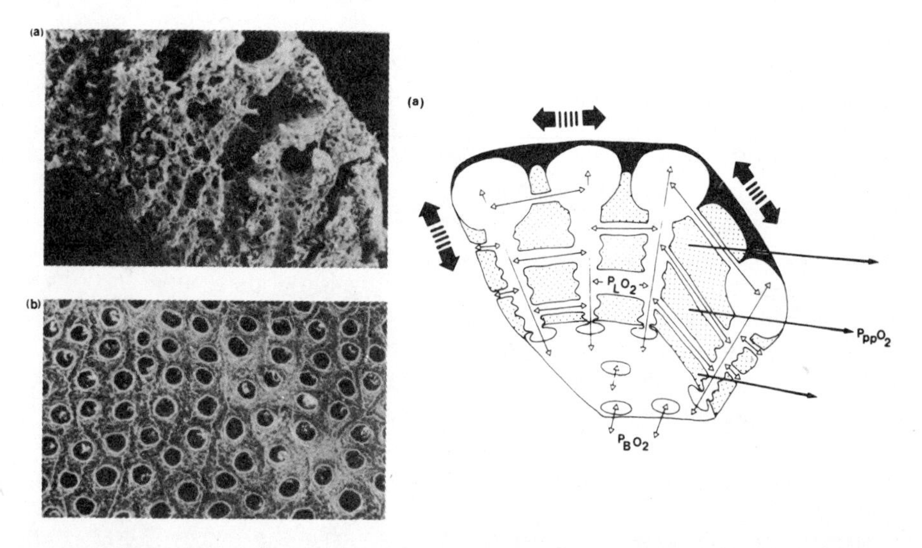

Figure 11.3: (Upper left) Scanning electron micrograph of a cut section of the branchial chamber lung of *Pseudothelphusa transversa*, showing the air channels leading from the branchial chamber, at lower left, toward the carapace, ending in large blind sacs. (Lower left) Same, viewed from the branchial chamber's inner surface. The scale bar at the lower left is 100μm. (Right) The lung, showing the arrangement of air channels and the striking similarity to the structure of lungs of birds and mammals. (From Innes et al., 1987, by permission, ©Pergamon Press.)

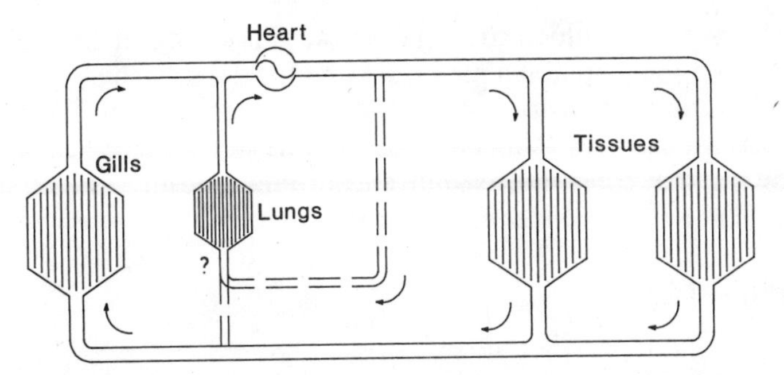

Figure 11.4: Circulatory pathways in an air-breathing crab. Since the gills can be removed and cauterized, there must be a bypass pathway for blood flow through the branchial chamber lungs. Whether the pathway is part of a portal system or has a direct supply from the heart is not known.

In a few air-breathing crabs the branchial chamber modifications have progressed much further, forming complex lungs. In the branchial chamber lining of the coconut crab, for example, the entire inner surface consists of a thick layer of spongy tufts (Fig. 11.2)(Harms, 1932; Cameron & Mecklenburg, 1973). While no measurements of total surface area are available for this animal, the result is clearly an effective aerial lung. Coconut crabs survive indefinitely after removal of all gills (Harms, 1932). Crabs of the Caribbean family *Pseudothelphusidae* have a rather different, but nonetheless elaborate lung constructed by invagination of the branchial epithelium (Fig. 11.3) rather than evagination as in the coconut crab. This lung is ventilated by slow expansion and contraction of the branchial chamber and appears to function as a ventilated pool lung, similar to those of the birds and mammals (Innes et al., 1987)(see Chapters 3, 13, and 14).

On the circulatory side, there are some modifications from the general crustacean plan (Innes et al., 1987). Coconut crabs have higher blood pressure than most crabs (Cameron & Mecklenburg, 1973) and a better developed vasculature, as shown in the somewhat stylized drawings by Harms (1932). There is an interesting question of whether the flow of blood can be diverted from gills to branchial chamber lining when switching from water- to air-breathing, as shown for *Holthuisana transversa* by Taylor and Greenaway (1984). The evidence so far is not overwhelming and depends upon application of microsphere techniques that have some uncertain assumptions. It is also interesting to note that there must be bypass shunts around the gills, at least in the coconut crab. Otherwise, when Harms (1932) removed the gills surgically, cauterizing the wounds, all circulation would have stopped. Oxygenation of the blood, then, is taking place in parallel circuit to the body circulation, rather than in series as in other crabs (Fig. 11.4). Similar conclusions have been drawn for some freshwater/terrestrial crabs by Innes et al. (1987) and Taylor and Greenaway (1979, 1984). The situation is apparently like that in reptiles and amphibians with partially divided hearts: The heart is pumping a mixture of oxygenated blood returning from the branchial epithelium (lung), deoxygenated blood returning from the systemic circuits, and perhaps poorly oxygenated blood coming from the gills during air-breathing. A further step toward the amphibian plan would be for the branchial epithelium (lungs) to receive a pulmonary arterial supply directly from the heart. There have been such arteries described (Vuillemin, 1963), but so far there has been no demonstration that they constitute a significant portion of the lung's blood supply (Taylor & Greenaway, 1984; Innes et al., 1987)

Respiratory Physiology

Ventilation

The flow of air through the lungs (gill chambers) of crabs is usually provided in the same manner as in aquatic ones: The beating action of the scaphognathite propels air out of the excurrent channels near the mouth, drawing air in through the Milne-Edwards openings and over the gas exchange surface. In some crabs ventilation may be supplemented by expansion and contraction of the branchial chamber (Innes et al., 1987). Rates of ventilation in air-breathing crabs are lower than those of water-breathers: The coconut crab pumps 74 ml kg^{-1} min^{-1} over the gills, compared to 555 ml kg^{-1} min^{-1} for a smaller blue crab (see Table 9.2). Extraction efficiency was considerably lower, however, averaging 5.8% (Cameron & Mecklenburg, 1973). Increases of almost 10-fold were possible, achieved almost entirely by increases in frequency at constant stroke volume. Notably absent were the periodic apneas and reversals so characteristic of ventilation in the aquatic crabs. Virtually any function attributed to these reversals, e.g. gill cleaning or flow augmentation, would not work in air anyway.

Ventilation in the coconut crab responds to hypoxia, though perhaps not as strongly as in aquatic crabs. At oxygen tensions below 50 torr ventilation rises sharply, with a fall in the extraction percentage and little change in oxygen con-

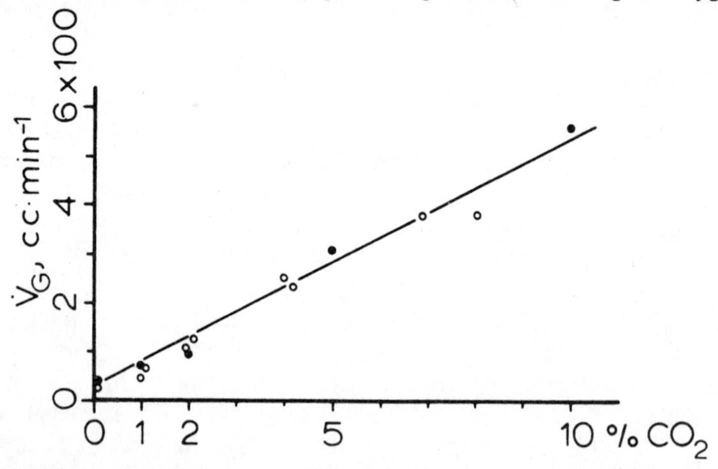

Figure 11.5: The ventilatory response to CO_2 in the air-breathing coconut crab, *Birgus latro*. (From Cameron & Mecklenburg, 1973, ©Elsevier North Holland, Inc.)

sumption. Interestingly, ventilation also responds strongly to CO_2 not only in *Birgus* (Fig. 11.5), but in *Gecarcinus* and *Cardisoma* as well (Cameron & Mecklenburg, 1973; Cameron, 1975b; Smatresk & Cameron, 1981). There is an almost linear response, which continues to CO_2 concentrations as high as 10% in air. A similar response is evoked by acid infusion but not by an iso-pH hypercapnic solution, implying that the response is primarily a pH effect (Smatresk & Cameron, 1981).

Several workers have remarked that aquatic crabs undergo an internal hypercapnia and resulting respiratory acidosis when emersed (Truchot, 1975b; DeFur et al., 1983). The long-term response to this hypercapnia is similar to what is seen during hypercapnic compensation in aquatic crabs. Turning the tables somewhat, when a primarily terrestrial crab such as *Cardisoma carnifex* is forcibly submerged and denied access to air, its limited gill surface area provokes an acute hypoxia. The hypoxia in turn stimulates ventilation, which leads to a respiratory alkalosis (Cameron, 1981b).

Since most of these terrestrial crabs live in hot climates, one might wonder how the animals prevent desiccation. The large, thin surface area used for gas exchange represents a potentially serious route of water loss. Part of the answer is behavioral, in that the animals tend to be nocturnal and/or confined to sheltered high-humidity microenvironments. During the dry season in the Marshall islands, for example, coconut crabs retreat into their burrows, seal them up, and remain there for several months (Gross, 1964). Part of the answer is also provided by water carried in the lung chambers. As first remarked by Milne-Edwards (1834, cited by Harms, 1932) and later by Semper (1878), Harms (1932) and von Raben (1934), some water is always present in the lung chamber, and this water is replenished by a modified leg in the anomuran crabs and by immersion in the brachyurans. One of its functions is clearly to humidify the lung chambers; whether its small volume is sufficient for any other purpose seems doubtful, but has been claimed by some (Wood & Randall, 1981a,b; Burnett & McMahon, 1987).

Circulation

Blood gases in the coconut crab are typical for crustaceans: Arterial oxygen tension averages about 80 torr and venous about 15 torr, for an A–V difference of roughly 50%. CO_2 tensions are considerably higher, however, with arterial P_{CO_2} averaging 6 to 7 torr and venous about 9 torr (Cameron & Mecklenburg, 1973; Smatresk & Cameron, 1981) and pH averaging 7.5 to 7.7 at 28°C. Cardiac output has not been measured directly in these animals, but should be in the same general range as for other crabs, based on blood gas data and total oxygen consumption.

Responses to Exercise

The capacity to exercise varies considerably in the terrestrial crabs, with some showing only short burst activity for escape, such as the ghost crab (*Ocypode quadrata*), and others such as the coconut crab, capable of sustained rapid walking

(Smatresk & Cameron, 1981). The pattern of changes in all parameters seems exactly like that described for the blue crab with the exception that the base deficit (excess acid load) in the blood during recovery exceeds the lactate concentration, in contrast to the other way around in aquatic crabs (see Fig. 6.17). The best explanation of this difference is that the terrestrial crabs are unable to temporarily shuttle excess H^+ ions to the external environment via ionic exchanges with seawater, so the excess H^+ remains in the circulation (Smatresk & Cameron, 1981). Some is probably buffered by the carbonates of the shell, judging from the increased circulating calcium during the recovery period (Henry et al., 1981; Smatresk & Cameron, 1981; Wood & Randall, 1981b).

That crabs have successfully invaded the land cannot be denied, but it must also be noted that their success is quite limited. They tend to be found near water, confined to humid environments. The coconut crab, for example, is distributed only on atolls and small islands of the Indo-Pacific, where it lacks competition from terrestrial vertebrates. None of these animals is able to reproduce or undergo larval development out of the water, although at least one species of fiddler crab, *Uca subcylindrica*, is able to rear young in water at the bottom of a burrow (Rabalais & Gore, 1985).

AIR-BREATHING FISH

General Ecology and Distribution

With a few exceptions, air-breathing fish are creatures of warm, swampy fresh or brackish waters where oxygen depletion is a frequent fact of existence. Although the most common air breathing organ (ABO) is some type of modified swimbladder, structures as diverse as the stomach, mouth, "labyrinth organ" and even the gills have been modified for the purpose. As with the crabs, one could probably lay out a series of fish ranging from no air-breathing ability to almost complete dependence upon air-breathing. Fish with no specialized ABO can occasionally take advantage of the increased oxygen availability of air: *Umbra limi* has been observed taking bubbles into its mouth under ice in "winterkill" lakes, for example (J. J. Magnuson, personal communication), and many fish will "gape" at the surface layer when oxygen concentrations are low (Burggren, 1982b). The ability to breathe air must clearly have been of great selective advantage, so it is not surprising that it occurs in many groups of fish distributed widely through the warm regions of the world.

Functional Morphology

Gills

As with the air-breathing crabs, there is a marked trend toward reduction of the total gill surface area in the air-breathing fish, reaching an extreme with the

African lungfish *Protopterus*, whose gill filaments are little more than mere nubs (Laurent & Dunel, 1980). Only a few of these fish actually emerge onto land, however, so in most there has not been the same trend toward thickening and stiffening of the gills – most still collapse together in air, becoming non-functional. Exceptions to this trend are fish such as the mudskippers, *Perioptbalmus*, whose gill chambers do act as a lung when the animal is on land.

The gill microvasculature is also strongly altered in many of the air-breathing fish. Lamellar channels may be much reduced, bypass shunts are more in evidence (Laurent, 1984), and in the gar there is a large basal channel in each secondary lamella that would direct much of the blood flow through a pathway with a very long diffusion path (Smatresk & Cameron, 1982a). Thus the reduction in functional surface area may be even greater than suggested by the external anatomy of the gills.

Air-Breathing Organs

Some of the most primitive air-breathers are the Holostean fish, a New World remnant group consisting of 10 extant species of gar (*Lepisosteus*) plus the bowfin, *Amia calva*. These fish have a moderately well vascularized and compartmentalized swimbladder-lung (Potter, 1927; Datta Munshi, 1976) connected to the esophagus by a pneumatic duct and ventilated by positive pressure generated by combined action of the buccopharyngeal basket (McMahon, 1969; Gans, 1970; Smatresk & Cameron, 1982a; Liem, 1984). Only the osteoglossids appear to have developed a muscular sheet underlying the swimbladder-lung that can generate negative filling pressures, i.e., aspiration breathing (Randall et al., 1978).

A number of other groups of fish also have modified swimbladder-lungs, including the South American erythrinids, osteoglossids, and several catfish groups (Johansen, 1970; Datta Munshi, 1976). The degree to which vascularization and elaboration of the internal surface area has been carried out is extremely variable, both within each phylogenetic group and among different groups. Despite some lungs being fairly well developed, the total respiratory surface area is only about $^1/25$ that of a mammal of comparable size (Hughes & Weibel, 1976).

Several other kinds of structures have evolved in fish for air-breathing, including very elaborate vascularization of the mouth cavity (electric eel) and extension of the opercular chambers to form a labyrinth organ (gourami). A few of these structures are shown in Fig. 11.6.

Circulatory Modifications

The principal routes of blood flow in various air-breathing fish have been described in Chapter 6 (see also Satchell, 1976). In all but a few of the lungfish, the lung "competes" with the rest of the body for blood flow. That is, all output from the heart goes first through the gills, then is apportioned between lung and systemic circuits, both of which return to the heart through veins. The heart, then, in all air-breathing fish is pumping a mixture of deoxygenated blood from the systemic circuit and oxygenated blood returning from the ABO. In gar, for example, the ABO is supplied by numerous short branches of the dorsal aorta. Only in some of the more advanced lungfish such as *Protopterus* does a separate afferent circulation begin to appear (Satchell, 1976; Laurent & Dunel, 1980). Specialized vessels from the posterior gill arches lead to a parallel pulmonary artery supplying the lung, and the heart is partially septate, possibly providing some separation of oxygenated and deoxygenated blood (Johansen, 1970; Ishimatsu & Itazawa, 1983).

There is presently little information regarding the control of blood flow to the ABOs of fish. In both bowfin and gar, there is a marked change in blood flow to the lung depending on whether the lung is ventilated or not (Johansen et al., 1970; Smatresk & Cameron, 1982a); this response appears to be under control of internally oriented oxygen chemoreceptors, but their exact location and neural circuitry are unknown (Smatresk et al., 1986).

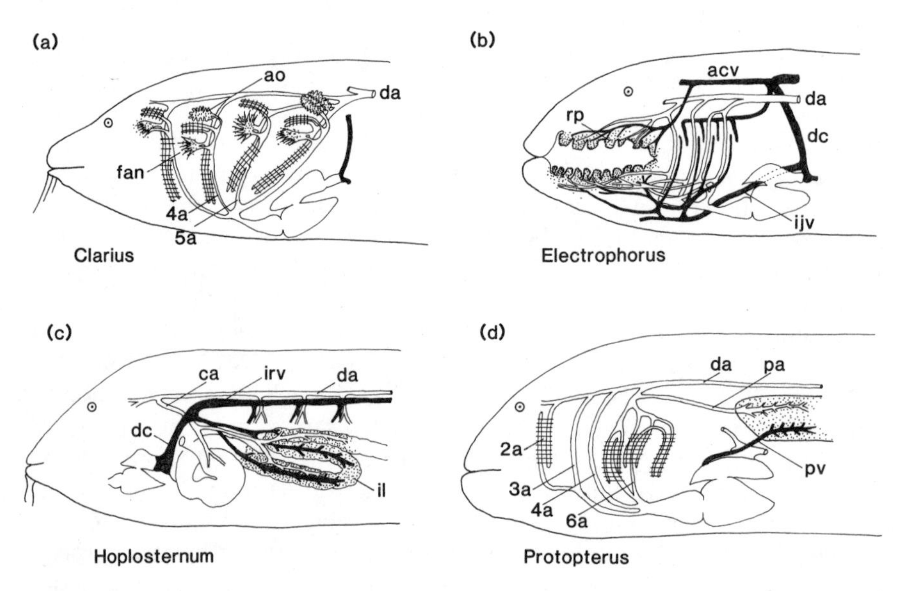

Figure 11.6: Some of the air-breathing organs (ABOs) found in different groups of fish. Returning veins that carry oxygenated blood are shown as solid black. (a) *Clarius* supplements its modified gills with fan-shaped extensions (*fans*) as well as with arborescent organs (*ao*) for both air and water breathing, and also has a vascularized buccal chamber. (b) The electric eel, *Electrophorus*, is a mouth breather, and has a vascularized lining which resembles the coconut crab's lung – highly tufted and irregular in shape with a high surface area. The gills are non-functional and are bypassed. (c) A intestinal breather, *Hoplosternum*, has a looping portion of the intestine that is thin and highly vascularized, supplied by the coeliac artery (*ca*). The gut is tidally ventilated by swallowing air, and the digestive portion of the gut is much reduced. (d) The lungfish, *Protopterus*, has highly modified blood vessels in the gills which act as shunts to the lung. The lung, a modified swimbladder, is supplied by a pulmonary artery (*pa*) and drained by a pulmonary vein (*pv*). Other symbols: *da*, dorsal aorta; *acv*, anterior cardinal vein; *dc, ductus cuvieri*; *ijv*, internal jugular vein; *irv*, interrenal vein; 2a – 6a, aortic arches. (Modified and redrawn after Satchell, 1976.)

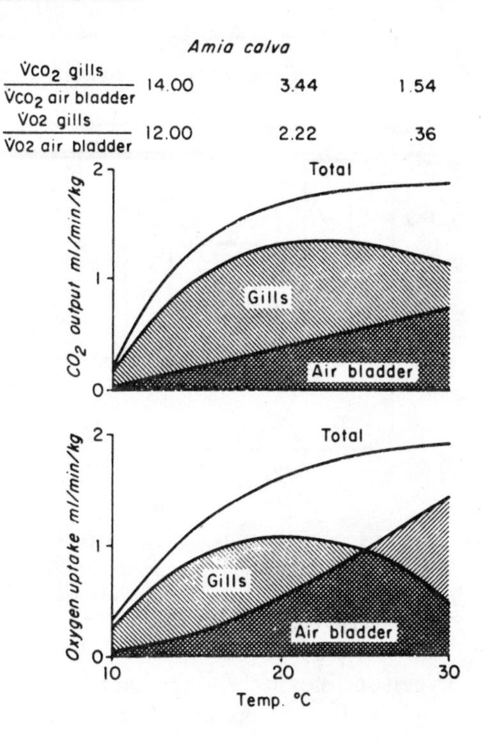

Figure 11.7: The proportion of oxygen and CO_2 exchanged by ABO and gills at different temperatures in the bowfin, *Amia calva*. (From Johansen et al., 1970, by permission, ©Elsevier North Holland, Inc.)

Respiratory Physiology

Sites of Gas Exchange

In nearly all of the air-breathing fish, the primary route for CO_2 excretion continues to be the gills, even when a large part of the oxygen demand is being met by the ABO (Fig. 11.7)(Johansen, 1970; Lenfant & Johansen, 1972). In Chapters 9 and 10 it was pointed out that the gills of water-breathers are typically hyperventilated with respect to CO_2, i.e., they have much greater capacity than is normally required. Obviously a considerable reduction of functional surface and flow can still be adequate for CO_2 excretion. At higher temperatures the fraction of CO_2 excreted by the lung rises somewhat but remains in the minority. Part of the low lung CO_2 excretion is simply due to the blood already having passed through the gills before entering the lung, thus ensuring a low blood-to-air gradient for CO_2 in the lung.

Many air-breathing fish also have reduced scales and well-vascularized skin and fins, so that a considerable portion of their gas exchange may take place cutaneously (Lenfant & Johansen, 1972; Romer, 1972).

Ventilation

Normal Ventilation. The frequency of air breaths can vary from more than 1 per minute to less than 1 per hour (Johansen et al., 1970; Smatresk & Cameron, 1982a). The frequency of air breaths is probably influenced by several factors. As

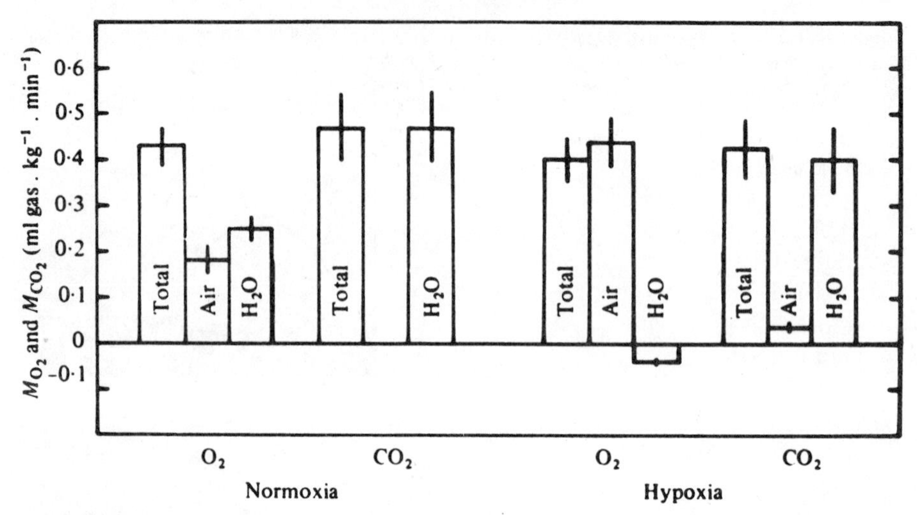

Figure 11.8: O_2 and CO_2 budget for spotted gar during normoxia and hypoxia. The bar marked "Air" indicates the amount exchanged by the lung and the bar marked H_2O the amount exchanged by the gill. Note that in hypoxia the gill allows a net O_2 loss. (From Smatresk & Cameron, 1982b, ©The Company of Biologists, Ltd.)

oxygen declines in the lung after a breath, the volume of the lung decreases since CO_2 is not being excreted into it in proportion (Johansen, 1970). Mechanoreceptors have now been described from the lungs of bowfin (*Amia calva*)(Milsom & Jones, 1985) and from the African lungfish (*Protopterus aethiopicus*)(DeLaney et al., 1983). They appear to include both rapidly adapting and slowly adapting stretch receptors (see Chapter 8) and provide a possible mechanism for triggering an air breath when the volume declines to some threshold. The mechanoreceptors may, however, function primarily in regulating the breath itself, i.e., providing feedback on the degree of lung filling achieved during inspiration and expiration. There is evidence of internal oxygen chemoreceptors that provide a stimulus for air breathing in the gar (Smatresk et al., 1986), and which receptor mode is dominant in controlling breathing frequency is not yet clear.

Response To Hypoxia. The typical response of a facultative air-breather to aquatic hypoxia Is a marked increase in lung ventilation and a reduction in gill ventilation (Willmer, 1934; Johansen et al., 1970; Smatresk & Cameron, 1982a). In the gar, lung ventilation volume increases more than 10-fold through increases in both frequency and tidal volume. There is some oxygen loss across the gills during severe hypoxia so that the lung is actually supplying more than the metabolic oxygen demand, but CO_2 excretion across the lung remains small (Fig. 11.8). During severe aquatic hypoxia it is interesting to ask why oxygen gained across the lung is not lost when the blood recirculates through the gills en route to the systemic circuit. The answer is mostly a matter of gradients: The oxygen partial pressure of air in the lung is about 150 torr, and deoxygenated blood might have a PO_2 of 15 torr. This 135 torr gradient strongly favors movement of oxygen from air to blood. In the gills, however, the ambient oxygen tension might be 0 to 5 torr, resulting in only a 10 to 15 torr gradient for loss. In addition, reduction of water flow and perhaps redirection of blood flow within the gill may impose an additional limitation on reverse oxygen exchange (Randall et al., 1981; Smatresk & Cameron, 1982a).

There is a strong interaction between air breathing and gill ventilation. For example, when air breathing is stimulated by aquatic hypoxia, gill ventilation is strongly reduced (Johansen et al., 1970; Burggren, 1979; Smatresk & Cameron, 1982a,b). The reduction appears to be a reflex inhibition mediated by oxygen chemoreceptors (Smatresk, personal communication), but as yet the pathways are not fully understood.

There is also a strong response of lung ventilation to aerial hypoxia, lung ventilation increasing 5-fold or so in the snakehead, *Channa argus*, as the aerial PO_2 declines from 155 to 40 torr (Glass et al., 1986). Of course aerial hypoxia is not normally encountered, but it does illustrate the action of oxygen chemoreceptors in reflexly stimulating lung ventilation.

Response to Hypercapnia. Some air-breathing fish have been reported to respond to aquatic hypercapnia by increasing lung ventilation. Burggren (1979) found sharp increases in *Trichogaster*, but Smatresk and Cameron (1982b) did not find any significant elevation in the spotted gar. Aerial hypercapnia also produced little change in lung ventilation in *Channa* (Glass et al., 1986). As with water-breathing fish, there is also the question of whether the responses seen in some fish are a direct result of a CO_2-pH effect, or an indirect effect of reduced oxygen content in the blood due to elevated CO_2.

Circulation

It would obviously be advantageous to have a mechanism for controlling the distribution of blood flow to the pulmonary and systemic circuits as the degree of air breathing changed. There is clear evidence from microsphere studies in gar (Smatresk & Cameron, 1981) and from studies of gas exchange in the bowfin (Randall et al., 1981) and other lungfish (Johansen et al., 1968) that blood flow to the lung increases significantly during hypoxia. Unfortunately the vascular anatomy of most of these fish makes it very difficult to make direct measurements of either total cardiac output or blood flow to the lung. Since the principal vein draining the lung in bowfin, for example, joins other large veins, samples withdrawn from it are likely to contain blood drawn retrograde from systemic returns and thus not be representative of the efferent lung flow (cf. Johansen et al., 1970; Randall et al., 1981).

Any increase in distribution of blood flow to the lung will require a corresponding increase in the total cardiac output in order to maintain a constant systemic flow. Randall et al. (1981) calculated that during hypoxia in *Amia* cardiac output increased 50%, but systemic flow remained constant. These changes are presumably accomplished by appropriate changes in vascular tone and capillary bed resistance in both lung and systemic circulation, but the control pathways are not presently known.

CONCLUSIONS

In this brief chapter the minimum modifications necessary to make the transition from water-breathing to air-breathing have been described. Both crabs and fish have been successful to a limited extent in exploiting the much greater oxygen reservoir contained in the atmosphere, but neither group provides us with an example of an animal that is fully terrestrial throughout its life cycle. In the case of the crabs some adults are fully terrestrial, but the larval forms require a brief aquatic existence. Among the fish, only a few manage to spend any appreciable time on land, though the limitations may be due more to support, locomotion and desiccation than to inadequacies of the respiratory system.

Despite dependence upon air for oxygen supply, the respiratory physiology of these animals is not greatly different than their fully aquatic cousins. Intriguing parallels do exist, however, between the progressive modifications seen in an increasingly aerial series and the evolutionary changes seen in higher groups. The gradual division of the pulmonary circulation into a separate circuit, which occurs to a considerable extent in lungfish and perhaps a little in the coconut crab, points to the evolution of the three- and four-chambered hearts and separate circulations seen in the higher vertebrates. The reduction in aerial ventilation begins to affect CO_2 tensions and probably spurred evolution of control mechanisms based on CO_2 and pH. Finally, the problems of presenting a large respiratory exchange surface without excessive desiccation are common to all air-breathing animals.

Chapter 12

Case Study: The Painted Turtle

(*Chrysemys picta*)

The painted turtle, *Chrysemys picta*, is the sole species of the genus, confined to North America and commonly divided into four or more intergrading races or subspecies. The painted turtle and the closely related red-eared turtle, *Pseudemys scripta*, are denizens of ponds and streams over most of the United States. These two genera are so close taxonomically that they seem to go through "lump and split" cycles, sometimes all included in *Chrysemys*, then again as separate genera.[1] They are primarily aquatic turtles, given to basking, and more or less omnivorous. Both species are known to hibernate at the bottoms of ponds in winter over the northern parts of their ranges, usually buried in leaves or trash. Both species grow rather quickly for turtles, reaching sexual maturity in as little as 1 to 3 years and attaining maximum sizes of about 180 mm (7 in.) for the painted turtle and perhaps 250 mm (10 in.) for the red-eared turtle (Carr, 1952).

Turtles are generally thought to have evolved from the early cotylosaurs sometime before the Triassic. The skulls still reflect that early reptilian plan, but the rest of the skeleton has undergone considerable modification. The dorsal carapace results from dermal bone plates that expand laterally from the mid-line, fusing with highly modified ribs. A ventral bony plate, the plastron, is formed from other dermal plates, resulting in an animal almost completely encased in bony armor. In the course of these developments, the pelvic and pectoral girdles joining the legs and the rest of the skeleton have been moved to a position actually inside the ribs.

There is a considerable diversity in respiratory physiology among the extant turtles, reflecting the equal diversity in habits. Some, such as the giant land tortoises, are sluggish overall, and so have relatively reduced respiratory systems, more geared to water conservation than for maximum gas exchange. Others, like the fast-swimming green turtle, reflect opposite needs and design. The painted and red-eared turtles represent a sort of middle between these two extremes. Owing to their abundance, low cost, and tractability in the laboratory, we know much more about these particular turtles than most others.

FUNCTIONAL MORPHOLOGY

Gas Exchangers

The respiratory tract in the turtle begins with the glottis, which opens to admit air that has entered the mouth from the internal nares. From the glottis air flows through the larynx (silent in most turtles) and then through the trachea, which branches into two bronchi, one to each lung. The lungs are not very different in appearance from those in lungfish, but the degree of compartmentalization and septation is somewhat greater, providing a larger respiratory exchange surface (Fig. 12.1).

Respiration is supplemented in many turtles at three other sites: the pharynx, the skin of the legs and abdomen, and the cloacal sacs. The epithelium lining the pharynx is highly vascularized in some species and can be used as a sort of water lung by rhythmically emptying and filling. Extensive foldings and vascularization of the skin also facilitate gas exchange in well-aerated water. The cloacal sacs are tidally ventilated through the anus and apparently subserve gas exchange. When metabolic rates are low, as in cold, well-aerated water, the auxiliary gas exchangers can suffice for all gas exchange, obviating lung ventilation.

Total lung surface area is highly variable in the turtles and rather tedious to measure due to the complex spherical geometry. The work of Tenney and Tenney (1970) indicated that a 200 gm painted turtle would have a total (inflated) lung volume of about 60 ml, a total respiratory surface area of 3000 mm^2, and a mean "alveolar" diameter of about 1.5 mm. The lungs were inflated in that study to a pressure of 25 cm H_2O, however, which may have led to an overestimate of their in vivo volume. Crawford et al. (1976) reported a lung volume for *P. scripta* of 160 ml kg^{-1}, about half as large. The total lung surface area of 15 mm^2 gm^{-1} is at least an order of magnitude lower than birds or mammals of comparable

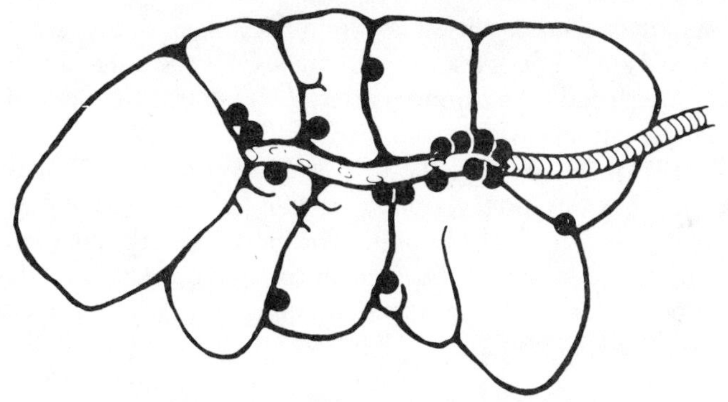

Figure 12.1: (Upper) Dry mount of the left lung of *Chrysemys picta* showing internal septation. (Lower) The approximate location of 17 pulmonary receptors located by punctate stimulation. (From Jones & Milsom, 1982, by permission, ©The Company of Biologists, Ltd.)

weight but is almost the same as that in the terrestrial coconut crab (12 $mm^2\ g^{-1}$, Chapter 11). Crawford et al. (1976) measured the diffusing capacity (DL_{CO_2}) using CO, and found a value of 0.081 ml (min × torr × kg)$^{-1}$, about one-tenth as large as the value for a dog lung but roughly proportional to the difference in metabolic rates.

Although the carapace and plastron in most turtles (including the painted turtle) forms a more or less rigid box, the lung is nevertheless ventilated in a manner similar to other vertebrates. Contractions of muscles in the leg pockets and beneath the lung expand it, reducing pressure and causing it to fill. Other groups of muscles aid in expiration by compressing the lung. Overall the metabolic cost of ventilation is fairly high compared to that for mammals; Kinney and White (1977) estimated it to be about 10% of total metabolism. The buccal force ventilation system used by air-breathing fishes and frogs is not found in turtles.

Circulatory System

Part of the interest in turtles from the comparative standpoint is that they represent a kind of in-between stage of partial separation of pulmonary and systemic circulations and thus stand midway between the fish and the so-called higher vertebrates (see Chapter 4). The heart, shown in Fig. 12.2, has two auricles. The right auricle receives return flow from the systemic circulation via one postcaval and two precaval veins which empty first into a sinus venosus. The left auricle receives its return flow from the pulmonary circuit via the pulmonary veins. Both auricles empty into the common ventricle, but the ventricle is partially divided into three chambers, the *cavum arteriosum, cavum venosum*, and *cavum pulmonale*. The positions of these cavities and the muscular ridges separating them has led to lively speculation on whether and to what degree functional separation of pulmonary and systemic circuits is possible during the normal contraction cycle of the heart (see below).

The circulation of the heart itself is also interesting and turns out to be more similar to fish and amphibians than to mammals. Only the outermost layer of the myocardium is perfused by a coronary arterial circulation, less than 10% of the total muscle mass (Juhasz-Nagy et al., 1963; Brady & Dubkin, 1964; Mackinnon & Heatwole, 1981). The bulk of the myocardium is nourished by the luminal blood, which invades the spongy, trabecular muscle mass during diastole and is squeezed out during systole. The limitation on cardiac energy metabolism is perhaps not quite so severe as in the fish, since the ventricle contains partially oxygenated blood returning from the pulmonary circuit, whereas the fish ventricle contains only venous blood.

BLOOD

Blood volumes of 8 to 9% of body weight and plasma volumes of about 7.5% were given by Thorson (1968) for various turtles including *Chrysemys*. Hemato-

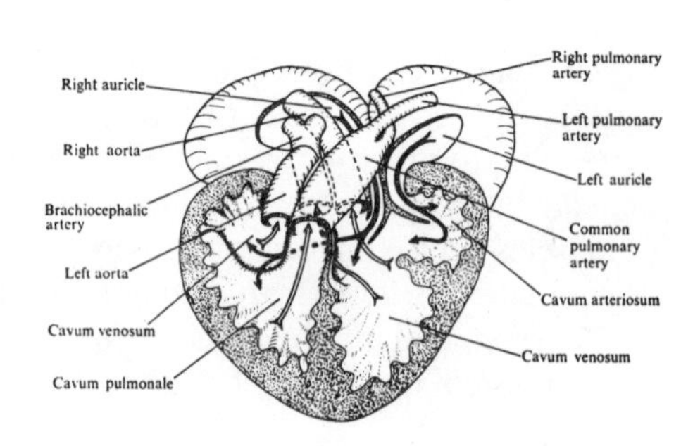

Figure 12.2: The turtle heart, shown in ventral aspect. The cavum pulmonale, which gives rise to the common pulmonary artery, lies ventral to the cavum venosum. All of the systemic arteries arise from the cavum venosum. The solid arrows indicate the gross movement of blood from the auricle into the incompletely divided ventricle. The open arrows indicate movement of blood from the ventricular chambers into the arterial arches. Arrows are not intended to illustrate the flow of separate blood streams through the ventricle. (From Shelton & Burggren, 1976, by permission, ©The Company of Biologists, Ltd.)

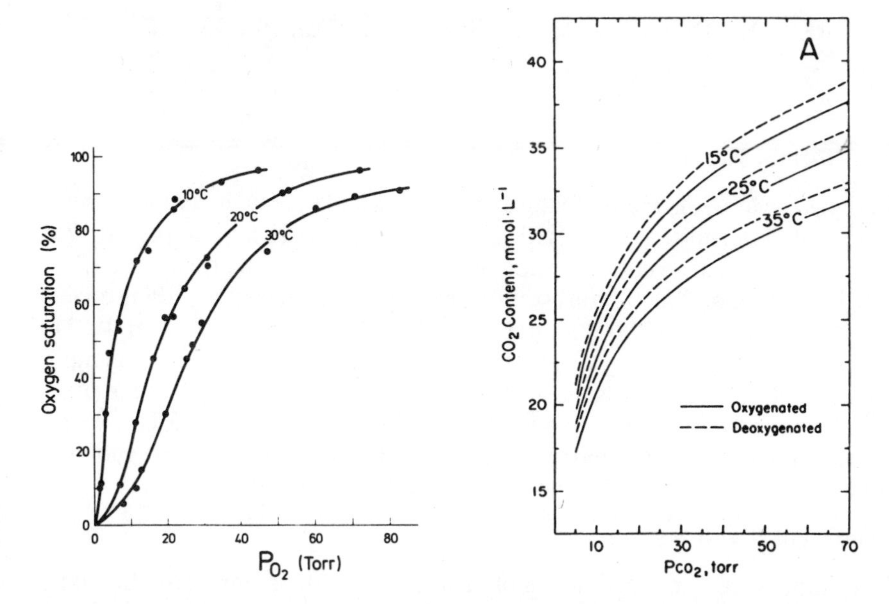

Figure 12.3: (Left) O_2 equilibrium curves for the turtle at P_{CO_2}s equal to the mean in vivo P_{CO_2} at each temperature. (From Glass et al., 1983, by permission, ©Springer-Verlag.) (Right) Mean oxygenated (solid lines) and deoxygenated (dashed lines) CO_2 dissociation curves for *P. scripta*. (From Weinstein et al., 1986, by permission, ©Elsevier North Holland, Inc.)

crit values for various turtles averaged 18 to 25%, with nucleated erythrocytes of about average size. The erythrocyte life-span is very long in the turtle, as it is in fish; Altland and Brace (1962) estimated it to be 600 to 800 days using ^{14}C-glycine incorporation into the proteins of newly forming cells.

Oxygen dissociation curves of turtle blood are pretty conventional; affinity declines fairly steeply as temperature and P_{CO_2} increase (Fig. 12.3)(Pough, 1980)(see Chapter 5). At 10°C and a P_{CO_2} of 16 torr, for example, the P_{50} is about 8 torr; and at 30°C and P_{CO_2} of 35 torr the P_{50} is about 28 torr (Maginnis et al., 1980; Nicol et al., 1983). Maginnis et al. (1980) demonstrated multiple hemoglobins in the erythrocytes of *Pseudemys scripta*, and showed that the resulting dissociation curve had a slightly different shape from that predicted with a single hemoglobin present. Pough (1980) reported the average oxygen capacity for 28 species of turtles as 7.6 vol% (approx. 3.4 mM L^{-1}) at an average hemoglobin concentration of 85 g L^{-} and hematocrit of 27%. Gaumer and Goodnight (1957) reported average hematocrit values of 27.5%, and average Hb concentrations of 101 g L^{-} for *Chrysemys picta*. On the whole, blood characteristics for *Chrysemys* and *Pseudemys* appear to differ only slightly.

The CO_2 combining curves for turtle blood are also quite conventional according to measurements by Weinstein et al. (1986)(Fig. 12.3), despite an earlier report of flatter than normal CO_2 dissociation curves in turtles (Ackerman & White, 1979). Both Bohr and Haldane effects in turtle blood are physiologically significant but not unusual.

RESTING GAS EXCHANGE AT VARYING TEMPERATURE

Ventilation

The Normal Pattern

Perhaps the most distinctive feature of reptilian, and especially chelonian ventilation is its intermittent or periodic nature. A typical breathing pattern for *Chrysemys* consists of a variable period of apnea, whether submerged or not, followed by a breathing "burst" consisting of several closely spaced breathing cycles (Fig. 12.4)(Glass et al., 1978; Burggren & Shelton, 1979; Funk & Milsom, 1987). It is therefore difficult to give any instantaneous breathing rates or minute volumes. Such measurements must be made over a long enough period of time to yield a representative value and must include measurements of time spent breathing and apneic. This fundamental periodicity imposes a similarly variable pattern on virtually all other variables of gas exchange in these animals.

Control of Breathing

In common with air-breathing amphibians, birds and mammals, the first line of defense appears to lie in the mechanoreceptors of the nares and pharynx. Sensory information from these receptors is carried by the trigeminal (V cranial) nerve, and forms a reflex arc inhibiting ventilation when water is present, thus preventing the animal from inhaling water into the lungs (Jones & Milsom, 1979, 1982). Pulmonary stretch receptors constitute a second principal class, producing sensory information of a more complex nature. Some of these receptors (SARs) have a tonic discharge, i.e., a steady firing rate dependent upon the degree of inflation (stretch) of the lung (Fig. 12.5). Some others (RAR's) have a phasic discharge pattern: They produce a burst when lung volume is rapidly changing but adapt quickly when lung volume is constant. Information from these

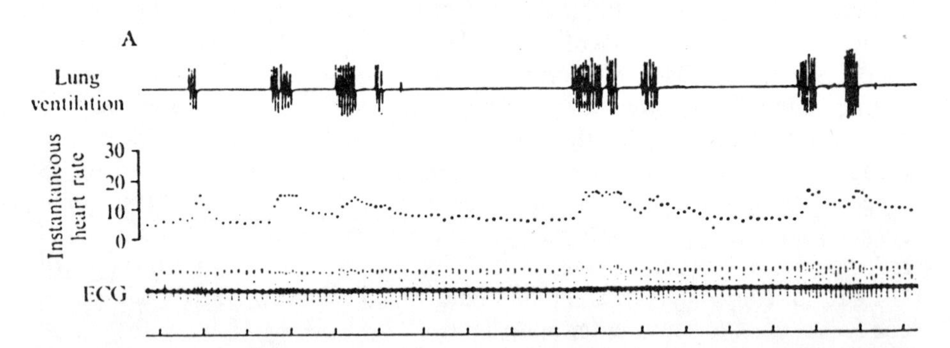

Figure 12.4: Intermittent breathing and heart rate. The uppermost trace shows measured breathing volume, the middle trace the heart rate, and the lower the EKG. Note the tachycardia accompanying each breathing burst. (From Burggren, 1975, by permission, ©The Company of Biologists, Ltd.)

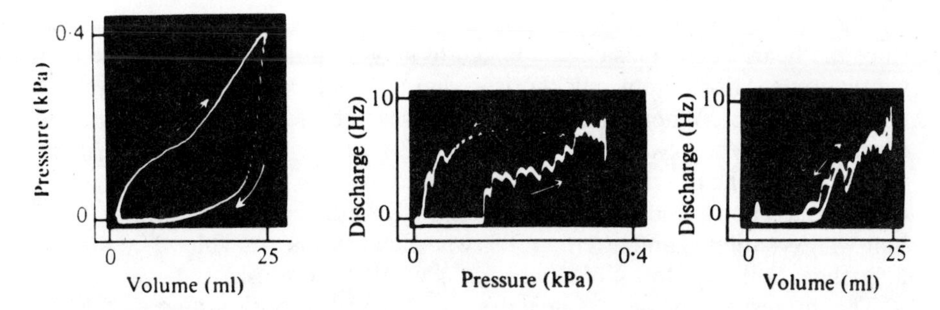

Figure 12.5: Pulmonary stretch receptor discharge in response to lung inflation and deflation. Left: a typical pressure-volume loop; center: frequency as a function of pressure, and right: frequency as a function of volume. (From Jones & Milsom, 1982, by permission.)

receptors is carried in the vagus and presumably provides the brain with feedback in regulating the mechanics of breathing (see Chapter 8).

In addition to their volume/stretch information, the pulmonary receptors also display varying degrees of CO_2 sensitivity. That is, for a given lung volume, the discharge frequency is decreased by increasing concentrations of CO_2 in the inspired air. If the discharge frequency is employed by the brain for assessing the degree of filling of the lung, then inflation to a given discharge frequency will lead to an increased tidal volume when hypercapnic. This situation is in fact observed in turtles, with ventilation volume increasing steadily at concentrations of CO_2 up to 6 to 8% (Jackson et al., 1974; Jones & Milsom, 1982; Funk & Milsom, 1987). The response to hypercapnia is not purely a function of lung receptors; the central nervous system may also serve a chemoreceptive function (Hitzig et al., 1985). The role of central versus peripheral receptors is still controversial, however, since the anoxic hyperventilation response is not affected by concurrent hypercapnia, indicating a predominance of peripheral chemoreceptor control (Wasser & Jackson, 1988)(see Chapter 8).

Changes in Ventilation with Temperature

Oxygen consumption rises in *Chrysemys* over the range of 10° to 30°C with a Q_{10} of approximately 2, but the earliest studies of ventilation over this temperature range showed virtually no change (Jackson, 1971; Jackson et al., 1974). More recent studies by Glass et al. (1985) and Funk and Milsom (1987) have shown that ventilation does increase at higher temperatures, but that it does not increase as much as oxygen consumption. The ratio of $\dot{V}E/\dot{M}O_2$, then, falls by about one-half from 10° to 30°C. Increased ventilation is accomplished by a combination of increases in the number of breaths per burst, a decrease in the apneic period between bursts, and an increase of tidal volume per breath. Of these parameters, the decrease in length of the apneic periods appears to contribute the most, as it does in air-breathing fish (see Chapter 11).

Circulation

Normal cardiac output in the turtle is as difficult to assess as the breathing rate, since the heart's activity is closely linked to lung breathing. During active lung ventilation cardiac output quickly rises, due to a reduction of vagal inhibition (Burggren, 1975). When lung ventilation ceases, the heart gradually slows, and cardiac output thus cycles through about a 2:1 activity ratio (Table 12.1). By various means the distribution of blood flow to the lung and to the systemic circulation has been found to vary correspondingly, with about 65% directed to the lung while ventilated and less than 50% when apneic. The total blood flow to the lung, then, drops by almost 65% when lung ventilation stops.

The interesting question of how this blood flow distribution is accomplished has generated considerable controversy. At one time it was thought that the muscular ridges between the ventricular chambers effected a functional separation of blood such that deoxygenated blood was directed primarily to the pulmonary circuit and oxygenated blood to the systemic circuit (White & Ross, 1966). A finding of different systolic pressures or of differential timing of systole among the various ventricular chambers would support such a scheme, but in fact no significant pressure differential among the ventricular chambers is found at any stage of the contraction cycle (Burggren, 1987, for review). The preponderance of evidence is that pulmonary/systemic distribution is controlled by relative changes in the peripheral resistance. The pulmonary vasculature has a number of sites at which control can be exerted including the major vessels as well as the lung microvasculature (Burggren, 1977, 1985).

The presence of substantial R–L and L–R shunts under virtually all conditions in the turtle heart has led to a view that the turtle heart represents a sort of inferior or incompletely evolved mammal heart. Such a view could be supported if it could be shown that the arrangement of flow in the turtle heart were in some way less advantageous to a turtle than the mammalian arrangement, but in fact the opposite may be so. Burggren (1987) reviewed five hypotheses that attribute functional advantages to shunting. Some of them seem to have little supporting

Table 12.1: Cardiac output parameters for a resting turtle at 20°C. From Burggren, 1987.

Parameter	During Lung Ventilation	During Apnea
$\dot{M}O_2$, ml kg^{-1} min–1	41.4	41.4
Heart rate, min^{-1}	23	11
$\dot{Q}$, ml kg^{-1} min^{-1}	57	27
Stroke volume, ml	2.48	2.45
% Directed to lungs	65	49
Metabolic cost of $\dot{Q}$ (%)	4.8	2.2

The metabolic cost saving of right-to-left (R–L) shunts during apnea is about 0.5%. From Burggren (1987), by permission.

evidence; the metabolic cost saving of shunts during apnea, for example, seems insignificant (Table 12.1). The shunts do allow, however, for blood to be diverted from the lung when ventilation is shut down, which may reduce plasma filtration (Burggren, 1982a) or assist in thermoregulation.

Gas Exchange

Normal Blood Values

A compilation of normal gas exchange parameters for the turtle at 10°, 20° and 30 °C is given in Table 12.2. Any data on blood gases must, of course, be examined carefully, since blood from various points in the circulatory system may be venous, arterial, or an admixture of both. Burggren and Shelton (1979), for example, gave average oxygen tension values from normoxic *P. scripta* at 20°C as: left pulmonary vein, 130 torr; left pulmonary artery, 23 torr; right aorta, 72 torr; left aorta, 60 torr; and anterior vena cava, 21 torr. These differences reflect not only whether blood is returning from lungs or systemic circuits but the degree of R–L and L–R shunt at the time of sampling.

Effects of Temperature

Although the study of Glass et al. (1983) showed little change in arterial saturation as temperature increased, other studies (Glass et al., 1985; Nicol et al., 1983; Wood et al., 1987) have shown decreases at the lowest and highest temperatures, partly due to the operation of the ventricular shunts. Wood et al. (1987) have suggested on the basis of model studies that the normal intracardiac shunts not only limit high temperature metabolism but have an important influence on hypoxic tolerance.

The data for blood P_{CO_2} at various temperatures are more uniform, although the studies of ventilation vary somewhat in predicting how the pattern is achieved. All the studies (Table 12.2) show increasing P_{CO_2} at higher temperatures, and all show a decreasing ratio of ventilation to oxygen consumption. There is considerable variation, however, in the relative changes of these two parameters: in Jackson et al.'s (1974) study ventilation hardly changed, whereas the other studies found roughly 4-fold increases in ventilation from 10° to 30°C. An examination of the data of Jackson et al. (1974) suggests that perhaps their 10 °C ventilation data were too high for some reason; their other values seem more or less in line with more recent work. The end result is the same in all, however, with increasing P_{CO_2}, roughly constant total CO_2 content (Robin, 1962; Nicol et al., 1983), and declining pH with temperature (see Chapter 6).

Efficiency

Sufficient data are on hand for the turtle to allow some fairly complete analyses of the performance of the gas exchangers. White and Bickler (1987) put together a model, using mostly data for *P. scripta*, which showed that the capacity-rate ratios (G_{vent}/G_{perf} in their notation) cluster around a value of 1 for oxygen, sug-

Table 12.2: Gas exchange parameters for a turtle at 10, 20 and 30°C.

Parameter	10°C	20°C	30°C	Source
$\dot{M}O_2$, ml kg^{-1} min^{-1}	0.16	0.87	1.55	Funk & Milsom, 1987
	0.19	0.57	1.68	White & Bickler, 1987
	0.38	0.70	1.30	Jackson et al., 1974
	0.20	0.61	1.29	Glass et al., 1985
$\dot{V}E$, ml kg^{-1} min^{-1}	7.0	21.8	31.0	Funk & Milsom, 1987
	10.4	20.3	39.6	White & Bickler, 1987
	29.8	23.8	30.6	Jackson et al., 1974
	6.0	17.5	26	Glass et al., 1985
$\dot{V}E/\dot{M}O_2$	44	25	20	Funk & Milsom, 1987
	54.7	35.6	23.6	White & Bickler, 1987
	76.8	33.0	22.7	Jackson et al., 1974
	36.5	30.0	21.3	Glass et al., 1985
Heart rate, min^{-1}	3.6	30	–	Jackson, 1987
	8.5	19	43	Kinney et al., 1977
$\dot{Q}$, ml kg^{-1} min^{-1}	8.5	26.7	84.5	White & Bickler, 1987
B.P., syst. art.	22	27	23	Kinney et al., 1977
	–	22	39	Jackson et al., 1987
PaO_2, torr	91	93	85	Glass et al., 1983
	30	63	60.5	Glass et al., 1985
	20	68.6	70.8	Nicol et al., 1983
PvO_2, mixed, torr	–	25	–	Wood et al., 1987
	–	20	–	Burggren & Shelton, 1979
$PaCO_2$, torr	20.8	27.5	29.9	Funk & Milsom, 1987
	18	22	32	Glass et al., 1983
	16	23	35	Nicol et al., 1983
CCO_2, art., mM L^{-1}	43.2	44.3	43.3	Nicol et al., 1983
pH, arterial	7.847	7.778	7.628	Nicol et al., 1983
	7.84	7.71	7.63	Funk & Milsom, 1987
	7.84	7.73	7.625	Glass et al., 1985

The data were compounded from various studies of *Chrysemys* and *Pseudemys* species. *Chrysemys picta*: Glass et al., 1983, 1985; Funk & Milsom, 1987; Jackson, 1987; Nicol et al., 1983. *Pseudemys scripta*: Burggren & Shelton, 1979; White & Bickler, 1987; Jackson et al., 1974. *Pseudemys floridana*: Kinney et al., 1977.

gesting a highly efficient exchanger (see Chapter 7). As temperature increases, the lung is progressively underventilated with respect to oxygen, possibly related to the regulation of CO_2 to maintain acid-base status (see below). The comparable values for CO_2 in their model are low but much more stable than those for O_2, again with implications for acid-base regulation. It is not clear, however, exactly how their model would be affected by including terms for cutaneous CO_2 exchange, nor have the effects of changes in the cutaneous component with temperature been studied.

Acid-Base Balance and Temperature

The most conspicuous feature of the turtle's acid-base regulation as temperature increases is the progressive decline in ventilation relative to oxygen consumption and the resultant increase in P_{CO_2} (Fig. 12.6). The result is declining pH with body temperature at roughly constant total CO_2 content. These data have fed the lively controversy over various paradigms of acid-base regulation and temperature (Cameron, 1984; Heis̆ler, 1984). When the existing data are summarized (Glass et al., 1985; Nicol et al., 1983), it becomes fairly clear that they fit neither the constant relative alkalinity hypothesis proposed by Rahn (1967) nor the imidazole alphastat hypothesis put forward by Reeves (1972). Both of these hypotheses make specific predictions about the value of the pH-temperature slope, the former based upon the dissociation properties of water, and the latter based on the dissociation properties of the imidazole moiety. Although it may be

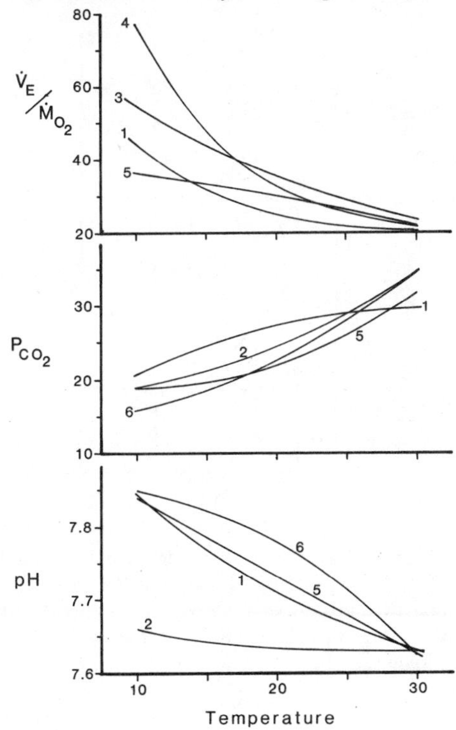

Figure 12.6: Acid-base parameters and temperature from six studies of *Chrysemys* and *Pseudemys*. 1: Funk & Milsom (1987); 2: Robin (1962); 3: White & Bickler (1987); 4: Jackson et al. (1974); 5: Glass et al. (1983); 6: Nicol et al. (1983).

difficult to specify an exact set of rejection criteria, the average for all the studies is –0.012, substantially lower than the slopes predicted by either hypothesis (Glass et al., 1985)(see Chapter 6).

It may be argued that intracellular, not extracellular acid-base status is the important parameter to be regulated, but unfortunately we have many fewer data on intracellular conditions. Malan et al. (1976) found tissue pH/temperature slopes for *P. scripta* ranging from –0.012 to –0.023, but their temperature slope for blood (–0.0207) was nearly twice as high as in all other studies. Interestingly, the cardiac muscle slope was significantly lower than other muscle tissues, similar to the findings for fish and crabs (see Chapters 6, 9 and 10). The pH-temperature relation of various turtle tissues appears to require further investigation.

DIVING, ANOXIA, AND HIBERNATION

The Normal Dive

Spontaneous dives by most turtles are relatively short, ranging from less than a minute to 15 or 20 minutes. Dives usually begin at the end of a breathing burst, when the air in the lung has been exchanged several times. During the dive the alveolar oxygen is gradually depleted, CO_2 builds up to a limited extent, and blood gases cycle in a corresponding manner (Burggren, 1988b)(Fig. 12.7). These cyclically varying blood gas tensions pose problems for deciding how respiration is

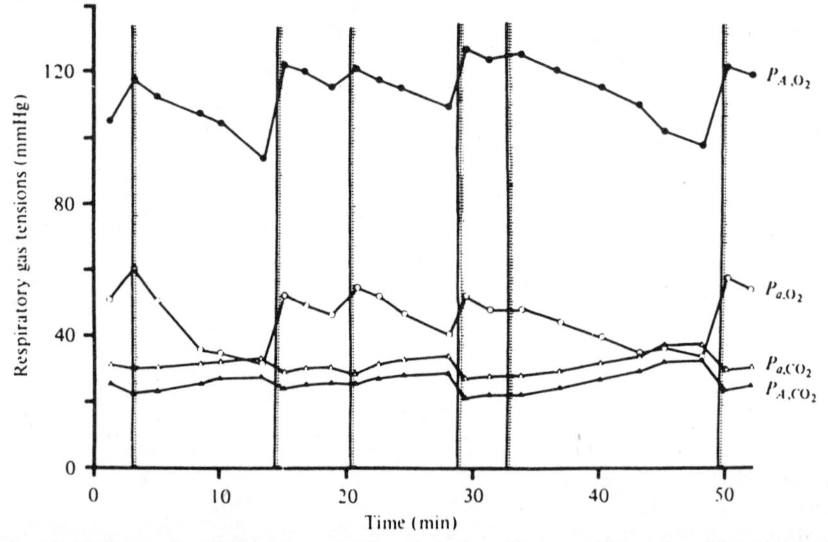

Figure 12.7: Fluctuations in lung O_2 and CO_2 gas tensions (P_{AO_2} and P_{ACO_2}) and femoral artery O_2 and CO_2 gas tensions (P_{aO_2} and P_{aCO_2}) during 1 hr of intermittent breathing in a freely diving, unrestrained 0.97 kg *P. scripta*. Brief bouts of surfacing and lung ventilation are indicated by the shaded vertical bars. (From Burggren & Shelton, 1979, by permission, ©The Company of Biologists, Ltd.)

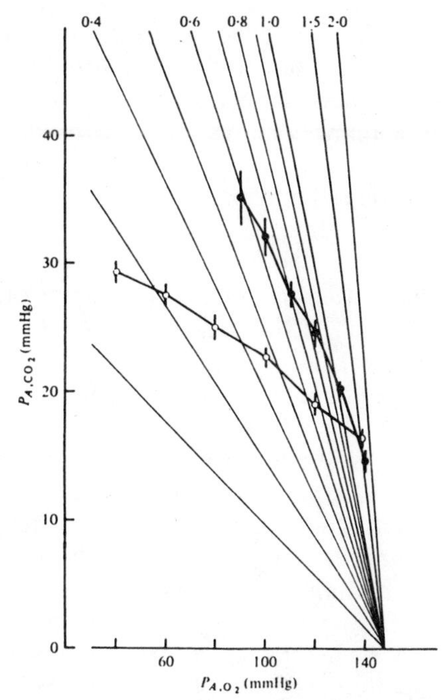

Figure 12.8: The pulmonary gas exchange quotient (R) for the lung during a dive in *P. scripta* (open circles) and *Testudo graeca* (solid circles). The numbered lines shown different constant values of R. (From Burggren & Shelton, 1979, by permission, ©The Company of Biologists, Ltd.)

regulated, since no simple set-point scheme appears to be satisfactorily adapted to intermittent breathing.

During the normal dive the pattern of CO_2 excretion into the lung is complex, the R ratio varying from over 1 to less than 0.4 (Fig. 12.8). The reasons for the variation are complex but include a progressive increase in cutaneous CO_2 loss as the PCO_2 of the blood increases and temporary storage of CO_2 in various tissues and body fluid pools (Lenfant et al., 1970; Burggren & Shelton, 1979). There is a significant build-up of CO_2 during experimental dives of 2 to 4 hours, with a resultant mixed acidosis (Jackson & Silverblatt, 1974). Blood pH can be depressed from 7.6 to 6.8, and lactate increases from the control value of 2.3 mM L^{-1} to over 28 mM L^{-1} at the end of the dive. Hyperventilation at the end of the dive eliminates the respiratory (CO_2) component of the acidosis within an hour or two, but the lactic acidosis takes many hours to eliminate. Although the experimental dives studied may be quite unrepresentative of normal behavior for these particular species, other turtles may have normal submergence periods of several days. The role of cutaneous respiration in determining normal dive behavior has not been systematically investigated but is likely to be quite important.

Although *Chrysemys* inhabits relatively shallow ponds and lakes, some of its marine relatives dive to considerable depth, posing additional problems for gas exchange. The Ridley sea turtle has been observed at a depth of 300 m, and leatherback sea turtles dive to over 1000 m depth (Kooyman, 1988). Since hydrostatic pressure increases roughly 1 atm per 10 m depth, most of the lung would probably be collapsed during such deep dives. Any remaining gas, however, would continue to exchange with blood, and nitrogen narcosis ("bends") would seem to be a serious problem for these turtles. At present nothing is known about either lung compression or gas exchange during such deep dives by turtles.

Anoxic Hibernation

Turtles of the *Pseudemys/Chrysemys* group naturally hibernate during winter at the bottoms of ponds. Since many of these ponds are shallow, rich in organic material, and likely to be snow-covered, oxygen depletion is a common condition (Robin et al., 1963; Belkin, 1968; Ultsch & Jackson, 1982). The turtles are thus placed in cold anoxia for periods up to several months. Laboratory studies of turtles under these circumstances have shown that they can survive for over 5 months, although considerable mortality begins to occur during rewarming after such long periods (Ultsch & Jackson, 1982). Increasing the temperature to 10°C dramatically shortens the time of survival to an average of only 17 days for *C. picta* (Ultsch et al., 1984), but if the water is reasonably well aerated survival is almost indefinitely prolonged.

During such long anoxic periods, oxygen stores are quickly exhausted, and anaerobic glycolysis appears to supply all of the needs of the much reduced metabolism (Gatten, 1987). As might be expected, plasma lactate concentrations rise steadily and pH falls, until after 12 weeks' anoxic submergence lactate may exceed 200 mM L^{-1} (Fig. 12.9)(Jackson & Heisler, 1982; Jackson & Ultsch, 1982; Jackson et al., 1984; Ultsch et al., 1984). Concomitantly both calcium and magnesium concentrations also rise to values as high as 120 mEq L^{-1} for calcium, leading to interesting questions about the maintenance of nerve conduction and

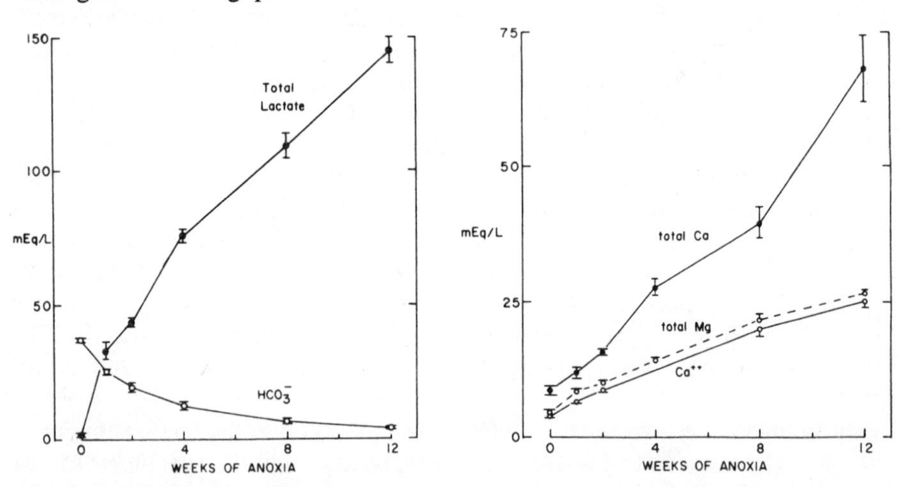

Figure 12.9: (Left) Plasma [HCO_3^-] and [lactate] for 12 weeks of anoxic submergence at 3°C. (Right) Total Ca^{2+}, total Mg^{2+}, and free ionized Ca^{2+} for 12 weeks of anoxic submergence. (From Jackson & Heisler, 1982, by permission, ©Elsevier North Holland, Inc.)

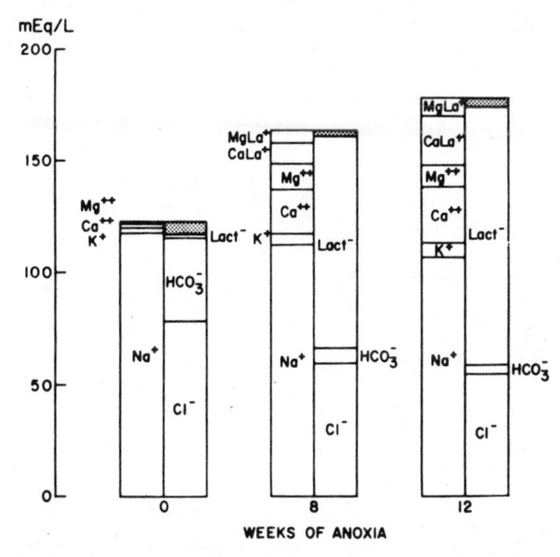

Figure 12.10: "Gamblegrams" from control turtles and those submerged for 8 and 12 weeks in anoxic water (see Fig. 12.9). Only by taking into account the formation of the ion pairs $CaLa^+$ and $MgLa^+$ can an electrolyte balance be calculated. (From Jackson & Heisler, 1982, by permission, ©Elsevier North Holland, Inc.)

muscle contraction (Yee & Jackson, 1984). By the end of a 12 week submergence period, the electrolyte status, both intra- and extracellular, is most peculiar, evidently including formation of significant quantities of a calcium-lactate ion pair (Fig. 12.10). Further investigations of how such conditions are tolerated should be quite interesting (see Jackson, 1987, for review).

EFFECTS OF HYPERCAPNIA

Compensation

The primary response to elevated CO_2 is an increase in ventilation (Glass et al., 1985; Funk & Milsom, 1987), which tends to reduce the alveolar-arterial P_{CO_2} difference. When high external concentrations of CO_2 are imposed experimentally, however, increased ventilation cannot reduce arterial P_{CO_2} below that of the ambient air, so other acid-base regulatory mechanisms are required. The responses of *Chrysemys picta* to breathing 5.7% CO_2 in air (P_{CO_2} = 42 torr) are shown in Fig. 12.11, which looks quite familiar. The initial response involves a largely passive respiratory acidosis, followed by slower compensation involving an increase in [HCO_3^-] and the corresponding changes in "strong ions." (Compare this pattern with those shown for the blue crab and the rainbow trout in Chapters 9 and 10.) It is not entirely clear how the ionic compensation occurred, however, since there was no difference between normal animals and those with

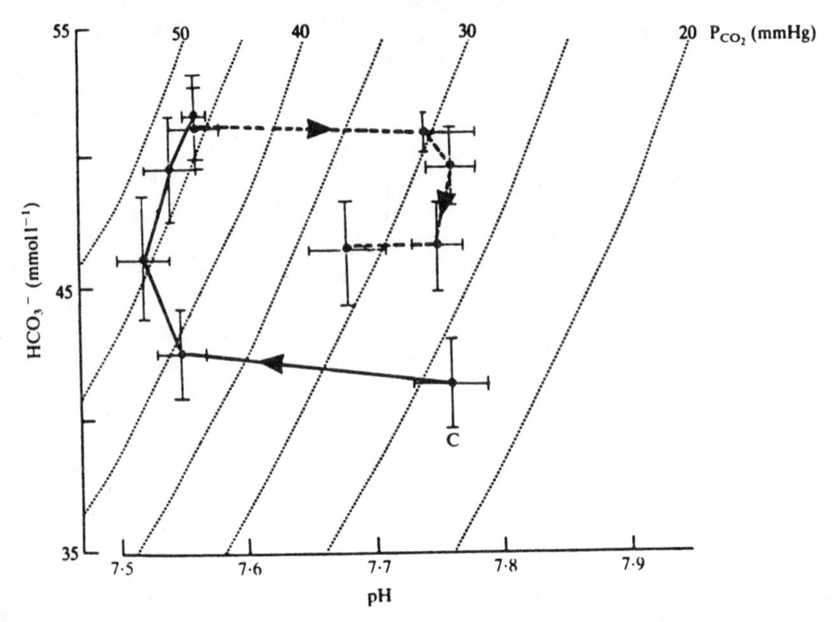

Figure 12.11: The time course of response to hypercapnia (P_{CO_2} of inspired air = 42 torr) in the turtle. The point C depicts values for control animals breathing air and the arrows follow the time course: 1, 5, 24, 48, and 72 hours breathing 5.7% CO_2, then 1, 5, 24, and 48 hours of recovery in air. (From Silver & Jackson, 1985, by permission, ©The company of Biologists, Ltd.)

urinary bladders removed; the urine made no contribution to compensation (Silver & Jackson, 1986). Because there are no gills to conduct ionic exchanges with the water, the changes must occur through a combination of cutaneous exchanges (probably small) and redistribution of ions between blood, tissue fluids, and bone. The significant increase in calcium indicates significant participation of bone in hypercapnic compensation, like land crabs, but unlike fish and aquatic crabs.

CONCLUSIONS

The turtle is a fine example of the modification of physiological processes to solve a particular set of environmental and behavioral problems. Although primarily an aerial animal, the extended times spent under water and the requirement in some species for winter hibernation has led to a respiratory system almost equally adapted to aerial or aquatic existence. Oxygen consumption is usually conducted by the lung, but can be supplemented or even replaced by cutaneous exchange. CO_2 excretion can occur by either route but normally is excreted mainly by skin. An additional large CO_2 capacitance allows considerable flexibility in the rate and timing of CO_2 release. The circulatory system is finely tuned to distribute blood preferentially to either lung or systemic vasculature, an option not open to the "higher" vertebrates with their more "advanced" four-chambered hearts.

It is therefore a mistake to look upon turtles as imperfectly evolved mammals. Their hundreds of millions of years of success in a wide variety of terrestrial and aquatic habitats is a tribute to the superiority of their design.

Footnotes

[1]The genus is currently called *Trachemys* by some taxonomists.

Chapter 13

Case Study: The Duck

When considering respiration in a bird, we have for the first time an animal that is homeothermic, depends entirely on a single, aerial respiratory organ, and has a metabolic rate about an order of magnitude higher than the groups considered previously. When choosing a particular bird to discuss, the duck was selected for several reasons. It is first of all common and familiar, has had considerable research done on it, and is capable of energetic flight and diving. The work on birds, however, has not been nearly so concentrated on a single species or a group of species as has the work on fish, crabs, and turtles. The work on embryonic development has been done mostly with the domestic chicken, flight energetics with pigeons and budgies, lung structure with a variety of birds, etc. Even among the ducks there are profound differences between the dabblers, which include domestic ducks derived from the mallard (*Anas platyrhynchos*) and the wild stock, and the diving ducks, such as scaup (*Aythya affinis*). Since the general characteristics of birds, as well as certain special capabilities, are to be illustrated, what is described here is a sort of "generic" duck, drawn from many studies of different species.

FUNCTIONAL MORPHOLOGY

Lungs and Air Sacs

Basic Plan of the Airways

The internal organs of a bird are compressed, with the lungs coming to lie in an anterodorsal position above the greatly enlarged sternum. The ribbed trachea branches at the level of the syrinx (a sound-producing organ) into two major bronchi, each leading to a complex consisting of a lung and several large air sacs (Figs. 3.9, 13.1). The lungs examined in previous chapters have had a more or less arboreal structure, the airways branching progressively and ending in many blind-ended sacs or alveoli. The lungs of fishes, amphibians, reptiles, and man all show

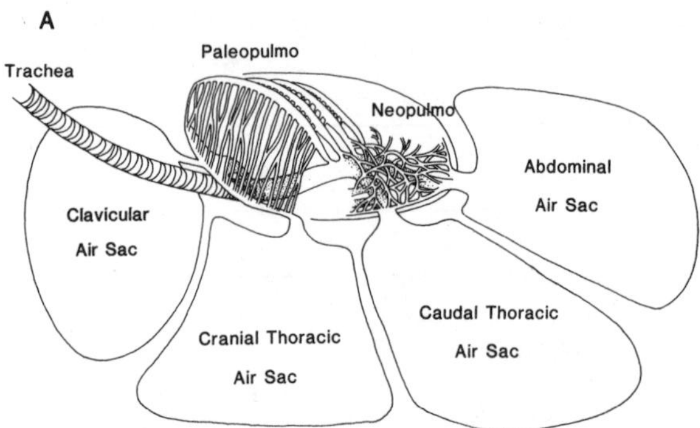

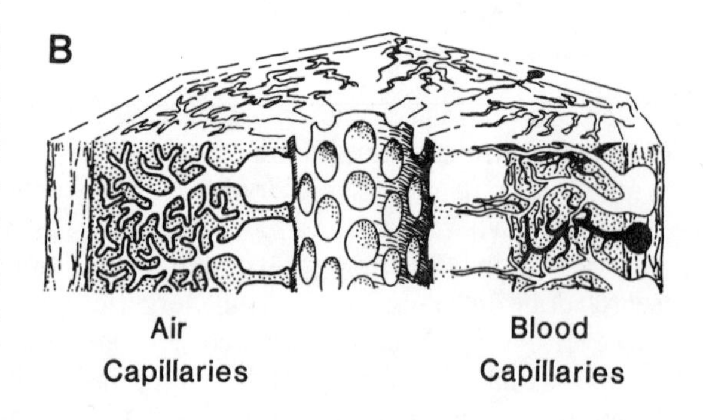

Figure 13.1: (A) The major components of the duck lung. The paleopulmo contains the bulk of the parabronchi, oriented in a parallel fashion; the neopulmo contains a smaller proportion of more randomly oriented parabronchi. (B) Schematic diagram of the parabronchi shown in (A). Air capillaires are shown on the left, blood capillaries on the right. (Both parts redrawn from originals supplied by D. R. Jones.)

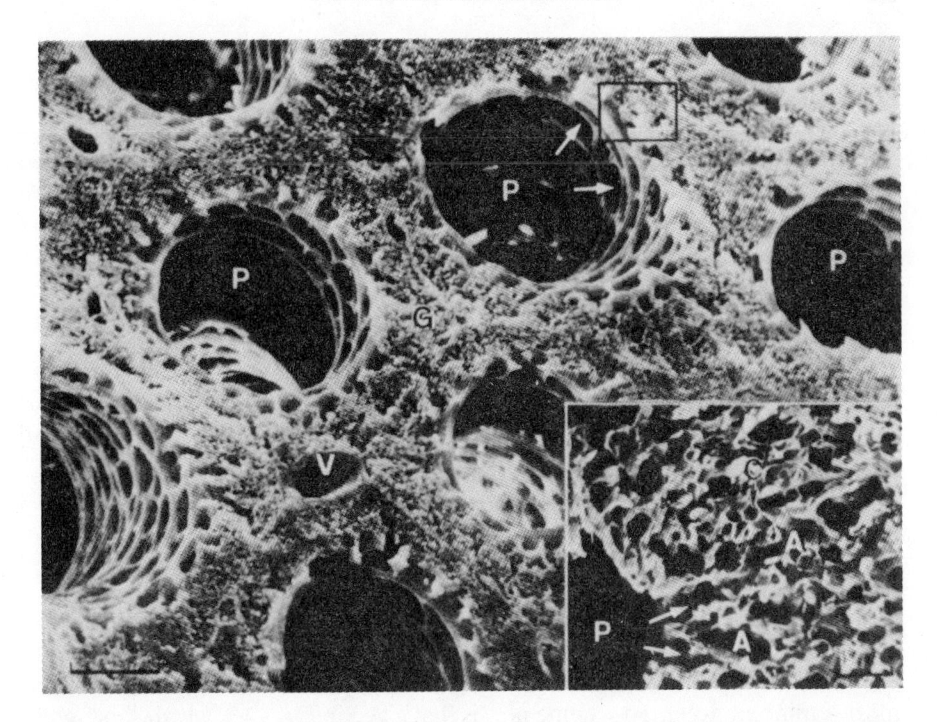

Figure 13.2: Scanning electron micrograph of a cross section of a bird lung. The walls of the cylindrical parabronchi (P) are marked by bundles of smooth muscle surrounding atria (arrows). Blood vessels (V) are embedded in the gas exchange tissue (G) between the parabronchi. The inset is a higher magnification of an area similar to that marked by a square, top left center. It shows that the gas exchange tissue between parabronchi is a dense meshwork of blood capillaries (C) and air capillaries (A). Scale marker, lower right = 200 μm. (Photograph by H. R. Duncker; from Weibel, 1984, by permission, ©Harvard Univ. Press.)

this structure and are tidally ventilated. In the bird, however, the organization of fine airways in the lungs is completely different – the smallest airways, called parabronchi, are flow-through tubes with many short air capillaries leading away from them (Figs. 13.1, 13.2). They connect on the afferent end with dorsobronchi and at their outflow with other collecting airways called ventrobronchi.

Patterns of Flow

Perhaps the most interesting and at the same time enigmatic feature of the bird lung is that the flow through the parabronchi is one-way and not tidal as in other lungs. Many workers, beginning with Dotterweich (1936) and reviewed by King (1966), Bretz and Schmidt-Nielsen (1971) and Banzett et al. (1987), surmised that the purpose of these tertiary airways (parabronchi) might be to conduct air flow through the gas exchanger in a one-way fashion. By studying the deposition of aerosols or fine particles, by sampling gas tensions, and by measuring flow in various experimental situations, several schemes for flow have been

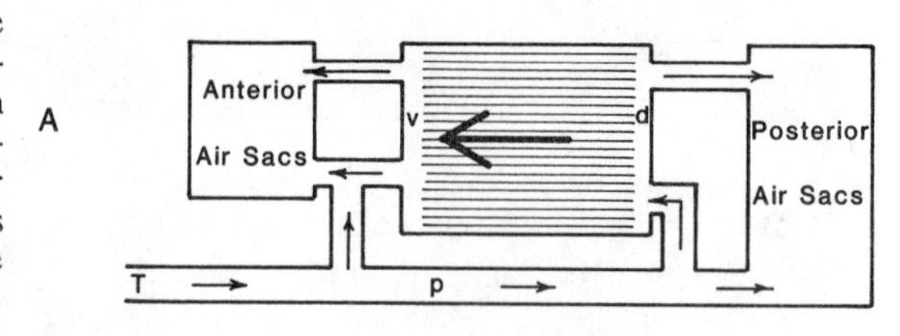

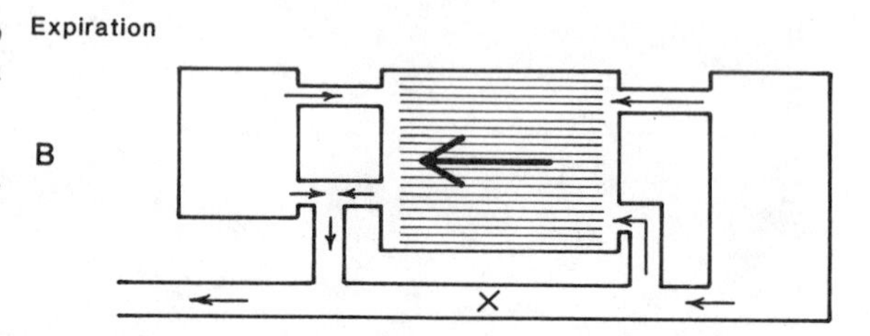

Figure 13.3: Air flow during inspiration and expiration in the lung and air sacs. During inspiration air flows through the trachea (T) and primary bronchi (p), filling both anterior and posterior air sacs. During expiration, flow is directed preferentially into the dorsobronchi (d) from the posterior air sacs, and there is little flow in the primary bronchi. The net flow through the parabronchi is from dorsobronchi to ventrobronchi (v) during both phases. See text for further explanation. (After Bretz & Schmidt–Nielsen, 1971, and Banzett *et al.*, 1987).

proposed. Since there have never been any anatomical structures described that could function as valves, the enigma has been to explain how flow could be channeled in the appropriate manner during both inspiration and expiration. The most commonly accepted scheme now is that given in Fig. 13.3, which shows the major air sacs and their connections with the lung.

During inspiration, air flows at a fairly high velocity in the primary bronchi directly into both the dorsobronchi of the lung and into the posterior air sacs. Only a little flow is directed toward the anterior air sacs (A, Figs. 3.9 and 13.3), since their openings branch off at an acute backward angle (Figs. 13.1 and Fig. 3.9, arrows). The resistance of the various airways is such that the anterior air sacs tend to fill more by drawing air from the parabronchi in the lungs and from the connecting airways. During expiration, air is apparently channeled from the posterior air sacs mostly through the secondary bronchi and parabronchi, since there is little outward flow at the distal end of the primary bronchus at this time (Bretz & Schmidt-Nielsen, 1971; Banzett et al., 1987; Keuthe, 1988; Wang et al., 1988). Thus not only is a one-way flow maintained throughout the ventilatory cycle, but in effect the ventilatory organs (air sacs) and the gas exchanger (lungs) have been functionally and anatomically separated. The lung itself is in fact quite rigid.

The "aerodynamic valving" exhibited by the bird lung has been shown to depend upon inertial forces of the gas stream combined with the particular orientation and size of the airway branches. Replacement of normal air with gas of lower density or experimental reduction of the gas stream velocity greatly reduces the effective valving (Banzett et al., 1987). Turbulence at airway constrictions and in bends is also thought to provide some flow rectification (Scheid et al., 1972), i.e., higher resistance to flow in one direction than the other.

This unique arrangement of air sacs and parabronchi gives the bird lung the characteristics of a cross-current exchanger, since most of the capillary circulation is in effect perpendicular to the direction of airflow (King, 1966; Abdalla & King, 1975). Actually, parts of capillaries in the lung are oriented countercurrent to the airflow, a fact not taken into account in model analyses of bird lung function. Theoretically, the bird lung should be able to achieve arterial oxygen tensions greater than expired oxygen tension, but there is so far no experimental proof that exchange is ever that efficient in birds.

Volume, Surface Area, and Diffusing Capacity

Although the total respiratory system volume is relatively large in the duck compared with other air-breathers, much of the volume, about 90%, is contained in the air sac system and major airways (Table 13.1). The data given are for domestic ducks; data from the goose, swan, chicken and other birds shows considerable variation in the proportions for each air sac, but the general pattern is the same (Duncker, 1972, 1974). In the lung itself, about two-thirds of the total volume

Table 13.1: Morphometric data for the respiratory system of the duck.

Parameter	Value	Source
Total respiratory volume	213 ml kg^{-1}	1
Total air sac volume	183 ml kg^{-1}	1
Interclavicular	45 ml kg^{-1}	1
Prethoracic	43 ml kg^{-1}	1
Postthoracic	57 ml kg^{-1}	1
Abdominal	37 ml kg^{-1}	1
Lung volume	22.3 ml kg^{-1}	2
Volume of parabronchi	15.5 ml kg^{-1}	3
Surface area:		
Per unit volume	200 mm^2/mm^3	4
Total	14.1 m^2	4
Blood:air capillary vol. ratio	0.83	4
Harmonic mean barrier thickness	0.29 μm	4
Contact time	0.9 sec	4
D_{LO_2}, ml min^{-1} $torr^{-1}$	2.28	5

1: Scheid et al. (1974).
2: Calculated from Duncker's (1972) data for goose and swan.
3: Calculated from the lung volume and Powell and Mazzone's (1983) fractional volume data.
4: From Powell & Mazone (1983).
5: From Burger et al. (1979).

consists of the parabronchi, the actual exchanging units, which is higher than other vertebrates' lungs. The total air-to-blood surface per unit volume is also quite high: the hummingbird has an exchange surface to volume ratio of over 400 mm^2/mm^3 (Fedde, 1986), whereas in a mouse of the same weight the ratio is 110 mm^2/mm^3 (Weibel et al., 1981).

The high proportion of air and blood capillaries in a unit lung volume means that the contact time for the erythrocytes in the exchanging capillaries is higher than might otherwise be predicted (Table 13.1). The average of 0.9 to 1.0 second appears to be adequate for complete gas exchange, although it is not certain how it might change with greatly increased cardiac output during flight.

Blood Vascular System

Hematology

The erythrocytes of birds are nucleated, oval, and appear to have a short life-span. The range of erythrocyte life-span for birds is roughly 30 to 60 days, with a figure of 39 to 42 days given for ducks by Brace and Altland (1956). The cells are of moderate size, averaging 10 to 15 μm in length; red cell numbers are generally in the range of 2 to 5 $\times$ 10^6 per mm^3 (Kisch, 1949; Jones & Johansen, 1972). Hematocrit, hemoglobin concentration, erythrocyte count, and total blood oxygen capacity tend to vary considerably with such factors as species, sex, condition, and age (see Table 13.2, below). Blood volume estimates vary widely, but the range of 5 to 13% of body weight is given by Jones and Johansen (1972) for birds generally. Determinations in the Pekin duck gave values of 9 to 10% of body weight (Hudson & Jones, 1986).

The Heart

Working from a sketch of the reptilian (particularly the crocodilian) heart, only a couple of modifications (closing the foramen of Panizza and moving the left aorta to the left ventricle) would be necessary to describe the basic plan of the bird heart (Fig. 13.4). Like the crocodile and the mammal, the atria and ventricles are completely separated in the adult, although the embryonic development is different in mammals and birds. The formation of the intraventricular septum is complete, and the major shunt connections found in the crocodile and other reptiles are absent. Thus the bird has a complete two-circuit circulation: The left ventricle supplies the systemic circuit and return flow enters the right atrium; the blood then enters the right ventricle for circulation to the lung and finally the return from the lung enters the left atrium to feed the left ventricle. The primitive sinus venosus is lacking in most birds, although evidence of its presence persists in some in the form of common connections to the right atrium. In most, however, several large veins enter the right atrium independently, including the anterior and posterior venae cavae and the coronary vein(s) (Jones & Johansen, 1972). The number and arrangement of heart valves is complex in birds and somewhat variable among different orders (Goodrich, 1930).

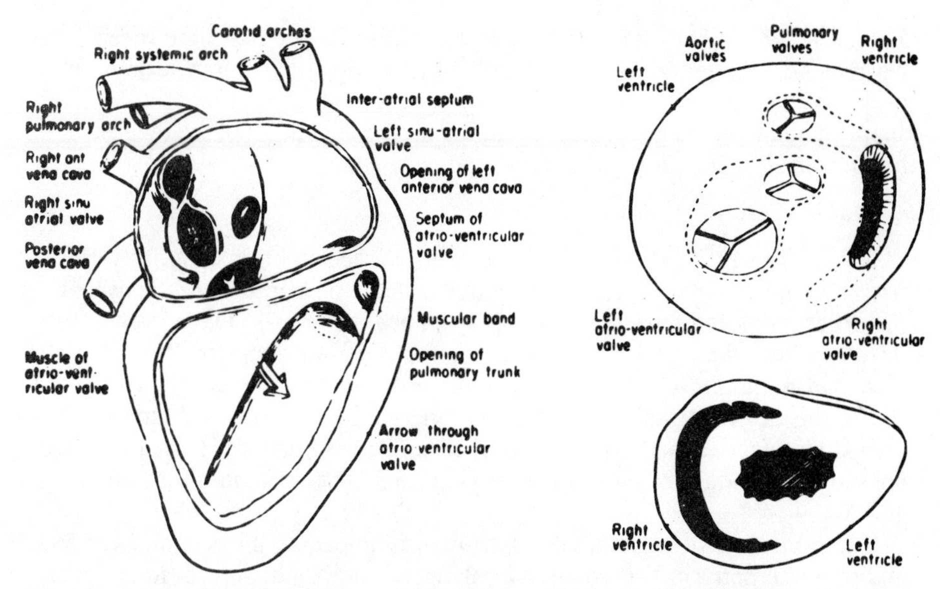

Figure 13.4: The bird heart. (Upper right) Section at the level of and parallel to the atrio-ventricular junction. (Lower right) Transverse section through the ventricles of the european crane, showing the right ventricle wrapped around the much more muscular left ventricle. (After Goodrich, from Jones & Johansen, 1972, by permission, ©Academic Press.)

The two major blood circuits have quite different characteristics: The systemic circuit, fed by the much more muscular left ventricle, has peak systolic pressures typically around 150 to 300 mm Hg; the pulmonary circuit has much lower pressure, with peak systolic pressures typically around 25 to 30 mm Hg. In the very thin-walled capillary circulation of the noncompliant lung, the higher pressures typical of the systemic circuit would no doubt produce unacceptably high rates of lymph filtration, reducing the effectiveness of the lung in gas exchange. An extensive lymphatic system has developed in the rest of the body to return plasma filtrate to the heart. The complete separation of systemic and pulmonary circulations, then, has an important advantage in allowing the gas exchanger to operate with a different pressure regimen, and possibly under independent pressure control.

In concert with the much higher relative metabolism and correspondingly higher circulation rates that have evolved in birds, there are changes in the vascular organization of the heart itself. The most notable is the reorganization of the muscle to consist almost entirely of compact, circumferentially arranged fibers with the disappearance of the spongy, trabecular myocardium. An extensive coronary circulation has developed, with two to four coronary arteries originating from the systemic aorta close to the left ventricle, branching out over the entire myocardial surface. The myocardial tissue is invested with a generous supply of nutritive capillaries whose drainage is gathered in a network of coronary veins emptying into the right atrium (Lindsay, 1967; review by Jones &

Johansen, 1972). Thus this highly energetic tissue for the first time receives an oxygenated, arterial blood supply, unlike the "lower vertebrates" examined in previous chapters. Judging from the consequences of interrupting the coronary supply in man, one might presume that the metabolic output of the myocardium has become dependent upon a high oxygen environment.

Major Vessels

The embryonic development of the major vessels in birds is rather variable from one species to another. The major vessels can be traced from the five pairs of aortic arches, but the sequence of appearance is complex. The systemic aorta develops from the right IVth arch, for example, rather than from the left IVth arch as in mammals. The Ist and IInd arches disappear completely, the IIIrd gives rise to the carotid arteries, and the pulmonary artery is derived from the VIth arch (Jones & Johansen, 1972). A number of different patterns of major vessels are seen in the adults of various species (Fig. 13.5), the duck most like that in Fig. 13.5A.

The properties of most major vessels in birds appear to allow analysis of flow and pressure patterns according to a relatively simple Windkessel model. That is, the vessel walls conform to a simple resistance-capacitance model and do not

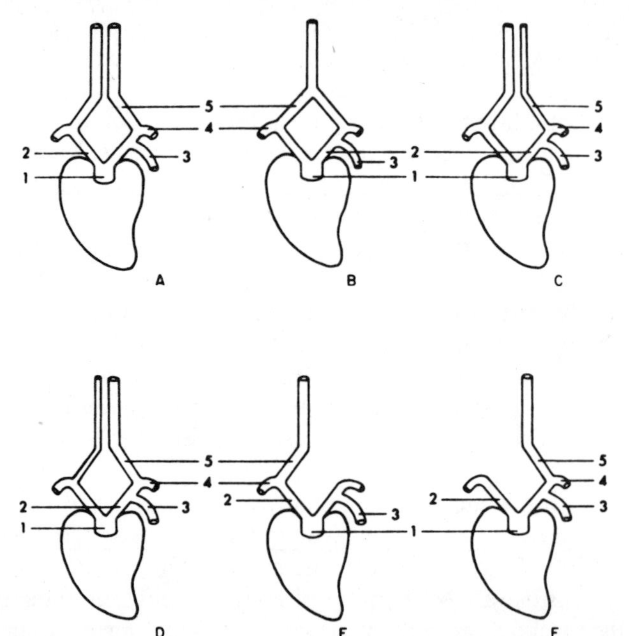

Figure 13.5: Dorsal views of the main arteries of birds in the region of the heart. (A) Most prevalent avian type. (B) Type found in *Botaurus*. (C) Type found in *Phoenicopterus*. (D) Type found in *Cacatua sulphurea*. (E) Type found in *Passeriformes*. (F) Type found in two species of *Eupodotis*. 1, aortic root; 2, innominate artery; 3, fourth right aortic arch; 4, subclavian artery; 5, carotid artery. (After Glenny, from Jones & Johansen, 1972, by permission, ©Academic Press.)

enhance harmonic components of the pressure wave. The principal effect of the arterial Windkessel is to convert a highly pulsatile ventricular output to a smoothed flow in the periphery. The transmission of the pulse wave down the aorta, however, shows amplification and distortion, implying a more complex elastic behavior.

OXYGEN BINDING CHARACTERISTICS OF THE BLOOD

Lutz (1980) reviewed a number of studies of the oxygen affinity of bird blood and reported P_{50} values for the duck ranging from 35 to 50 torr at an average P_{CO_2} of 35 torr with an average body temperature of 41°C. The best estimate of P_{50} from these studies is probably 43 torr, but there may be genuine differences due to species or variety of the duck and changes in organic phosphates in the erythrocytes. Isaacks and Harkness (1980), for example, reported considerable amounts of inositol phosphate (ITP), adenosine phosphates (ATP, ADP), 2,3-diphosphoglycerate (2,3-DPG), and other phosphorylated metabolic intermediates from duck erythrocytes; the amounts appear to depend on age and sex, and ITP in particular may vary in an adaptive manner depending upon acclimation conditions. Whether the small shifts in the oxygen dissociation curve that result are functionally significant in respiration has not been demonstrated. Most birds show a typical Bohr shift with increasing P_{CO_2}, and there also appears to be a direct CO_2 effect (independent of pH), presumably from carbamate formation (Lutz, 1980).

GAS EXCHANGE AT REST

Ventilation

Since the actual exchange volume of the lung, i.e., the parabronchi and their associated air capillaries, occupies a relatively small percentage of the total volume of the lungs and air sacs (Duncker, 1972, 1974), it might be expected that there is at least the potential for a large dead space. Rapid and shallow ventilation, for example, would be relatively inefficient compared to a slower, deeper breathing pattern. Data from ducks and other species of birds confirms that the predominant pattern in normal ventilation is a relatively low frequency of breathing with a large tidal volume (Table 13.2). When birds become hyperthermic they do pant, but at least some workers have reported that the rapid, shallow ventilation associated with panting is simply superposed upon the normal pattern of slower and deeper breathing, satisfying both thermoregulatory and respiratory needs (Ramirez & Bernstein, 1976). Since the aerodynamic valve action inherent in the bird's airway system also depends upon inertia and flow velocity (Banzett et al., 1987; Wang et al., 1988), changes in ventilation might be more effectively accomplished by changes in tidal volume than by changes in frequency.

Table 13.2: Gas exchange parameters for a duck with a body temperature of 41°C.

Parameter	Value	Source
W, weight, kg	2	Assumed
$\dot{M}O_2$, ml kg^{-1} min^{-1}	23.1	1
$\dot{M}CO_2$, ml kg^{-1} min^{-1}	15.7	1
$\dot{V}_E$, ventilation volume, ml kg^{-1} min^{-1}	324	1
V_f, ventilation frequency, min^{-1}	10.5	1
V_T, tidal volume, ml kg^{-1}	31.9	1
V_B, blood volume, ml kg^{-1}	46.3	1
$\dot{Q}$, cardiac output, ml kg^{-1} min^{-1}	391	1
fH, heart beat rate, min^{-1}	130	1
SV, stroke volume, ml kg^{-1}	3.0	1
Mean circulation time, sec	7.5	1
$\dot{V}_E/\dot{M}O_2$, Air convection requirement, O_2	14.0	Calculated
$\dot{V}_E/\dot{M}CO_2$, Air convection requirement, CO_2	20.6	Calculated
$\dot{V}_E/\dot{Q}$, Ventilation-perfusion ratio	0.83	Calculated
CRR, Capacity-rate ratio	0.99	Calculated
EPR, Exchange potential ratio	0.70	Calculated
Hct, Hematocrit, %	39	2
αBO_2, Blood O_2 capacity, mM L^{-1}	7.63	2
PaO_2, Art. oxygen tension, torr	96.1	1
	82	3
PvO_2, Mixed venous O_2 tension, torr	55.9	1
$PaCO_2$, Art. CO_2 tension, torr	35.9	1
	38	3
$PvCO_2$, Mixed ven. CO_2 tension, torr	42.6	1
pH_a, arterial pH	7.46	1
	7.49	2
pH_v, venous pH	7.42	1
P_{50}, Half saturation tension of blood	43	4
CaO_2, Art. oxygen content, mM L^{-1}	7.01	Calculated
CvO_2, Venous oxygen content, mM L^{-1}	4.20	Calculated
P_{IO_2}, Inspired oxygen tension, torr	145	1
P_{EO_2}, Expired oxygen tension, torr	100	1
P_{ECO_2}, Expired CO_2 tension, torr	34	1
%E, Percent O_2 extraction in lung	31	Calculated

Data compounded from various studies of domestic duck varieties.
1: Jones & Holeton, (1972a).
2: Holle et al. (1977).
3: Kawashiro & Scheid (1975).
4: Lutz (1980).

Perfusion

With the complete double circulation, lung perfusion and systemic perfusion must be equal, although the cardiac output is now only that pumped by one ventricle, not the sum of both. For resting ducks a value of 400 ml min^{-1} kg^{-1} is typical (Table 13.2). As mentioned above, the systemic blood pressure tends to be high in birds in general; for the duck 150 to 200 mm Hg appears to be typical of systolic values, with a pulse pressure of around 40 mm Hg (Jones & Purves, 1970a; Butler & Jones, 1971). Changes of cardiac output are achieved almost entirely by changes in frequency, however, with a fairly constant stroke output (Jones & Holeton, 1972b). Given a blood volume of 10% body weight, the mean circulation time is approximately 15 seconds in the systemic circuit.

For the lung capillaries the transit time is less, probably always less than 1 second. Under experimental conditions pulmonary transit times as short as 0.3 second have been observed (Henry & Fedde, 1970), and during flight exercise they are undoubtedly very short. Obviously gas exchange must take place very rapidly in the bird lung, which corresponds well with alveolar walls that are the thinnest of any animal group, and with short mean diffusion paths (Abdalla & King, 1975). Even so, arterial oxygenation may not be completely achieved at the highest rates of blood flow (Jones & Johansen, 1972).

Efficiency Indices

From representative data for the resting duck, the various efficiency indices defined in previous chapters have been calculated in Table 13.2. Despite a different respiratory medium and very different structure, the ventilation-perfusion and capacity-rate ratios for the bird lung are very close to values found for crabs, fish, and reptiles. The requirements for efficient design have tended to produce a very similar result by different means. Various model studies have shown that the bird lung has efficiency advantages over the mixed-pool lung of man, for example, but is not as efficient as the counter-current gills of fish (Piiper, 1982). Some modification of these models may be appropriate, however, in view of the partially counter-current arrangement actually found in parts of the lung circulation (Abdalla & King, 1975).

Normal Acid-Base Status

Arterial and venous pH and P_{CO_2} values for resting ducks are given in Table 13.2. The steady-state concentration of CO_2 in ducks is regulated by a combination of ventilatory control and renal bicarbonate resorption, as is the case in most terrestrial animals. Since birds are normally homeothermic, the complication of regulating the acid-base status over a wide environmental temperature range is removed. Many birds allow blood to cool as it flows through the feet (Scholander, 1958; Steen & Steen, 1965), but the closed-system behavior of blood at constant

CO_2 content will ensure that appropriate pH-temperature relations are maintained (see Chapter 6).

The rapid lung transit mentioned in connection with oxygen exchange is also a factor in CO_2 excretion. As in other animals, the principal reservoir of CO_2 in the blood is plasma HCO_3^-, which must be dehydrated by entering the erythrocyte. Rates of reaction for the Cl^-/HCO_3^- exchange across the erythrocyte membrane and the subsequent dehydration and diffusion processes have been measured; technical difficulties with these measurements prevent giving a precise figure, but overall half-times between 0.1 and 0.5 second suggest that complete CO_2 equilibrium may not always be achieved during the lung transit (Lückner, 1939; Piiper, 1969).

EFFECTS OF AMBIENT GASES

Changes in Ambient Oxygen

The response of ducks to inhaling air with reduced P_{O_2} is to increase ventilation by increasing both tidal volume and breathing frequency (Jones & Purves, 1970b; Butler & Taylor, 1973). These responses are mediated by the carotid body chemoreceptors; denervation of these receptors abolishes nearly all the ventilatory response to hypoxia (Jones & Purves, 1970b). There is also a reduction in ventilation in response to acute hyperoxia, although the subsequent increase in arterial P_{CO_2} re-stimulates ventilation, masking the response. When ducks are exposed to hypercapnic hypoxia while being force-ventilated in air, a marked bradycardia develops, whereas in freely ventilating ducks bradycardia is masked by large increases in ventilation attributable both to the hypoxia and the hypercapnia, and the cardiac output rises (Butler & Taylor, 1973). Bradycardia also develops during aerial apnea, so the bradycardia associated with diving (see below) can be attributed as much to hypercapnic hypoxia as to any direct reflex.

Changes in CO_2

Hypercapnia also elicits a ventilatory response: Jones & Purves (1970b) found a doubling of ventilation when the arterial P_{CO_2} was increased from the normal value of 40 torr to 55 torr, and Butler and Taylor (1973) found a 2.7-fold increase in ventilation upon increasing arterial P_{CO_2} from 36 to 43 torr, with no change in heart rate or cardiac output. The hypercapnic response is mediated primarily by central chemoreceptors, with a smaller contribution by carotid body receptors (see Chapter 8). Jones and Purves (1970b) found only slight changes in the rate of response in denervated ducks, and Milsom et al. (1981) showed that the central drive was much greater than any peripheral source. The intrapulmonary chemoreceptors do not appear to affect ventilatory drive during hypercapnia but do alter the pattern of breathing somewhat (Milsom et al., 1981).

CARDIOVASCULAR AND RESPIRATORY ADAPTATIONS FOR DIVING

Cardiovascular Responses

The responses of ducks to diving depends upon whether they are dabblers or divers, whether the dive is voluntary or forced, and to some extent whether they have been trained (Jones et al., 1988). This is hardly surprising, since the behavior and anatomy of dabblers and divers is quite different. Dabbling ducks such as the mallard and pintail typically immerse their heads (usually for feeding) for periods of only 3 to 6 seconds, followed by a several-second interval of emersion. The cardiovascular responses to these brief immersions are negligible (Butler & Woakes, 1979). Dabbling ducks trained to make shallow voluntary dives do show a reduction in heart rate to about 250 per min, regardless of the pre-dive rate (which is usually higher) (Furilla & Jones, 1987). During involuntary dives (i.e., forced immersion of the head), the responses are quite different: during the first 30 to 60 seconds of the dive, heart rate declines to only a fraction of the normal value (Andersen, 1963; Johansen, 1964; Butler & Jones, 1968), cardiac output declines proportionally (Jones & Holeton, 1972b), and there is a redistribution of blood flow to favor the head and critical organs (Johansen, 1964; Folkow et al., 1966; review by Jones & Johansen, 1972.

Although the responses of true diving ducks such as the redhead and scaup to involuntary dives is similar to those of dabblers, in voluntary dives the picture is quite different. About 5 sec before the dive there is an increase in heart rate, and upon submergence the heart rate drops almost instantaneously. During a prolonged dive the heart rate gradually recovers to nearly the predive rate, and finally just before emergence there is an anticipatory increase in the heart rate (Fig.

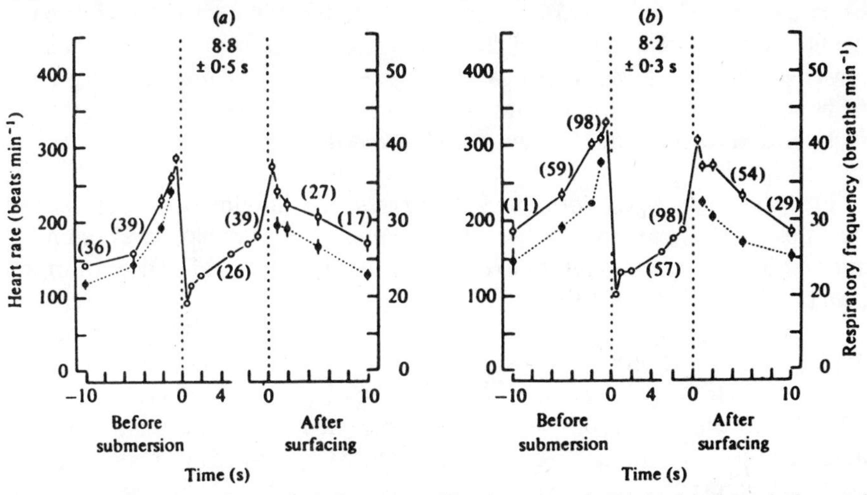

Figure 13.6: Heart rate (open circles) and breathing frequency (solid dots) before, during, and after spontaneous dives of free-swimming ducks. The dotted vertical lines show the duration of the dive. (From Butler & Woakes, 1979, by permission, ©The Company of Biologists, Ltd.)

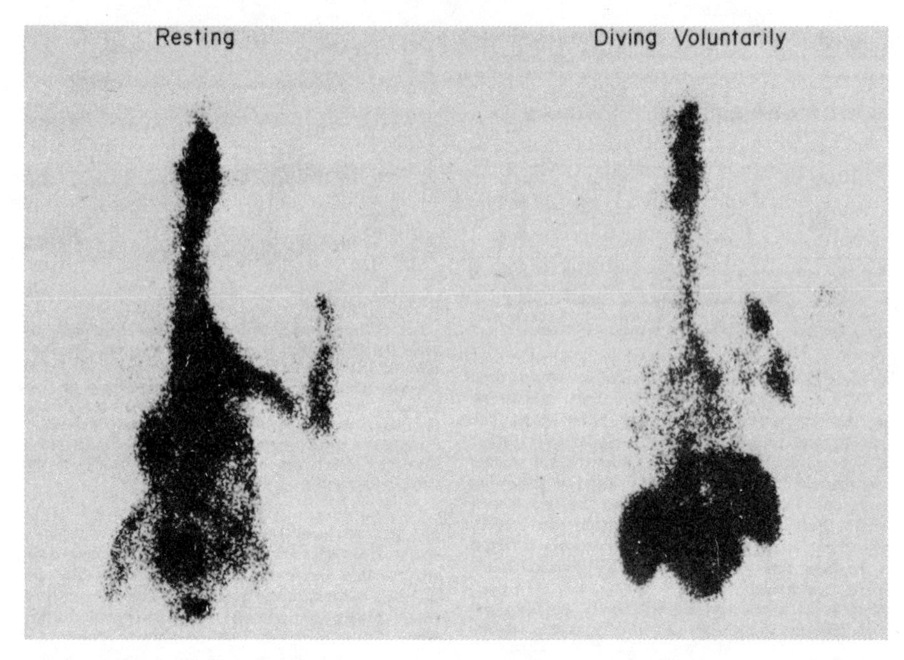

Figure 13.7: Gamma camera photograph of blood flow distribution in the scaup at rest (left) and during a voluntary dive (right). The pictures were obtained after a remotely triggered infusion pump injected technetium-99 bound to particles that were trapped in the capillary circulation. The pictures show the marked redistribution of blood flow to favor the leg muscles during the dive. (From Jones et al., 1988, by permission.)

13.6)(Butler & Woakes, 1979). Despite the bradycardia there does not appear to be much reduction in total cardiac output, consistent with the high energy demand during voluntary diving. The duck must not only swim vigorously but must overcome the effects of buoyancy (Jones et al., 1988). During the dive there is, however, a dramatic redistribution of blood flow away from the pectoral muscles and favoring the leg muscles involved in swimming (Heieis & Jones, 1988)(Fig. 13.7).

From data on oxygen stores, metabolic rates, and duration of dives, it is clear that most dives are aerobic. Oxygen stores would last for perhaps up to 1 minute, and respiratory variables return to normal within 10 seconds after the end of the dive (Butler & Woakes, 1979).

Implications for Control

Since diving must be accompanied by hypercapnic hypoxia, a combination that has been shown to strongly stimulate ventilation (Butler & Taylor, 1973), other control centers must override the ventilatory drive. Some of the reflex inhibition of ventilation and heart rate has been shown to be attributable to airway receptors ("defense receptors" of Jones & Milsom, 1982) near the bill and glottis of the duck (Butler & Jones, 1968; Bamford & Jones, 1974; Blix et al., 1976). Peripheral chemoreceptors of the carotid bodies apparently maintain the ventilatory drive, but effects of this drive are overruled during the dive (Jones & Purves, 1970b). The clear evidence of anticipation of the dive, however, with an attendant increase in heart rate (Butler & Woakes, 1979) strongly implies coordination of the diving responses in the brain. At present we do not know much about how coordination is achieved (Jones et al., 1988).

GAS EXCHANGE IN FLIGHT

Birds are above all designed for flight, a capability shared only by insects and bats. Perhaps the two most critical design requirements for flight are light weight and a high power output, although almost every aspect of birds' design is affected. Studying the physiology of flight is, of course, technically very tricky, and as a result we have data for only a handful of species. Tucker (1966, 1968, 1972) has studied the house sparrow, budgerigar, and laughing gull, Bernstein et al. (1973) have worked on the fish crow, and several authors have studied pigeons (Aulie, 1972; Pennycuick, 1975; Butler et al., 1977). Since there have not been any studies of ducks, Butler et al.'s data on pigeon will be used to illustrate the sorts of changes that occur in respiratory physiology during flight.

The limits of flight are quite interesting in themselves. Birds can travel farther, faster, and for longer than any other group of animals. Maximum speeds of over 100 mph (160 kph) have been reported for various swifts and falcons. Many birds undertake nonstop transoceanic migrations, and certain seabird species do not touch land for nearly a year and a half between breeding seasons. Altitude records, however, seem to capture the imagination more than others; the record is held by a vulture, which met its end in a collision with an aircraft at an altitude of 37,000 ft (11,300 m)! More routinely, some birds nest above 6000 m, and many migrate over the Himalayas at altitudes of 7000 to 9000 m (see reviews by Tucker, 1972; Scheid & Piiper, 1986).

Energetics of Flight

Flying is hard work. The maximal sustained energy expenditure during flight is around 10 times the (already high) resting level and about 3 times as high as maximal metabolic rates of mammals of comparable size (Tucker, 1971, 1972). Oxygen consumption reaches maximal rates within the first minute of flight and is maintained at this high level throughout (Fig. 13.8). That these rates of oxygen consumption are maintained at great altitudes is even more remarkable; the oxygen tension of the atmosphere at 8000 m is only 55 torr, about one-thirdof the sea-level value. These studies yield some clues as to why human muscle-powered flight has proved so elusive — we have had to design flying vehicles of almost no weight and to strain the limits of aerobic performance to achieve even the slowest low-altitude flight under the most favorable wind conditions (Welch, 1988).

Breathing in Flight

The air sacs not only contribute to the ventilatory system but actually invade other parts of the body including the hollow wing bones, thus lightening and streamlining the structure in critical places. The major flight muscles, the pectorals, are attached to the enlarged sternum, so it is reasonable to suppose that there would be some interaction between flight muscle movements and ventilatory movements. In fact there is no particular pattern – sometimes breathing and wing movements are coordinated 1:1, sometimes with other ratios, and sometimes not at all (Tucker, 1972; Scheid & Piiper, 1986). Part of the reason for this lack of coordination is probably the relation between ventilation frequency and tidal

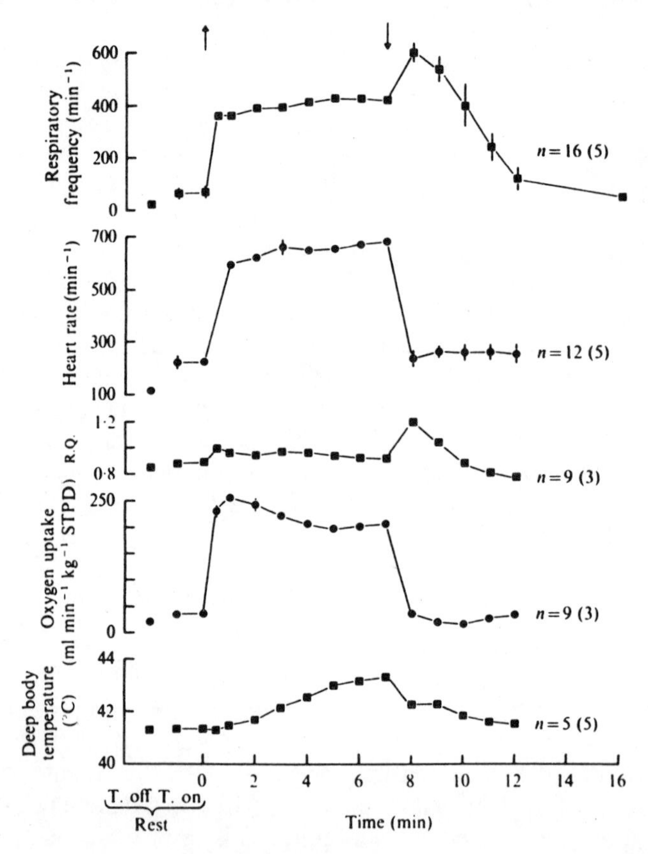

Figure 13.8: Changes in respiratory rate, heart rate, RQ, oxygen consumption, and body temperature in pigeons before, during and after flying at 10 m sec^{-1} for 7 minutes. Vertical lines represent ± 1 s.e., and the number of observations is shown at the right. The values were determined in a wind tunnel. (From Butler et al., 1977, by permission, ©The Company of Biologists, Ltd.)

Table 13.3: Mean values (± s.e.) of respiratory and cardiac variables measured in pigeons under various conditions.

Parameter	Resting Tunnel Off	Resting Tunnel On	Flying at $10\ m\ sec^{-1}$
Heart rate, min^{-1}	115 ± 2	222 ± 20	670 ± 14
Mean arterial pressure, mm Hg	142 ± 6	150 ± 5	147 ± 7
Mean venous pressure, mm Hg	1.2	–	2.5
Respiratory frequency, min^{-1}	19.7 ± 1.6	59.4 ± 15.9	411 ± 8.8
Oxygen uptake, $ml\ kg^{-1}\ min^{-1}$	20.3 ± 0.7	34.7 ± 2.3	200 ± 5.9
CO_2 production, $ml\ kg^{-1}\ min^{-1}$	17.2 ± 1.4	30.3 ± 2.8	184 ± 6.4
RQ	0.85	0.87	0.92
Pa_{O_2}, mm Hg	87 ± 2	–	95 ± 1
Ca_{O_2}, $mM\ L^{-1}$	6.74 ± 0.67	–	6.12 ± 0.54
pHa, arterial pH	7.43 ± 0.02	–	7.36 ± 0.02
Pv_{O_2}, mixed venous P_{O_2}, mm Hg	57 ± 2	–	42 ± 5
Cv_{O_2}, mixed venous O_2, $mM\ L^{-1}$	4.69 ± 0.58	–	2.41 ± 0.27
pH_v, mixed venous pH	7.36 ± 0.01	–	7.24 ± 0.03
Pa_{CO_2}, mm Hg	27 ± 0.6	–	16 ± 0.6
Pv_{CO_2}, mm Hg	35 ± 0.4	–	32 ± 2
Hematocrit, %	42 ± 1.3	–	41 ± 1.5
Lactic acid, $mM\ L^{-1}$	1.0 ± 0.3	–	6.64 ± 2.38
Body temperature, °C	41.3 ± 0.1	–	43.3 ± 0.2

The three conditions were: at rest with the wind tunnel turned off; with it turned on; and after 6 min of steady, level flight at $10\ m\ sec^{-1}$. Average weight = 442 gm; ambient temperature = 25.5°C. (From Butler et al., 1977, by permission.)

volume alluded to earlier. Particularly in smaller birds, whose wingbeat frequencies are much higher, rapid and shallow ventilation in time with the wingbeat would probably lead to inefficient ventilation of the lungs, moving air mostly back and forth in the dead spaces (Tucker, 1972).

There is, however, an enormous increase in ventilation during sustained flight, more than 20-fold higher than the resting rates (Table 13.3). Since there is only a 10-fold increase in oxygen consumption, there is evidently a reduction in the lung's efficiency, possibly due to either increased functional dead space or incomplete equilibration in the alveolar capillaries. Hart and Roy (1966) found little change in tidal volume during flight, so the assumption that ventilation volume increased in proportion to breathing frequency seems justified. Some authors have speculated that the disproportionate increase in ventilation serves to dissipate excess heat generated during flight (Zeuthen, 1942).

Cardiovascular Responses to Flight

From the data on oxygen consumption (Table 13.3), heart rate (also Fig. 13.8) and blood gases, cardiac output and stroke volume may be calculated for both resting and flying birds. The resting cardiac output of 555 ml kg^{-1} min^{-1} increased 4.4-fold during flight, and the A–V O_2 difference also increased by a factor of 1.8 (Butler et al., 1977). These two parameters together account for an 8-fold increase in oxygen transport, somewhat less than the 10-fold increase observed, but the discrepancy was attributed to variance in the resting data under different experimental conditions. Most of the increased cardiac output was due to increased heart rate, which is often synchronized with the flight muscles (Aulie, 1971, 1972).

The greatly increased ventilation led to a reduction in arterial P_{CO_2} from 27 to 16 torr, but the increase that would normally be expected in arterial pH was partially offset by an increase in lactic acid in the blood (Table 13.3). A marked hypocapnia seems characteristic of flight and exerts a beneficial effect on blood oxygenation. In some birds arterial P_{CO_2} as low as 10 torr has been recorded, and with the higher ventilation required at high altitude, even lower values may be achieved. The progressive reduction in P_{CO_2} shifts the oxygen dissociation curve progressively left, i.e., toward higher affinity, assisting in oxygen loading (Fig. 5.7)(Tucker, 1972). In Butler et al.'s (1977) study of the pigeon, arterial oxygen actually rose a little during flight.

Water Balance in Flight

Each time air is taken into the lung and expelled again, a certain amount of water vapor is lost, since the alveolar environment is saturated at a relatively high body temperature. For long flights such as migrations, then, water loss may be a critical limiting factor. Metabolism of fat, a major fuel reserve for long flights, generates about 5 gm of water for each liter of O_2 consumed. Tucker (1972) has calculated that rates of evaporative water loss would need to be about 7 gm of water per liter of O_2 consumed if the bird has water reserves equal to 15% of its weight and metabolizes 25% of its body mass as fat. Measured rates of evaporative water loss are only a little higher, but the difference could be due to additional stresses of experimental situations. Many birds have counter-current heat exchangers in the nasal passages, which greatly reduce water loss (Jackson & Schmidt-Nielsen, 1964).

GAS EXCHANGE IN THE EGG

Structure of the Avian Egg

As the avian embryo develops it makes a gradual transition from the aquatic environment of the egg fluids to the aerial environment in which it will later live. It does it by developing a lung-like exchanger within the egg that consists of a gradually increasing air space and a highly vascularized chorionic membrane. The chorioallantoic artery carries deoxygenated blood to this exchange surface, where it exchanges gases and returns to the embryo via a chorioallantoic vein (Fig. 13.9).

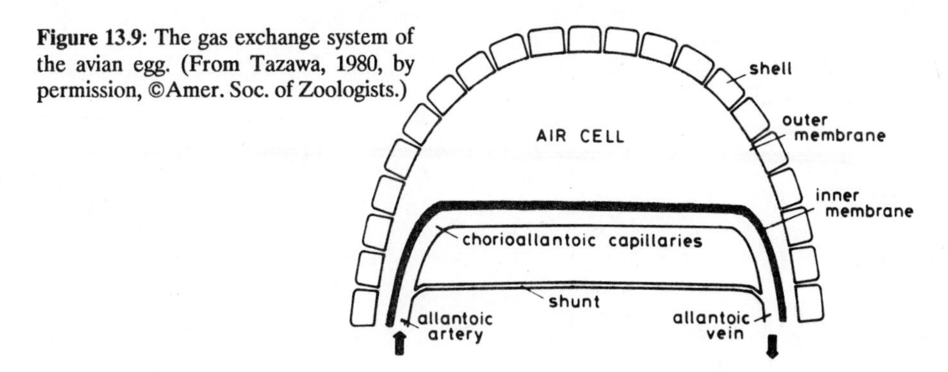

Figure 13.9: The gas exchange system of the avian egg. (From Tazawa, 1980, by permission, ©Amer. Soc. of Zoologists.)

With patience and cleverness, techniques have been developed for sampling blood and gases from developing embryos without interrupting the normal course of development (Tazawa, 1971, 1980).

The mineralized shell is penetrated by many small pores (Tyler, 1964) whose total cross-sectional area (in the domestic chicken) is about 2 mm^2 (Wangensteen, 1972). These fine pores provide a diffusive pathway for inward movement of oxygen and outward movement of both CO_2 and water vapor. The

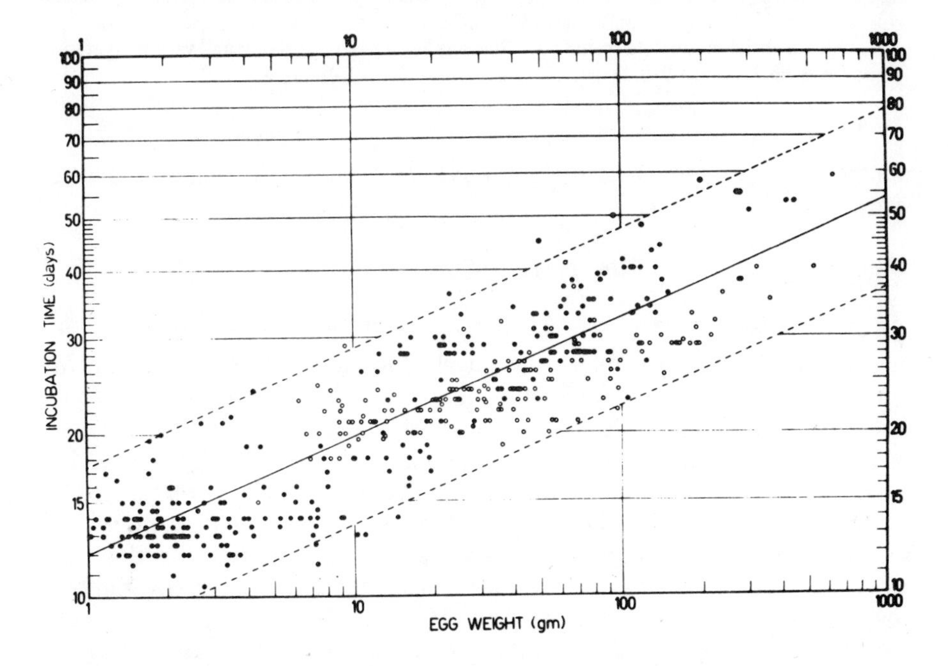

Figure 13.10: Allometric relation between incubation time and egg weight. Within the overall relationship, shown by the regression line (solid line) and ± 2 s.e. error limits (dashed lines), there is considerable variation. (From Hoyt, 1980, by permission, ©Amer. Soc. of Zoologists.)

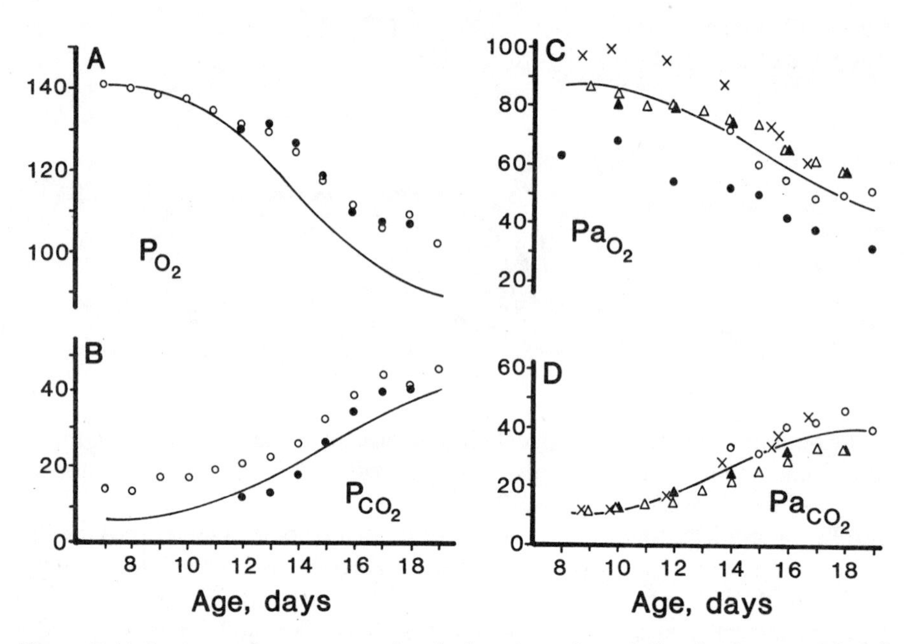

Figure 13.11: Summary of gas concentrations in the avian embryo during development. The left panel shows air cell values from three studies (solid line, open and closed circles), and the right panel shows arterial blood data from five studies of the chicken (solid line and various symbols). (Redrawn from Tazawa, 1980, by permission, ©Amer. Soc. of Zoologists.)

number and size of pores is actually quite variable but bears a relationship to egg size and incubation time.

Water Loss, Incubation Time, and Egg Size

As any farmer knows, the air space enlarges as incubation progresses, and the egg also loses weight. Weight loss is equivalent to water loss (Ar & Rahn, 1978), but results from a complex interaction between evaporative loss and metabolic water generation. Data collected from a large number of bird species shows that incubation time is generally related to the initial weight of the egg (Fig. 13.10), although the range of incubation times is only about 10 to 70 days with egg weights ranging from 1 to nearly 1000 gm (Ar & Rahn, 1978; Hoyt, 1980). Compared at the same egg weight, however, there is still a large variance in incubation time. Hoyt (1980) has shown that this variance is nicely accounted for by covariance in water conductance of the shell, and Ar and Rahn (1980) argue convincingly that the pore area of egg shells is adjusted to provide for constant hydration of the egg throughout development. Put another way, if the pore area were much larger or smaller, water loss would be either too fast or too slow to allow normal development. The effects of changes in pore area can be mimicked to a certain extent by

Figure 13.12: Summary of the oxygen gradients in an 18 day chick embryo. (From Wangensteen, 1972, by permission, ©Elsevier North Holland, Inc.)

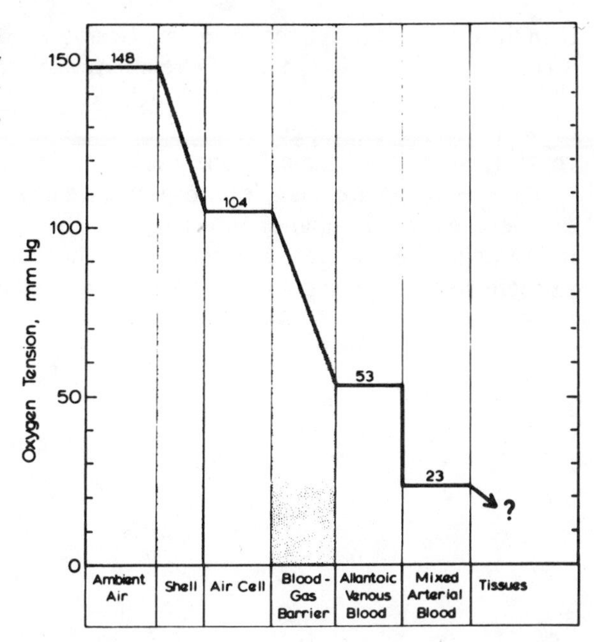

manipulating the humidity of the egg's environment; high mortality results from either too high or too low humidity (Romanoff, 1960).

Gas Exchange of the Embryo

As the development of the embryo progresses and its metabolic rate increases, a progressively steeper gradient for gas diffusion develops. Blood gas data taken at various stages show a progressive decline in air space and arterial oxygen, and corresponding increases in CO_2 tensions (Fig. 13.11). Resistance to diffusive exchange is present primarily at the shell/air space boundary and at the air space/chorionic membrane interface, resulting in gas gradients late in development as shown in Fig. 13.12. There is a progressive decline in O_2 tension in the air cell to about 100 torr at hatching, combined with a progressive increase in CO_2 tension to values near those of the adult.

The oxygen dissociation curve of the blood also changes during development. In the chicken the P_{50} is around 70 torr at 8 days of development, and declines to about 30 torr at 18 days (see review by Tazawa, 1980). Most of this change is due to the increasing P_{CO_2} during this time, but there are also changes in intra-erythrocytic phosphates that contribute to the P_{50} change (Mission & Freeman, 1972; Isaacks & Harkness, 1980).

The changes in gas exchange requirements during development are very closely linked to the water balance of the egg. In one study, the natural variability in water conductance was utilized to study the interrelationship. Those with relatively high water conductance had higher P_{O_2}, lower P_{CO_2} and higher pH values than the ones with lower conductance (Tazawa et al., 1983). Interestingly, there was some compensation so that the relations were not exactly as predicted. Possibly the reduced oxygen permeation in the less conductive eggs depressed metabolic rate and prolonged incubation. There was also some acid-base compensation so that the pH slope was flatter than expected.

The overall picture, then, is for eggs to have a fairly precisely controlled relation between size, incubation time and porosity. The progressive water loss and increases in gas exchange as hatching approaches result in constant hydration and a progressive change in gas tensions toward the adult values.

Chapter 14

Case Study: Man

With Notes on Other Mammals

In many respects, man is not a particularly good example of mammalian respiration, since we do not excel at diving, like the seal, at flying, like the bat, or at hibernating, like the marmot. We are not especially capable athletes by the standards of the animal world, nor do we exemplify extreme tolerance to temperature, altitude, or deviations from the standard atmosphere. Nevertheless, we do have a more or less typical mammalian respiratory system, and the importance of understanding our own physiology is obvious. What follows is a description of the standard man plus some examples of special mammalian physiology where appropriate.

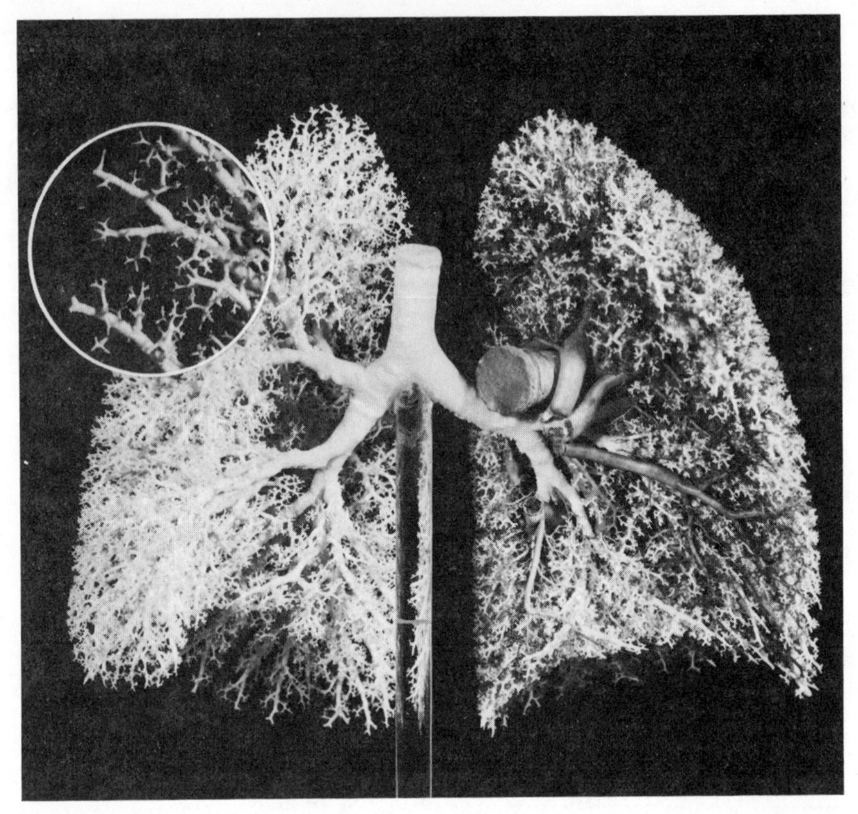

Figure 14.1: Cast of human lung airways. (From Weibel, 1984, by permission, ©Harvard Univ. Press.)

FUNCTIONAL MORPHOLOGY

Lung

In man we see a return to the line of arboreal lungs similar to those developed in lungfish, amphibians, and reptiles. The paired lungs of man are usually divided into three lobes on the right and two lobes on the left, although there is considerable variability among individuals (Krahl, 1964). Casts made of the airways show a complex branching network occupying 80% of the lung volume (Fig. 14.1). The single trachea branches into two primary bronchi, and they in turn bifurcate repeatedly, forming the various bronchioles, alveolar ducts, and finally alveolar sacs and the alveoli themselves. Each level of branching can be assigned a generation or order number, and certain useful analyses of the lung structure can be carried out with models of this dichotomy. In general, the first 16 generations or orders of division constitute the ever-finer bronchioles, which serve for air convection and distribution to the more distal portions of the airways. They are often termed the "conducting zone" (Fig. 14.2). At about order 16 a transition zone

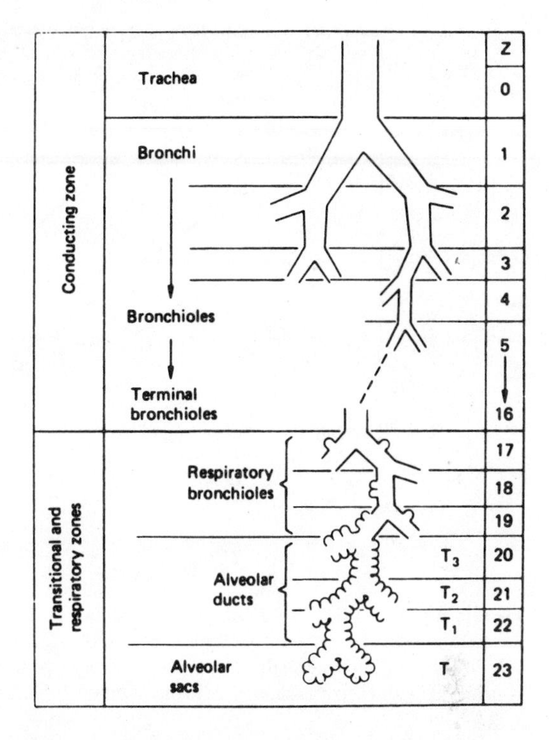

Figure 14.2: Organization of airway branching and zones of the lung. The number, Z, at right indicates the branching order or generation. (From Weibel, 1963, by permission, ©Springer-Verlag.)

begins, including the respiratory (finest) bronchioles and portions of the alveolar ducts. Finally, the terminal portions of the airways, which are lined with exchanging capillaries, constitute the respiratory zone, at approximately levels 20 through 23 in Fig. 14.2.

The order of division turns out to be a good predictor of several functional properties of the airways, including diameter, cross-sectional area, and airflow velocity. At each level of division the airway diameter decreases in a regular fashion, obeying the equation:

$$d(z) = d(0) \times 2^{-z/3} \qquad \text{(Eq. 14.1)}$$

where d(z) is the diameter at order z, and d(0) is the initial (tracheal) diameter. This equation yields the straight line relation of Fig. 14.3, and fits the hydrodynamic relation predicted for flow with minimum energy loss in branching tube systems (Weibel, 1984). As the average airway diameter decreases, the number of airways in each order increases by a power of 2, so that the total cross-sectional airway area increases in an exponential fashion (Fig. 14.3). As the total ventilatory flow distributes itself further down the airway orders, the velocity slows to a point where the diffusional velocity of oxygen is greater than the convective flow rate. The regular relation between airway diameter and branching order breaks down in this diffusional exchange zone, and the airway diameter stays about constant. Constant airway diameter happens to favor most efficient diffusional flow in a branched system. The mechanical design of the lung, then,

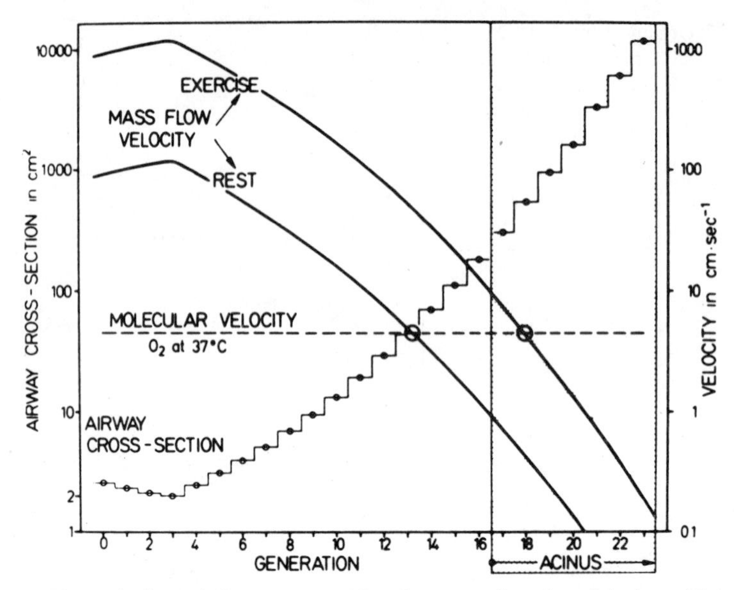

Figure 14.3: Flow velocity and airway cross sectional area as a function of the branching order or generation in the human lung. At about generation 13 at rest and 18 during exercise, the diffusional velocity of oxygen exceeds the convective velocity due to ventilation, and from there on the lung is primarily a diffusive exchanger. (From Weibel, 1984, by permission, ©Harvard Univ. Press.)

appears to be nicely optimized for convective flow in the proximal airway regions and diffusional flow in the distal alveolar regions.

Total Lung Surface and Volume

The complex spherical geometry and high distensibility of the human lung makes morphological measurement of volume and surface area a tricky task. Nonetheless many estimates of lung surface and volume have been made, with considerable variation in the results. The most recent estimates of Weibel (1984)(Table 14.1), using careful fixation conditions and EM sections to estimate the contribution of alveolar roughness and folding to total area, yield an estimate about 40% higher than earlier work by the same author (Weibel, 1964). The static lung volume is of questionable significance, however, since lung volume and particularly alveolar volume change continually during the ventilation cycle and are strongly influenced by inflation pressure (Agostoni & Mead, 1964). In functional (i.e., physiological) studies several different volumes are defined: The total lung capacity (TLC) is the volume of the entire respiratory system at maximum inhalation; residual volume (RV) is the gas remaining after maximum exhalation; functional residual volume (FRC) is the gas remaining after a normal exhalation; and tidal volume (VT) is the volume of a breath, or average of a series of breaths. Some portion of the gas inhaled with each breath merely fills the major airways, contributing nothing to gas exchange; this portion is the anatomical dead space, and for man this volume is about 170 ml (Table 14.1).

Table 14.1: Morphometric data for the human lung. Data for a standard 70 kg man.

Parameter	Value	Source
Total lung volume (TLC)	5500 ml	1
	4340 ml	2
Anatomical dead space (ADC)	170 ml	2
Transition zone	1500 ml	2
Respiratory zone	3150 ml	2
Total number of alveoli	296×10^6	1
Total alveolar surface area	143 m^2	2
Total alveolar capillary surface	126 m^2	2
Capillary volume	213 ml	2
Air/blood barrier, harmonic mean	0.62 μm	2
DLO_2, Morphometric, rest	137 ml O_2 min^{-1} $torr^{-1}$	2
Physiological, rest	30.2	
	38.3	
Physiological, exercise	100.2	

1: Dunnill (1962).
2: Weibel (1984).
3: Dejours (1975).

There is a good match between the total alveolar air surface and the total alveolar capillary surface; and the barrier between the air and the blood is very thin. The barrier is of uneven thickness, however, with the thinner parts contributing proportionally more to gas exchange than thicker parts. It is therefore most useful to compute a harmonic mean thickness of the barrier, rather than an arithmetic mean. For man the harmonic mean thickness is about 0.6 μm, roughly twice as thick as the barrier in the bird lung (Weibel, 1972, 1973, 1984)(see Chapter 13). From the capillary volume and total blood flow (cardiac output) a contact or transit time of about 0.9 second can be computed, more than adequate time for equilibration between blood and air under resting conditions.

Area and volume of the lung are primarily of interest insofar as they indicate the capacity of the lung to exchange gases. This capacity can be estimated in two ways, either by calculations of transfer based on morphometric data or from physiological tests. The former method is based upon calculating the gas transfer for a given set of diffusion coefficients, barrier thicknesses, temperature, etc. In the latter, the most common method is to allow a subject to inhale a small quantity of carbon monoxide, which binds tightly to hemoglobin, allowing analysis based upon washout of the nonexchanging portion of the gas. Morphometric measurements of the diffusing capacity were first made by Hufner almost a century ago (Hufner, 1897), and when correct diffusion coefficients for tissue are substituted in his calculations the estimate is very close to the most recent ones (Table 14.1). These morphometric measurements usually exceed physiological measurements, probably due to a number of factors. At rest equilibration probably occurs in only a fraction of the contact time, for example, which leads to an underestimate of the true capacity. Since the units of the diffusing capacity are amount of oxygen transferred per unit partial pressure gradient per time, a mean alveolar-blood partial pressure gradient can be calculated using the measured oxygen consumption rate. For resting man the gradient turns out to be about 8 torr for oxygen, a value consistent with measurements. During heavy exercise, however, the predicted gradient rises to about 80 torr, an unreasonably high value, which must mean that the functional diffusing capacity of the lung is not fixed but can increase under various conditions (Forster, 1964; Dejours, 1975; Forster & Crandall, 1976; Weibel, 1984).

Mechanics of Ventilation

The human lung is ventilated by aspiration, i.e., a reduction of pressure within the chest cavity by the combined action of the diaphragm and intercostal muscles of the rib cage. Since the airway diameter is large relative to the normal flow rates, the pressure required to drive ventilation is small, only about 1 cm H_2O. The even branching of the airways seems to ensure fairly even ventilation of the different regions of the lung, although contraction of the smooth muscles of the bronchiole walls can significantly alter this distribution.

Emptying of the lung (expiration) is achieved partly by the external intercostal muscles but primarily by the elastic rebound of the relaxed diaphragm and of the lung itself. Elastic forces within the lung constitute a balance between the tension exerted by intrapulmonary connective tissue fibers and that exerted by surface tension of the alveolar walls. Since the surface of the alveolus is wet, it can be analyzed as a bubble of comparable size. As bubbles get smaller, the surface tension increases, tending to further reduce the size. The alveolus, then, would appear to be inherently unstable and should collapse. That it does not is due to opposing mechanisms, one of which is simply the tension exerted by other alveoli through the fibrous interconnections (Weibel, 1984). The other important fac-

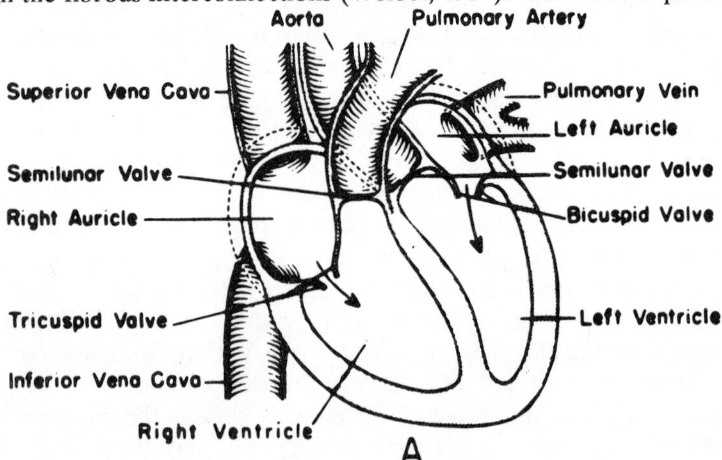

Figure 14.4: The human heart. The right ventricle is shown disproportionately large. (From Florey, 1966, by permission, ©W. B. Saunders, Inc.)

tor in maintaining alveolar integrity is a surfactant layer present on the alveolar surface. Small bubbles observed in foams or in solutions grow smaller and smaller: As surface tension increases, the gas tension increases inside the bubble, leading to diffusive loss of the bubble volume. The loss in turn reduces bubble volume, further destabilizing the bubble, which rapidly disappears. In froths made from lung washings, however, Pattle (1955, 1976) observed that the bubbles were stable, leading him to suppose that there must be a material present that reduced the surface tension progressively as bubble size decreased. This material has since been identified as a phospholipid monolayer (Clements et al., 1961) on the alveolar surface. Compression of the monolayer (e.g., as an alveolus contracts) increases the effect of the surfactant, balancing the forces of wall tension and luminal pressure. Without the surfactants, lungs would collapse and could not be reinflated, a condition seen in some premature infants and in various diseases.

Vascular System

The Heart

The human heart has probably been studied more than any other organ in the biological kingdom, and the relevant literature is immense. The literature includes two volumes of the early *Handbook of Physiology*, three volumes of the more recent edition, many books, and a number of journals specifically dealing with various aspects of the heart and its physiology. What follows here, then, is only an abbreviated general description of its form and function, in order to compare it with the animals studied in previous chapters and to illustrate general trends.

Like the crocodilian reptiles and birds, the human heart is four-chambered with complete separation of the pulmonary and systemic circuits. While superficially resembling the reptile and bird hearts, there are differences in the way the separation arises during development that suggest closer ties with a common amniote ancestor rather than in a line with reptiles and birds (Goodrich, 1930). The right atrium and ventricle are much less muscular than the left (Fig. 14.4). Although the output of both right and left pumps is necessarily the same, the pressure in the pulmonary circuit is only about one-tenth that of the systemic circuit. The work load of the "right heart" is therefore correspondingly less, matching its smaller muscle mass.

Blood beginning in the right atrium flows to the lungs via a semilunar valve and the pulmonary artery. As in other higher vertebrates, the arterial vasculature in the pulmonary circuit carries deoxygenated blood and the veins oxygenated blood. The pulmonary artery branches progressively in a manner parallel to airway branching, terminating in the many millions of alveolar capillaries. They in turn gather into a pulmonary venous drainage, eventually returning to the heart via the pulmonary vein. Blood entering the left atrium from the pulmonary vein progresses to the ventricle during diastole via the bicuspid valve. During systole this valve closes and blood is forcefully ejected into the systemic aorta, from which it is distributed to all parts of the body. Return of systemic blood to the right atrium via the venae cavae completes the circuit.

Events of the Cardiac Cycle

In the region of the right atrium, near where the superior and inferior venae cavae join the atrial wall, there is a specialized patch of tissue known as the sino-atrial (SA) node. It has been known for a long time that isolated pieces of myocardium could maintain rhythmic contraction but that the pacemaker region of the SA node provides the synchronizing signal for the entire heart. The depolarization wave begun at the SA node spreads at about 1 m sec^{-1} through both atria, triggering atrial contraction and diastolic filling of the ventricles (Fig. 14.5). This depolarization wave can be seen as the "P wave" in a conventional electrocardiogram (ECG) recorded at the body surface and also corresponds to the modest pressure wave in the atria. Electrical communication between the atria and the ventricles is controlled by another specialized tissue region, the atrioventricular (AV) node, which lies at the base of the interatrial septum at the top of the ventricles. Conduction of the P wave through this node is very slow, around 0.1 m sec^{-1}, so the AV node acts not only to couple the atria and ventricles but to delay the latter by the appropriate amount of time. Once through the AV node, the excitation wave travels rapidly through the syncytial myocardium and specialized Purkinje fibers, generating the large ECG spike known as the QRS complex. The repolarization of the atria occurs during ventricular contraction, but ventricular repolarization at the end of systole shows up as the T wave in an ECG tracing. Atrial depolarization takes about 80 msec to spread and lasts about 150 msec; ventricular depolarization starts approximately 80 msec after the end of the atrial depolarization and lasts about 300 msec (Scher & Spach, 1979). The muscle fibers in each part of the heart contract during depolarization (systole) and relax during repolarization (diastole). The relation of these electrical events to mechanical and pressure events in the heart is shown in Fig. 14.5.

Coronary Blood Supply

As in the birds, nearly all of the myocardial tissue receives an arterial blood supply via a well-developed coronary circulation. While this is true in the adult, the early fetal heart does not have a coronary supply and appears in section like that of the amphibians and fish, i.e., spongy and trabecular, with luminal blood perfusing the myocardial muscle (van Mierop, 1979). The coronary blood supply is of great interest to medicine, since even temporary interruptions of it lead

Figure 14.5: (Facing Page) Events of the cardiac cycle. Various pressures, valve events, heart sounds, and other measurements of the cardiac cycle are plotted as a function of time (abscissa). The top half depicts the appropriate pressures in the aorta, left ventricle, pulmonary artery, right ventricle, and right and left atria. Valve opening (O) and closing (C) is indicated for the mitral (M), tricuspid (T), pulmonary (P), and aortic (A) valves. The locations of the four heart sounds (S_1–S_4) and systolic ejection clicks are noted. The volume curve of the left ventricle shows a slight increment near end-diastole due to left atrial contraction. This is followed by an abrupt reduc-

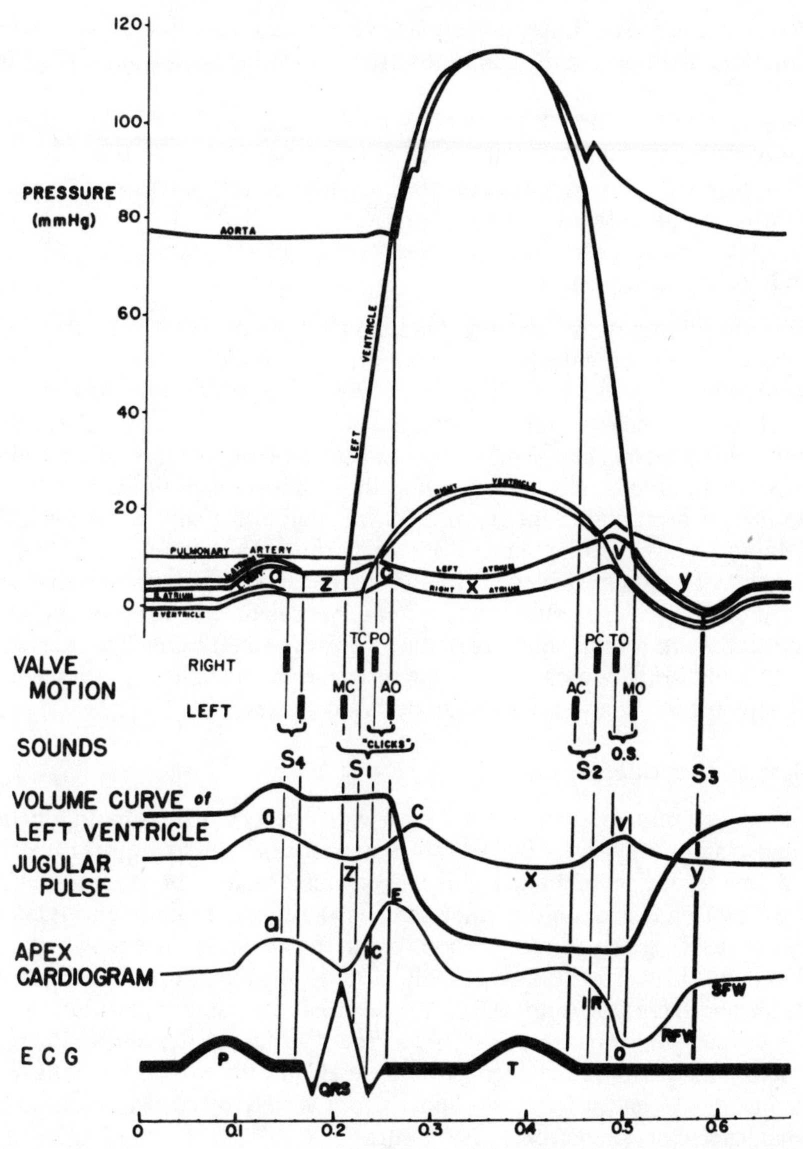

tion in volume during systole with subsequent filling during the following diastolic cardiac cycle. The designations in the jugular pulse and right and left atrial pressures are as follows: the *a* wave is due to atrial contraction followed by a *z* wave of relaxation. The *c* wave of the jugular pulse represents bulging back of the tricuspid valve into the right atrium during systole. This is followed by the *x* descent as the base of the heart moves downward during systole. The *v* wave represents continued atrial filling against a closed tricuspid valve, followed by a *y* descent after tricuspid valve opening. The characteristic P wave, QRS complex, and T wave of the electrocardiogram are shown near the bottom for timing purposes. (Original diagram from J. W. Hurst & R. B. Logue, *The Heart* (3rd ed.), McGraw–Hill; used by permission. Legend explanation from Parmley & Talbot, 1979, by permission.)

to severe reductions in heart performance and permanent tissue damage from anoxia, "heart attacks" in the common parlance (see Berne & Rubio, 1979, for a review). The coronary arteries arise from the aorta, spread over the entire heart surface, and serve a dense network of capillaries that invade the heart muscle. In much of the ventricular mass, the ratio of capillaries to muscle fibers is nearly 1:1. Drainage of the coronary capillaries is gathered into a system of coronary veins which drain back into the right atrium.

Pulmonary Vasculature

Some allusion has already been made to the pulmonary vasculature in connection with describing the lung (above). From the standpoint of the cardiovascular system, however, it is of some interest to compare it to the systemic circulation or to the incompletely separated circulations that we have examined in previous chapters. The tissue it serves, i.e., the lung, is more or less homogeneous, both functionally and anatomically, so it is perhaps not surprising that there is little provision in the pulmonary vasculature for control of blood flow distribution. The circuit is of low resistance, perfused by a high-output, low-pressure pump. The low resistance is important, because imposing the same pressures in the lung as are found in the systemic vasculature would probably cause serious problems for maintenance of the alveolar geometry as well as causing excessive fluid filtration rates. Most of the resistance in the pulmonary vasculature is attributable to the alveolar capillaries (Weibel, 1984).

Systemic Vasculature

Blood flow from the aorta must serve a wide variety of tissues, both functionally and anatomically diverse. The brain, for instance, seems to require a constant blood flow regardless of mental activity, body temperature, or other conditions. Blood flow to the skin, on the other hand, is extremely responsive to temperature, as muscles are to exercise. Blood flow to the kidney is extremely high relative to either its weight or its metabolic activity, obviously for the purposes of filtration and excretion. A priori it seems clear that the systemic vasculature must possess the capability for modifying blood flow distribution through a wide range and in response to rapidly changing conditions. Contractile smooth muscle in the walls of arteries and arterioles responds to a variety of neural, humoral and mechanical factors (Bevan et al., 1980; Burnstock, 1980; Johnson, 1980), with the largest resistance provided by the arterioles.

Since the systemic vasculature represents an arboreal network similar to the airway network of the lung, its volume increases with the order of branching. The diagrammatic analysis of Rushmer (1965)(Fig. 4.18) gives an idea of the changes in vessel cross-sectional area and flow velocity. Note that the steepest velocity drop comes in the neighborhood of the arterioles, and that the high-pressure artery system represents a potential energy reservoir to provide a quick blood flow response to peripheral vasodilation.

Pressure in the arterial system is not static, of course, but has a considerable pulse pressure imposed by ventricular ejection (Fig. 14.5). The elastic properties of the arteries provide considerable damping (Windkessel) action, but there are some peculiar anomalies of arterial pressure whose interpretation has been controversial. If the systolic pulse pressure is measured at increasing distance from the heart, the peak pressure actually rises, and the waveform changes. The changes result from a combination of reflection of pressure waves from branch points, frequency-dependent damping, and frequency-dependent differences in wave propagation velocity (Rushmer, 1965).

BLOOD

Hematology

Human blood is not too different from that of other vertebrates, except perhaps in having nonnucleated erythrocytes with a relatively short life-span, around 120 days (Berlin et al., 1959). The hematocrit is relatively high, as is the hemoglobin concentration (Table 14.2). The erythrocytes are of fairly small size and are rather readily deformable as they travel through capillaries (see Chapter 5).

Oxygen Carriage by Hemoglobin

The oxyhemoglobin dissociation curves of man are quite conventional; i.e., they have a sigmoid shape indicating cooperativity among binding sites, a moderate Bohr shift, and a moderate Haldane effect (Table 14.2). The P_{50} at normal values for pH (7.40) and P_{CO_2} (40 torr) is 27 torr (see Chapter 5).

Human blood is considerably influenced by concentrations of phosphorylated metabolic intermediates in the erythrocyte. Increasing 2,3-DPG in a range of 0.2 to 1.0 mM L^{-1}, for example, almost doubles the P_{50} (Benesch & Benesch, 1967; Chanutin & Curnish, 1967; Eaton et al., 1970). Although changes in intraerythro-

Table 14.2: Average blood values for 70 kg adult men and women.

Parameter	Men	Women
Hematocrit, %	47	42
Hemoglobin, g dL^{-1}	16	14
RBC count, $10^6/mm^3$	5.4	4.8
Mean corpuscular Hb, μg	2.96	2.92
RBC diameter, μm	8	
Blood oxygen capacity, mM L^{-1}	9.73	8.48
P_{50} at pH = 7.4, torr	27	
$\Delta \log P_{50}/pH$*	–0.51	
Blood Volume, L	1.6	

Data from Altman & Dittmer (1974) except *Dejours (1975).

cytic phosphates can be demonstrated in response to altitude acclimation and after birth, the in vivo shifts are not very large and their functional significance is not entirely clear (Eaton et al., 1969).

GAS EXCHANGE AT REST

Ventilation

The normal 70 kg "standard man" at rest consumes about 300 ml O_2 per minute, breathes over 7 L per minute, and has an RQ of 0.85 (Table 14.3). From the difference between inspired and expired air we can calculate that extraction of oxygen from the lung is only about 21% efficient, and that about 25 L of air must be moved in and out of the lungs to obtain 1 L of oxygen. Notice that the alveolar O_2 tension is lower than the expired tension, which reflects the diffusion gradient between the ventilated and exchanging volumes of the lung (see above). The alveolar-arterial oxygen gradient matches earlier calculations of the diffusing capacity and sets an upper limit on oxygen loading.

The normal exchange parameters allow for a considerable "venous reserve." That is, venous blood is still about 65% saturated, which allows tissue extraction to increase substantially for acute increases in demand. Since these values are for mixed venous return, however, they obscure the regional differences in oxygen extraction. Blood flowing through the kidney, far in excess of metabolic needs, only has about 8 to 10% of its oxygen utilized, whereas for skeletal muscle and brain the figure is about 50%, and for heart muscle 66% (Farhi & Rahn, 1960).

The values for alveolar and arterial CO_2 bear a fixed relation with the oxygen partial pressures, as worked out long ago by Fenn et al. (1946). To use a simple example, if the partial pressure of alveolar O_2 drops by 50 torr and the RQ is 0.8, the rise in alveolar CO_2 will be 40 torr because the capacitance ("solubility") of the two gases in air is the same. In Table 14.3, the inspired-expired oxygen difference is 32 torr, and the inspired-expired CO_2 difference 26 torr, a ratio of 0.81, which is not significantly different from the observed RQ.

Perfusion

The total cardiac output for man (Table 14.3) provides for close matching of ventilation and perfusion in the lung. The $\dot{V}_A/\dot{Q}$ ratio of 1.31 and the almost identical oxygen contents at normal air saturation yield a capacity-rate (or conductance) ratio very near 1. The exchange potential ratio (see Chapter 7) is 0.27, however, reflecting the lower overall efficiency of the mixed pool lung compared to either the counter-current fish gill (see Chapter 9) or the cross-current avian lung (see Chapter 13).

Normal pulse pressure and heart beat rate are highly variable among individuals; young people and those more physically fit tend to have slower resting

Table 14.3: Gas exchange parameters for a resting 70 kg human.

Parameter	Value	Source
Weight, kg	70	Assumed
$\dot{M}O_2$, ml kg^{-1} min^{-1}	4.04	1
$\dot{M}O_2$, μM kg^{-1} min^{-1}	180	Calculated
$\dot{M}O_2$, ml min^{-1}	283	Calculated
RQ	0.85	2
$\dot{M}CO_2$, ml kg^{-1} min^{-1}	3.43	Calculated
$\dot{M}CO_2$, μM kg^{-1} min^{-1}	153	Calculated
$\dot{V}E$, L min^{-1}	7.60	1
$\dot{V}E/\dot{M}O_2$, L mM^{-1}	0.60	Calculated
%E from lung	19.1	3
$\dot{Q}$, L min^{-1}	5.81	3
Heart rate, min^{-1}	65	2
Stroke volume, ml	89	Calculated
Blood pressure, mm Hg	120/80	3
P_{IO_2}, torr	150	1
P_{EO_2}, torr	118	1
P_{ECO_2}, torr	26	4
P_{AO_2}, torr	102	4
P_{ACO_2}, torr	39	4
Pa_{O_2}, torr	93	2
Pa_{CO_2}, torr	39	1
Pv_{O_2}, torr	38	2
Pv_{CO_2}, torr	44	4
Ca_{O_2}, mM L^{-1}	8.48	2
Cv_{O_2}, mM L^{-1}	6.56	2
Ca_{CO_2}, mM L^{-1}	21.34	5
Cv_{CO_2}, mM L^{-1}	23.3	5
pH_a	7.40	5
pH_v	7.47	5
β, mM pH^{-1}	−29	2
P_{50} at pH 7.40	26.8	2

1: Torrance *et al.*, 1970
2: Altman & Dittmer, 1974
3: Ekelund & Holmgren, 1964
4: Dejours, 1975
5: Davenport, 1974

heart beats, and old age or disease can substantially increase blood pressure. Blood pressure is usually measured in the brachial artery below the shoulder, but at locations between that site and the heart the maximum pulse pressure may reach over 160 mm Hg due to the enhancement effects discussed above. Since the major pressure drop occurs across the smaller arteries and arterioles, pulse pressure is strongly damped by the time blood enters the capillaries. Capillary flow is slow (see Fig. 4.18) and steady.

Normal Acid-Base Status

It is a certainty that soon after the glass pH electrode was invented in 1909, the pH of human blood and plasma was measured, and the relatively narrow range of about 7.35 to 7.45 was established. There was a tendency to overgeneralize, however, and it was more than half a century later before it was realized that 7.4 was not necessarily the "right" pH for all animals at all temperatures (see Chapter 6). Still, healthy humans do regulate their acid-base status within narrow limits, and the large statistical data base has allowed fairly precise clinical definitions of a variety of disturbances of diagnostic importance (Hills, 1973). A conventional pH-bicarbonate diagram was shown in Chapter 6 (Fig. 6.14), along with identification of the principal types of acid-base disturbance.

As with other animals, the blood is studied as a reflection of the acid-base status of the whole animal (human), even though the intracellular fluids make up a far larger, better buffered, and probably more critical compartment. Compensation of acid-base disturbances consists of a complex interplay between ventilation, renal H^+ excretion, and extra/intracellular fluid exchanges (Hills, 1973), plus minor contributions from such processes as bone buffering (Bettice & Gamble, 1975; Poyart et al., 1975a,b; Bettice, 1984).

GAS EXCHANGE DURING EXERCISE

In humans of about average conditioning, oxygen consumption can rise approximately 8- to 10-fold during heavy exercise (Table 14.4). The primary metabolic fuel for short-term exercise is carbohydrate, which tends to increase the RQ somewhat, but in much longer-term exercise a progressively larger portion of the energy is provided by fat, leading to a secondary decline in the RQ toward 0.7. During the onset, maintenance, and recovery from exercise, even though rates

Table 14.4: The respiratory responses to heavy, non-steady-state exercise in a 70 kg standard man.

Parameter	Rest	Exercise	Source
$\dot{M}O_2$, ml min^{-1}	278	2400	1
RQ	0.85	0.9 –1.0	1
$\dot{V}E$, L min^{-1}	7.1	68.1	1
PE_{O2}, torr	116	118	1
%E, air	23	21	1
$\dot{Q}$, L min^{-1}	5.81	50.2	Calculated; 2
Heart rate, min^{-1}	76	166	1
Stroke volume, ml	76	302	Calculated
Pa_{CO2}, torr	38	35	2
pH_a	7.44	7.38	2
[Lactate], mM L^{-1}	1	12	2

1: Ekelund & Holmgren (1964).
2: Wasserman et al. (1979).

of oxygen and carbon dioxide transport are constantly changing by an order of magnitude, a remarkable constancy of blood gases and acid-base status is observed (Wasserman et al., 1979). Thus we should expect to see that the changes in ventilation and cardiac output are very closely matched to the changes in metabolic rate (Table 14.4). If, for example, ventilation failed to rise as much as metabolism, a quite pronounced respiratory acidosis would occur and if the increase were more than metabolism, an alkalosis would be expected. Actual data show that over long periods of heavy exercise there is indeed some acidosis, but it is the result of increasing lactic acid production by muscle that is partially compensated by increases in the ratio of ventilation to O_2 consumption (see Chapter 8).

The data for expired gases are also interesting: The extraction efficiency of the lung does not change, even though the rate of gas transport has increased by an order of magnitude. Evidently the alveolar-arterial gradients are not very different during exercise than during rest, which matches the earlier speculations about changes in the diffusing capacity. More efficient ventilation and capillary recruitment must balance the increased gas transport requirement.

Cardiac output rises in linear proportion to the increase in ventilation, mostly by increases in stroke output, with a smaller increase in rate. Most of the increase in stroke volume takes place via increased end diastolic volume, so the resulting contraction ejects a larger volume with higher efficiency (Khouri et al., 1965).

RESPONSES TO CHANGES IN AMBIENT GASES

Hypoxia and Altitude

The physiology of man at high altitude (and its attendant hypoxia) became of urgent interest during World War II, when nonpressurized aircraft were reaching ever-higher altitudes. Studies have continued more recently with mountaineers in various expeditions and in high-altitude natives of the Andes, motivated in part, perhaps, by a sense of adventure as well as by scientific curiosity.

The primary effect of high altitude, of course, is a decrease in the partial pressure of oxygen (see Chapter 13). Since high altitude natives have approximately the same weight-specific oxygen consumption (Torrance et al., 1970), they must have proportionately increased ventilation in order to compensate. At an altitude of 5400 m, where P_{IO_2} is only 93 torr, arterial oxygen tension has declined to less than 50 torr, and the hemoglobin is only about 75% saturated. Venous saturation is little changed, however, and the reduced A–V difference is partially compensated for by an increase in the oxygen-carrying capacity. Long-term exposure to high altitude stimulates production of erythrocytes, so Hct and Hb are elevated as much as 30–40% in high-altitude residents. There is also some rightward shift in the oxygen dissociation curve caused by increased 2,3-DPG, but this does not appear to be of adaptive value, since it further adds to the difficulty of loading oxygen in the lungs (Eaton et al., 1974; review by Bouverot, 1985).

At progressively higher altitudes, arterial saturation begins to "slide" down the dissociation curve until at some point the oxygen demand cannot be met. For practical purposes, the limit for humans is about 7000 m (19,000 ft), but aviators are required to use supplemental oxygen at altitudes above 4400 m (14,500 ft) (Anonymous, 1987). Prolonged exposure to hypoxia more severe than that can cause loss of consciousness. Exposure to milder hypoxia, however, may also cause permanent damage to the eye and brain, and both motor coordination and judgement are significantly impaired with only slight hypoxia.

Hypercapnia

External or environmental hypercapnia is a very unusual situation for air-breathing animals. Perhaps in caves or deep mines it is occasionally encountered, but hypercapnia is most often simply an experimental tool for probing the respiratory responses that would be appropriate to internally elevated CO_2. In Chapter 8 the relation between ventilatory drive and CO_2 was discussed, pointing out the sensitivity of both pulmonary stretch receptors, regulating the depth of breathing, and central nervous system chemosensors to changes in P_{CO_2} and, indirectly, pH. These reflex controls of ventilation provide for short-term adjustment of the arterial P_{CO_2}, effecting pH homeostasis in the process. When P_{CO_2} is forced high, either by breathing high P_{CO_2} ambient air or by various disease states, renal compensation becomes more important in long-term adjustment (Pitts, 1973). Alterations in the degree of HCO_3^- resorption from renal filtrate and/or the rate of H^+ secretion in the kidney tubules affects the plasma $[HCO_3^-]$ and electrolyte status. These mechanisms are not fundamentally different from those seen in other animals (see Chapter 6). Compared with water-breathers the ventilatory P_{CO_2} control is much better, and renal mechanisms are less important in shorter-term adjustments.

COMPARATIVE ASPECTS OF MAMMALIAN RESPIRATION

Diving Mammals

In some interesting ways, the research on diving mammals has parallelled the work on diving birds, with some of the same pitfalls arising. That is, most of the studies were performed in laboratory situations with forcibly submerged animals, and based on their responses to these highly unnatural conditions, a paradigm was constructed for the "diving reflex." The story was roughly that there was an immediate reflex inhibition of breathing (naturally!) mediated by airway and nares receptors, a pronounced bradycardia, redistribution of blood flow to heart, brain and other critical organs, and high tolerance of anaerobic acidosis

(Scholander, 1940; review by Butler, 1982). In freely diving animals, a different pattern was observed, more like that found in spontaneously diving ducks (see Chapter 13). Most dives by free-ranging mammals are of fairly short duration, although in Weddell seals dives may exceed 1 hour (Kooyman et al., 1971). During these dives, heart rate is often increased, rather than decreased, and blood gas and acid-base data indicate that the animals remain fully aerobic throughout most of their dives (Elsner, 1965, 1969; Zapol et al., 1979; Butler, 1982).

In the case of the Weddell seals, whose dives may go below 300 m and last for an average of 15 minutes (Fig. 14.6), it is of interest to wonder why nitrogen narcosis does not occur. Compression of the considerable volume of nitrogen in the lung at the beginning of the dive would enhance entry of the highly soluble gas into the blood and would seem to pose a risk of bubble formation upon emergence. Scholander (1940) proposed that the alveoli and small airways collapse during deeper dives, limiting the gas space mainly to the conducting (i.e., nonexchanging) zone of the lung. Subsequent work has shown that the smaller airways are reinforced by cartilage and remain open during deep dives; the alveoli may then collapse, forcing air into the major airways (Kooyman et al., 1970, 1971; Kooyman, 1988). The ribs of seals and other marine mammals are flexible, allowing almost complete collapse of the chest during the dive.

Size and Respiratory Scaling

Mammals vary in body mass by almost eight orders of magnitude: The smallest is apparently the Etruscan shrew at about 2 grams, and the largest the blue whale at approximately 160 million grams (160 metric tons). Little physiological work has been done on the very largest mammals, but from shrew to domestic cow we

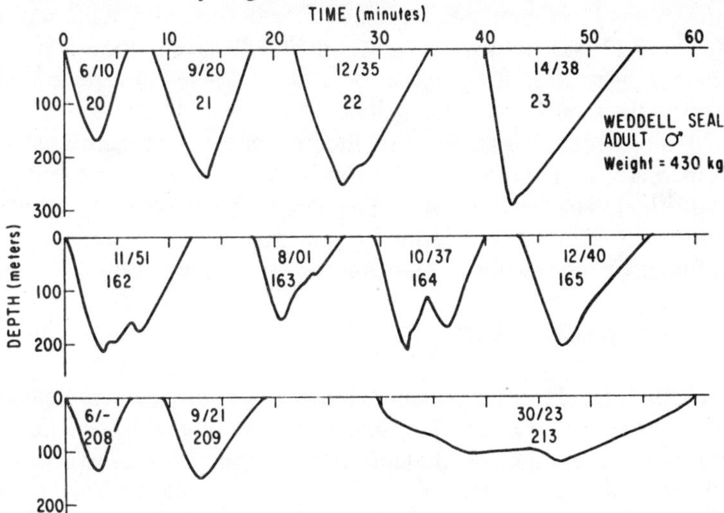

Figure 14.6: Depth and duration of dives in a Weddell seal. The length of the dive is given within the profile of each dive. (From Kooyman et al., 1971, by permission, ©Elsevier North Holland.)

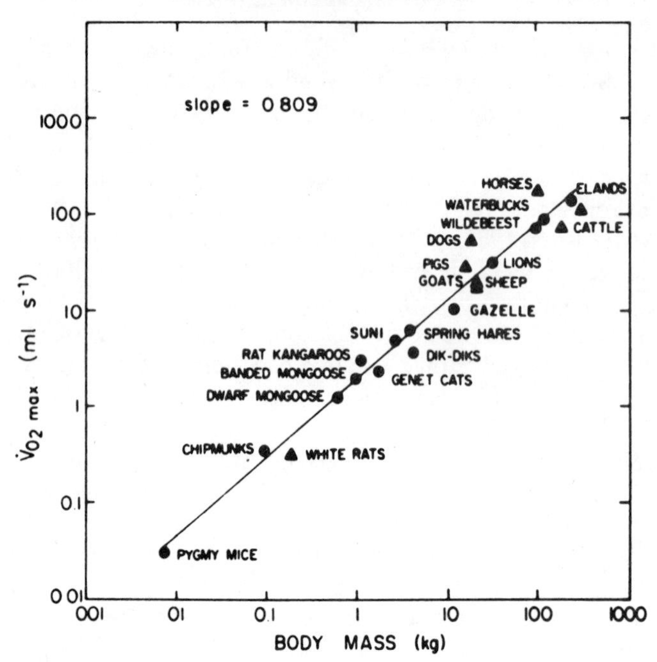

Figure 14.7: Scaling of $Vo_{2,max}$ versus body weight for a variety of wild and domestic mammals. (From Taylor et al., 1981, by permission, ©Elsevier North Holland, Inc.)

have over five orders of magnitude for comparisons. The relation between body mass and either resting or maximal oxygen consumption fits the equation:

$$\dot{V}O_{2,max} = aW^b \qquad \text{(Eq. 14.2)}$$

where the value of b is about 0.8 (Fig. 14.7). The 14 species of mammals studied by Taylor et al. (1981) yielded a mean slope of 0.81. Oddly enough, the diffusing capacity, lung volume, lung capillary volume, and alveolar surface area when compared on the same basis all have slopes of about 1.0 (Gehr et al., 1981)(Fig. 14.8). This slope seems to suggest that the lung capacity of increasingly larger animals increases more than is required for gas exchange, an apparent paradox discussed at length by Weibel (1984). That it is probably not real is shown partly by increasing alveolar-to-arterial oxygen gradients in larger animals, the reasons for which are several and not completely agreed upon.

Sleep, Torpor, and Hibernation

Changes in body temperature in mammals are associated with profound alternations in both metabolism and respiration. Sleep, torpor, and hibernation "... lie on a physiological continuum with sleep, in which energy is conserved in proportion to the reduction in metabolism and T_b [body temperature]: 1 or 2°C in sleep, 5 – 20°C in shallow torpor, and 14 – 35°C in hibernation" (Bickler, 1984). Metabolic energy expenditure is typically reduced by a factor of 30 or more (Morrison

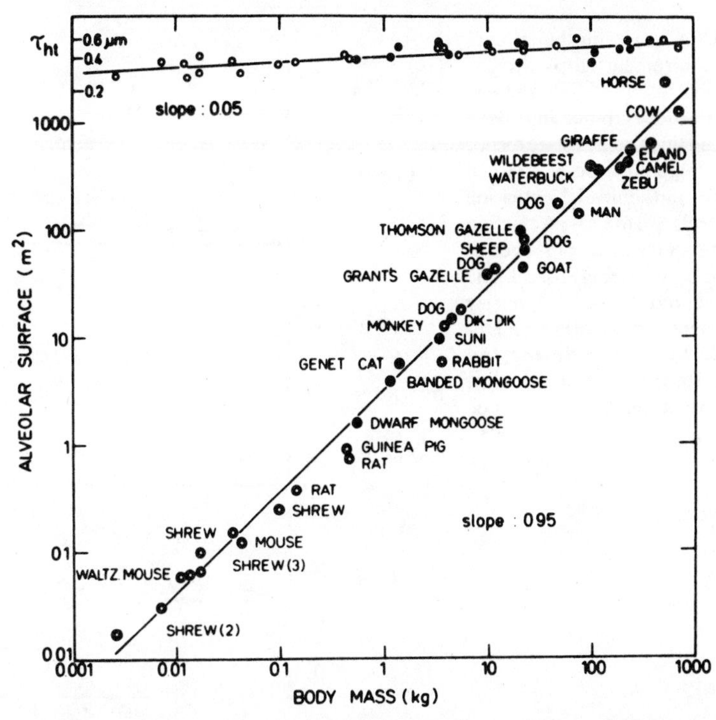

Figure 14.8: Scaling of total alveolar surface (lower) and τ (harmonic mean thickness; upper) versus body weight for wild and domestic mammals. (From Gehr et al., 1981, by permission, ©Elsevier North Holland, Inc.)

& Galster, 1975), leading to an interesting controversy over whether this steep a reduction is simply a Q_{10} effect of temperature or some additional biochemical inhibition is brought into play. For small hibernators the depression appears to be greater than would be expected from temperature alone, whereas in large hibernators and animals undergoing daily torpor temperature alone appears to account adequately for the depression (cf. Geiser, 1988; Malan, 1988).

Unlike clinical hypothermia, from which man is unable to spontaneously rewarm, sleep and torpor represent physiologically regulated states. In addition to the obvious change in body temperature, these states are characterized by considerable CO_2 retention, a relative acidosis, and reduced thermal sensitivity of the hypothalamic centers responsible for temperature regulation (Heller, 1988). Entry into either torpor or hibernation is characterized by a reduction in ventilation relative to CO_2 production (Fig. 14.9)(Bickler, 1984), leading to retention of CO_2, increased P_{CO_2}, and respiratory acidosis. The relatively reduced ventilation is regulated during the steady-state torpor or hibernation bout, then increases markedly with the first stages of arousal (Malan, 1973; Malan et al., 1973)(Fig. 14.9). Malan (1988) has argued that the acidosis may be important in effecting metabolic reduction beyond that expected from temperature, and there is some biochemical evidence in support of this idea (Hand & Somero, 1983b;

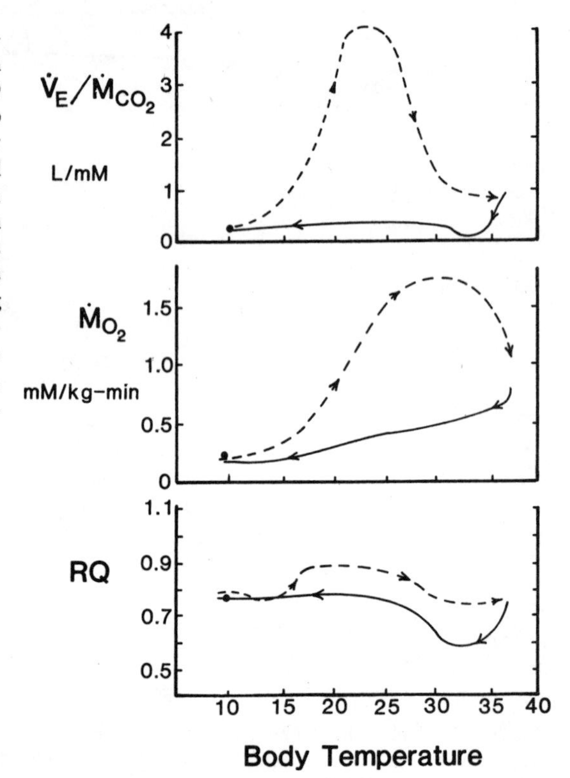

Figure 14.9: Changes in the air convection requirement (top), oxygen consumption rate (middle), and RQ (bottom) during entrance into deep torpor (solid line), steady state torpor (solid circle), and arousal (dashed line) in a ground squirrel. Ventilation is reduced relative to CO_2 production early in the entry to torpor, then is sharply increased during arousal. The result is a net retention of CO_2 during entry and steady–state torpor, and removal of the excess during the arousal. (Redrawn from Bickler, 1984, by permission, ©Amer. Physiol. Soc.)

review by Malan, 1988). The acidosis is relative in the sense that pH actually rises when Tb falls from 37° to 10°C, but much less than would be expected from the usual slope for poikilotherms (see Chapter 6)(Goodrich, 1973); Malan et al. (1985) found an increase of pH of less than 0.05, whereas a $\Delta pH/\Delta T^{\circ}$ of – 0.017 would have increased pH by 0.46.

The decreases in body temperature, the acidosis, and the retention of CO_2 may all be related through the CO_2 sensitivity of the hypothalamic centers. Increased CO_2 may lead to reductions in body temperature and in the threshold for shivering in normothermic mammals, and has been shown to have direct inhibitory effects on the hypothalamus (Wunnenberg & Baltruschat, 1982; Heller, 1988). Cause and effect are somewhat difficult to determine in the sleep/hibernation phenomenon, but the subject provides an interesting opportunity to compare the physiology of poikilotherms, strict homeotherms, and various groups of heterotherms.

SUMMARY

The mammalian respiratory system is apparently an efficient and adaptable design. With only minor modifications it serves athletes, jaguars, whales, camels, and shrews: large, slothful, quick, aquatic, desert and diminutive animals. While it is not capable of such high efficiency as the counter-current gills of fishes, a similar design for an aerial lung would no doubt lead to unacceptably high rates of water loss. The much higher diffusive velocity of oxygen in air compared to water allows the exchanging region of the lung to be designed for diffusive transfer rather than convective transfer, with attendant savings in energy and water. By controlling the extent of ventilation and the degree of capillary perfusion in the lung, a wide range of oxygen transport can be accomplished by the same structure, achieving a good compromise under all conditions. The respiratory system of man reflects our biological position as a generalized, rather than specialized creature. Although outdone by other animals in maximal activity or water conservation or some other factor, our system serves a wide variety of environmental and physiological situations.

Appendix

Computer Programs

The computer programs are supplied on the accompanying 5¼ inch floppy diskette. It is a 360K, double density, two-sided standard IBM PC format disk, and all the programs will run on IBM PC, XT, AT, and most compatibles. The programs also run on the Tandy 2000, a semi-compatible based on the 80186 processor, as well as on newer Tandy machines. The BASIC programs run under the IBM-supplied BASICA and should run all right using GW-BASIC supplied on most "clones." The PASCAL programs were all written and compiled on the IBM using Borland International's ©Turbo Pascal version 3.0 or 4.0. These programs will also run on the Tandy 2000, nearly four times faster than on the IBM PC-XT.

DAVGM.* Davenport Diagram Programs in BASIC, Screen Output

- Reference: Chapter 6

There are two files on the diskette which pertain to this program:

- DAVGM.BAS The BASIC program
- DAVGRAM.DOC A short text file on how to run the program

The first of these files is in the "tokenized" format used by BASIC, so to get a listing, you need to enter BASIC, load the program, and use the LLIST command. The second file is in ASCII text format, and may be listed using the DOS TYPE or PRINT commands, or with any word processor.

1. The program DAVGM.BAS is used to produce Davenport diagrams on the screen. To run it, enter BASIC (usually by typing BASIC↵ at the DOS prompt), then RUN "DAVGM.BAS"↵ at the BASIC "OK" prompt. [Text to be typed on the computer is shown in SMALL TYPE, and the ↵ symbol is used to indicate the Enter or Return key.] The program will prompt you for a number of parameters, some of them the usual ones such as the pH limits over which you want the plot, and the HCO_3^- limits. In order to properly calculate the solubility and pK′ using Heisler's (1984) formulae, it will also ask for ionic strength (entered as a decimal fraction), sodium concentration (entered in molar, e.g., 0.150), and the total protein concentration in gm L^{-1}. The latter two variables have only a small influence on the numbers computed, so if you have not measured them, try 0.15 and 5.

The uncompiled version is fairly slow to generate the P_{CO_2} isopleths. Overall program execution time is controlled by statement 740, which sets the step size for calculation. The default is 0.01 pH units, but changing it to 0.05, for example, will generate points that may be connected to form the isolines. Making the step much smaller, e.g., 0.002 pH units, will generate unbroken lines, but will take a long time. If the set of conditions is one you are going to use to analyze many data sets, generate the plot without putting on a buffer line or plotting points, and save it as a template. You can then recall it and use it without having to wait for the isopleths to be generated each time.

The buffer slope must be entered as a negative number.

2. This program can be speeded up considerably by compiling it with one of the popular BASIC compilers on the market. The procedure for compiling the program varies from one compiler to the next, but if the program is used frequently, it will probably be well worth the effort.

3. The plots generated can be printed from the screen using the Prt-Scrn key combination from the keyboard. Before doing this, however, you need to issue the appropriate GRAPHICS command from the operating system. For IBM graphics printers, this is simply GRAPHICS↵, but for some other printers there are minor variations; the IBM color printer requires GRAPHICS COLOR8, for example. All of this is discussed in the DOS manual.

DAVPLOT.* Davenport diagrams in PASCAL for

Houston Instruments and Hewlett-Packard Plotters.

- Reference: Chapter 6

There are five files on the disk for this program:

- DAVPLOT.EXE The compiled object code
- DAVPLOT.DOC A short text file of instructions
- DAVHI1.BAT Two batch files for setting up the serial port
- DAVHI2.BAT for Houston Instruments plotters.
- DAVHP1.BAT Batch file for Hewlett-Packard plotters.

All of these files except the .EXE file may be listed either from DOS using the TYPE or PRINT commands or by retrieving them with a word processor.

1. You must have either a Houston Instruments or Hewlett-Packard plotter to use this program. It was written for the Houston Instruments DMP-40, the or the Hewlett-Packard 7470 or 7475A, but since all the H.I. plotters share the same language, it should run on any of them.

2. The plotter must be connected either to COM1: or COM2: and the MODE commands must be issued to reroute the line printer output to the serial port and to set up the right parameters for the serial port. This is the purpose of the two short batch files. If you have your H.I. plotter connected to COM1:, type DAVPLOT1↵ to run the program, DAVPLOT2↵ if COM2:. For the H-P, type DAVHP1↵ for serial port #1.

3. Before running the program, turn on the plotter, reset it, and load it with paper.

4. The program operates much like the DAVGM program in BASIC. It is self-prompting, and again, if you have not measured Na concentration or protein, just guess, as it will cause only small errors.

When you get to the point-plotting routine, you can select several symbols for plotting, select the symbol size (H.I. only), and add an optional text label next to each point. You may have problems with points at the far right edge of the diagram – the plotter ignores anything beyond the plot margin.

5. Output for the H-P plotters may also be printed on the H-P LaserJet printers using a program called LaserPlotter (©1986, Insight Development Corp.). You must use the CAPTURE utility to put the plotter output into a file, then the main program to print it on the LaserJet. Version 1.3 of this program was written for the H-P 7470, which did not have the rotation function. The plots will have the top third cut off, so make the Y-axis longer than you need.

6. The program was written and compiled with Turbo Pascal (© Borland Intl.) version 4.0, and source code is available from the author. Several of the procedures can be used for other programs you might want to write, and modifications can be made if the program is recompiled with any changes.

GILLEX.* Simulation Model of Gas Exchange in the Trout Gill

- Reference: Chapter 7

There are four files on the diskette relating to this model:

- GILLEX.EXE — The executable PASCAL object code
- GILLEX.DOC — A short text file of instructions
- GILLEX87.COM — Object code compiled for an 8x087 math coprocessor
- GILLEXGR.EXE — Object code for a graphics version.

The .DOC file may be listed in the usual ways (see above). No special equipment is required to run the first program; just type GILLEX↵ at the DOS prompt to run the standard version, and GILLEX87↵ for the version compiled to support the 8087 math coprocessor.

Be advised that this program takes a lot of time. It is intensively "compute-bound" because of the small time step and iterative calculation routines. The presence of the math coprocessor, higher clock speeds than the standard PC-XT, or other processors will make an enormous improvement in the execution time. The time step is fixed at 0.001 second. The screen output has an inner loop counter, so 100 times through this loop simulates 0.1 second of "real" time. The model generally takes a few "real" seconds to stabilize, so count on 10 to 50 times the amount of time taken for 100 times through the inner loop. The table below shows the approximate times for the two program versions to complete 100 iterations on various equipment:

	Co-Processor	GILLEX	GILLEX87
IBM PC-XT, 4.77 MHz	8087	16'21"	1'55"
Tandy 2000	None	6'10"	
IBM PC-AT, 8 MHz	80287	4'55"	1'04"
AT Clone, 12 MHz	80287		0'59"

The times required for 20 seconds of "real" simulation range from more than 5 hours on the standard XT without math coprocessor to less than 20 minutes for the 12 MHz AT clone with 80287 installed.

In all of the program versions, the prompts ask for various parameters, but the default values for trout may be accepted by simply pressing ENTER at each prompt. The derivation of the value for each parameter is explained in Chapter 7, and most of the data may be found in Chapter 10. Some variables such as the time step size and the diffusion coefficient are fixed in the program, but only minor modifications to the source code (see below) are required to either make them variables, or change the values used. After any changes, the program must be recompiled.

The output screen shows the values for each compartment (down the screen) at the end of each 100 iterations (0.1 second total). The columns, in order, are compartment number, oxygen content of the blood in ml L^{-1}, partial pressure of oxygen in the blood in torr, partial pressure of the water in torr, and oxygen content of the water in ml L^{-1}. In order to decide whether the program has stabilized, watch the output at the end of each 100 iterations, and see if there is any significant change from the previously displayed values. Alternately, press any key during the calculation and the program will stop when 100 is reached; at this point you may print the screen, change parameters and continue, or stop.

The Ctrl-Break option is disabled during the run in order to speed up program execution. You can stop the run only by pressing a key and waiting for the inner loop count to reach 100 or by removing any floppy disks from the drive and activating the computer reboot (either with a switch or by pressing Ctrl-Alt-Del together).

Source code for these programs is also available on the source code disk from the author at the address at the end of this section.

HILLPLOT.BAS Hill Plots on Screen

•Reference: Chapter 5

The program HILLPLOT.BAS asks you to input data pairs of oxygen saturation and partial pressure. The percent saturation should be entered as percentages, i.e., 50 for half saturation, not 0.5. Partial pressures are entered in torr. When all numbers have been entered, enter a final number pair 999,999. The program then plots the pairs of data you entered, computes the regression line through the points, plots the line, prints the slope and intercept at the upper right corner of the screen, and stops. You can print the screen with the Prt-Scrn key command from the keyboard.

As with the DAVGM program, you must first issue the appropriate GRAPHICS command from the DOS prompt in order for your printer to print the graphics correctly (see above).

To obtain a complete listing of this program (or the DAVGM.BAS program), go to BASIC, load the program, and do LLIST↵.

TEMPTABL.COM Temperature-dependent variables table generator.

•Reference: Chapter 1

This program takes the place of the usual appendix tables found in many books. With the program, you can calculate the value of solubility, pK, etc., for oxygen, carbon dioxide, and ammonia as well as the vapor pressure of water under a variety of salinity and temperature conditions. You have a wide choice of units, and considerable flexibility in the output. The program is self-prompting.

Single values or complete tables can be generated either on screen or on your printer. The screen tables can also be printed using the Prt-Scrn option.

To run the program, simply type TEMPTABL↵ from the DOS prompt.

UNITS.COM Unit conversion program.

•Reference: Chapter 1

Rather than providing bulky tables or formulas, this program is provided for converting a wide variety of units into other units. There are sections for pressure units, volume, length, mass, etc. All parts of the program run from a master menu, and either single values or tables can be generated on screen or on your line printer.

Some of the menus are longer than the screen. Just keep pressing the down arrow and the menu will scroll up, displaying more choices. From the top line the up arrow will bring the bottom of the menu into view.

To run the program, simply type UNITS↵ from the DOS prompt. Source code is provided in case you want to add your own esoteric units to this table.

ALLREFS.TXT

All of the literature citations from the book are here in the form of an alphabetized, ASCII text file. These can be used for any sort of database reference, or literature citation program. A single-spaced printout will take over 30 pages. Beware that all codes such as underlining, bold, italic, super- or subscripts will be missing, and some foreign characters such as umlauted letters may be changed or missing.

Source Code:

The source code on 5¼ inch floppy disk along with a complete listing for all of the PASCAL programs is available from the author for $15. Please send requests to: J. N. Cameron, P. O. Box 742, Port Aransas, TX 78373.

References

Abdalla, M. A., and A. S. King. 1975. The functional anatomy of the pulmonary circulation of the domestic fowl. Respir. Physiol. 23:267-290.

Ackerman, R. A., and F. N. White. 1979. Cyclic carbon dioxide exchange in the turtle *Pseudemys scripta*. Physiol. Zool. 52:378-389.

Agostoni, E., and J. Mead. 1964. Statics of the respiratory system. In W. O. Fenn and H. Rahn, eds., *Handbook of Physiology*, Section 3, Vol. I. Bethesda: Amer. Physiol. Soc. pp. 387-409.

Albers, C. 1970. Acid-base balance. In W. S. Hoar and D. J. Randall, eds. *Fish Physiology*, Vol. IV. New York: Academic Press. pp. 173-208.

Aldridge, J. B., and J. N. Cameron. 1979. CO_2 exchange in the blue crab, *Callinectes sapidus* (Rathbun). J. exp. Zool. 207:321-328.

Aldridge, J. B., and J. N. Cameron. 1982. Gill morphometry in the blue crab, *Callinectes sapidus* Rathbun (Decapoda Brachyura). Crustaceana 43:297-305.

Alexander, R. McN. 1967. *Functional Design in Fishes*. London: Hutchinson University Library. 160 pp.

Altland, P. D., and K. C. Brace. 1962. Red cell life span in the turtle and toad. Am. J. Physiol. 203:1188-1190.

Altman, P. L., and D. S. Dittmer. 1974. *Biology data book*, 2nd edition, Vol. III. Bethesda: FASEB. 2123pp.

Andersen, H. T. 1963. Factors determining the circulatory adjustments to diving. I. Water immersion. Acta Physiol. Scand. 58:173-185.

Andrews, E. B., and P. M. Taylor. 1988. Fine structure, mechanism of heart function and haemodynamics in the prosobranch gastropod mollusc *Littorina littorea* (L.). J. Comp. Physiol. B 158:247-262.

Anonymous. 1987. *Airman's Information Manual.* U.S. Fed. Aviation Admin. 192pp.

Ar, A., and H. Rahn. 1978. Interdependence of gas conductance, incubation length and weight of the avian egg. In J. Piiper, ed., *Respiratory Function in Birds, Adult and Embryonic*. Berlin: Springer-Verlag. pp. 227-238.

Ar, A., and H. Rahn. 1980. Water in the avian egg: overall budget of incubation. Am. Zool. 20:373-384.

Arp, A. J., J. J. Childress and C. R. Fisher, Jr. 1984. Metabolic and blood gas transport characteristics of the hydrothermal vent bivalve *Calyptogena magnifica*. Physiol. Zool. 57:648-662.

Aulie, A. 1971. Coordination between the activity of the heart and the flight muscles in small birds. Comp. Biochem. Physiol. 38A:91-98.

Aulie, A. 1972. Co-ordination between the activity of the heart and pectoral muscles during flight in the pigeon. Comp. Biochem. Physiol. 41A:43-48.

Austin, J. H., F. W. Sunderman and J. B. Camack. 1927. The electrolyte composition and the pH of serum of a poikilothermous animal at different temperatures. J. Biol. Chem. 72:677-685.

Ballantijn, C. M. 1969. Functional anatomy and movement coordination of the respiratory pump of the carp (*Cyprinus carpio* L.). J. Exp. Biol. 50:547-567.

Ballantijn, C. M. 1982. Neural control of respiration in fishes and mammals. In Proc. 3rd Congr. Eur. Soc. Comp. Physiol. Biochem. Oxford: Pergamon. pp. 127-140.

Ballantijn, C. M., and G. M. Alink. 1977. Identification of respiratory motor neurons in the carp and determination of their firing characteristics and interconnections. Brain Res. 136:261-276.

Ballantijn, C. M., and J. L. Roberts. 1976. Neural control and proprioceptive load matching in reflex respiratory movements of fishes. Fed. Proc. 35:1983-1991.

Bamford, O. S. 1974. Oxygen reception in the rainbow trout (*Salmo gairdneri*). Comp. Biochem. Physiol. 48A:69-76.

Bamford, O. S., and D. R. Jones. 1974. On the initiation of apnoea and some cardiovascular responses to submergence in ducks. Respir. Physiol. 22:199-216.

Banzett, R. B., J. P. Butler, C. S. Nations, G. M. Barnas, J. L. Lehr, and J. H. Jones. 1987. Inspiratory aerodynamic valving in goose lungs depends on gas density and velocity. Respir. Physiol. 70:287-300.

Barnes, R. D. 1980. *Invertebrate Zoology*, 4th edition. Philadelphia:Saunders College 1089pp.

Barrett, D. J., and E. W. Taylor. 1984. Changes in heart rate during progressive hypoxia in the dogfish *Scyliorhinus canicula* L.: evidence for a venous oxygen receptor. Comp. Biochem. Physiol. 78A:697-703.

Barrett, D. J., and E. W. Taylor. 1985. The location of cardiac vagal preganglionic neurones in the brain stem of the dogfish *Scyliorhinus canicula*. J. Exp. Biol. 117:449-458.

Bartels, H. K., P. Dejours , R. H. Kellogg, and J. Mead. 1973. Glossary on respiration and gas exchange. J. Appl. Physiol. 34:549-558.

Batterton, C. V., and J. N. Cameron. 1978. Characteristics of resting ventilation and response to hypoxia, hypercapnia and emersion in the blue crab *Callinectes sapidus* (Rathbun). J. exp. Zool. 203:403-418.

Bauer, C. 1974. On the respiratory function of haemoglobin. Rev. Physiol. Biochem. Pharmacol. 70:1-31.

Beamish, F. W. H. 1978. Swimming capacity. In W. S. Hoar and D. J. Randall, eds., *Fish Physiology*, Vol. VII. New York: Academic Press. pp. 101-187.

Beamish, F. W. H., and P. S. Mookherji. 1964. Respiration of fishes with special emphasis on standard oxygen consumption. I. Influence of weight and temperature on respiration of goldfish, *Carassium auratus* L. Can. J. Zool. 42:161-175.

Belkin, D. A. 1968. Aquatic respiration and underwater survival of two freshwater turtle species. Respir. Physiol. 4:1-14.

Benesch, R., and R. E. Benesch. 1967. The effect of organic phosphates from the human erythrocyte on the allosteric properties of hemoglobin. Biochem. Biophys. Res. Commun. 26:162-167.

Berlin, N. I., T. A. Waldman and S. M. Weissman. 1959. Life span of red blood cells. Physiol. Rev. 39:577-616.

Berne, R. M. (Ed.) 1979. *Handbook of Physiology. The Cardiovascular System. The Heart.* Bethesda: Amer. Physiological Soc., Sect. 2, Vol. I.

Berne, R. M., and R. Rubio. 1979. Coronary circulation. In *Handbook of Physiology. The Cardiovascular System. Sect. 2, Vol. I: The Heart*. Bethesda: Amer. Physiol. Soc. pp. 873-952.

Bernstein, M. H., S. P. Thomas, and K. Schmidt-Nielsen. 1973. Power input during flight in the fish crow, *Corvus ossifragus*. J. Exp. Biol. 58:401-410.

Bert, P. 1867. Sur le physiologie de la seiche (*Sepia officinalis* L.). C. R. Acad. Sci. Paris 65:300-303.

Bettex-Galland, M., and G. M. Hughes. 1973. Contractile filamentous material in the pillar cells of fish gills. J. Cell. Sci. 13:359-370.

Bettice, J. A. 1984. Skeletal carbon dioxide stores during metabolic acidosis. Am. J. Physiol. 247:F326-F330.

Bettice, J. A., and J. L. Gamble, Jr. 1975. Skeletal buffering of acute metabolic acidosis. Am. J. Physiol. 229:1619-1624.

Beutler, E. 1968. *Hereditary Disorders of Erythrocyte Metabolism*. New York: Grune and Stratton.

Bevan, J. A., R. D. Bevan, and S. P. Duckles. 1980. Adrenergic regulation of vascular smooth muscle. In D. F. Bohr, A. P. Somlyo, and H. V. Sparks, Jr., eds., *Handbook of Physiology, Sect. 2: The Cardiovascular System, Vol. II: Vascular Smooth Muscle*. Bethesda: Amer. Physiol. Soc. pp. 515-566.

Biber, T. U. L., and T. L. Mullen. 1976. Saturation kinetics of sodium efflux accross isolated frog skin. Am. J. Physiol. 231:995-1001.

Bickler, P. E. 1984. CO_2 balance of a heterothermic rodent: comparison of sleep, torpor and awake states. Am. J. Physiol. 246:R49-R55.

Black, E. C., G. T. Manning, and K. Hayashi. 1966. Changes in levels of hemoglobin oxygen, carbon dioxide, pyruvate, and lactate in venous blood of rainbow trout (*Salmo gairdneri*) during and following severe muscular activity. J. Fish Res. Bd. Canada. 23:783-795.

Blix, A. S., A. Rettedal, and K. A. Stokkan. 1976. On the elicitation of the diving responses in ducks. Acta Physiol. Scand. 98:478-483.

Boggs, D. F., and D. L. Kilgore, Jr. 1980. Ventilatory response to CO_2 in a burrow-dwelling bird. Fed. Proc. 39:1061(Abstr.).

Bohr, C., K. A. Hasselbalch, and A. Krogh. 1904. Ueber einen in biologischer Beziehung wichtigen Einfluss, den die Kohlensäurespannung des Blutes aur dessen Suaerstoffbindung übt. Skand. Arch. Physiol. 16:402-412.

Boland, E. J., and K. R. Olson. 1979. Vascular organization of the catfish gill filament. Cell. Tissue Res. 198:487-500.

Bonaventura, J., and C. Bonaventura. 1980. Hemocyanins: Relationships in their structure, function and assembly. Amer. Zool. 20:7-17.

Bone, Q. 1966. On the function of the two types of myotomal muscle fibre in elasmobranch fish. J. Mar. Biol. Assoc. U.K. 46:321-349.

Booth, C. E., B. R. McMahon, P. L. DeFur, and P. R. H. Wilkes. 1984. Acid-base regulation during exercise and recovery in the blue crab, *Callinectes sapidus*. Respir. Physiol. 58:359-376.

Booth, C. E., B. R. McMahon, and A. W. Pinder. 1982. Oxygen uptake and the potentiating effects of increased hemolymph lactate on oxygen transport during exercise in the blue crab, *Callinectes sapidus*. J. Comp. Physiol. B 148:111-121.

Booth, J. H. 1979. The effects of oxygen supply, epinephrine, and acetylcholine on the distribution of blood flow in trout gills. J. Exp. Biol. 83:31-39.

Borradaille, L. A., F. A. Potts, L. E. S. Eastham and J. T. Saunders. 1963. *The Invertebrata.* 4th ed., revised by G. A. Kerkut. Cambridge Univ. Press. 820pp.

Botelho, L. H., S. H. Friend, J. B. Matthew, L. D. Lehman, G. I. Hanania, and F. R. N. Gurd. 1978. Proton nuclear magnetic resonance study of histidine ionizations in myoglobins of various species: Comparison of observed and computed pK values. Biochemistry 17(24):5197-5205.

Bourne, G. B., and J. R. Redmond. 1977. Haemodynamics in the pink abalone, *Haliotis corrugata*. I. Pressure relations and pressure gradients in intact animals. J. exp. Zool. 200:9-16.

Boutilier, R. G., M. L. Glass and N. Heisler. 1986. The relative distribution of pulmocutaneous blood flow in *Rana catesbiana*: effects of pulmonary or cutaneous hypoxia. J. Exp. Biol. 126:33-40.

Boutilier, R. G., and N. Heisler. 1988. Acid-base regulation and blood gases in the anuran amphibian, *Bufo marinus*, during environmental hypercapnia. J. Exp. Biol. 134:79-98.

Boutilier, R. G., T. A. Heming and G. K. Iwama. 1984. Physico-chemical parameters for use in fish respiratory physiology. In W. S. Hoar and D. J. Randall, eds., *Fish Physiology,* Vol. XA. New York: Academic Press. pp. 403-430.

Boutilier, R. G., and D. P. Toews. 1977. The effect of progressive hypoxia on respiration in the toad *Bufo marinus*. J. Exp. Biol. 68:99-107.

Bouverot, P. 1985. Adaptation to altitude-hypoxia in vertebrates. NewYork/Berlin:Springer-Verlag. 176pp.

Bouverot, P., D. Douguet, and P. Sbert. 1979. Role of the arterial chemoreceptors in ventilatory and circulatory adjustments to hypoxia in awake Pekin ducks. J. Comp. Physiol. 133:177-186.

Brace, K., and P. D. Altland. 1956. Life span of the duck and chicken erythrocyte as determined with C^{14}. Proc. Soc. Exp. Biol. Med. 92:615.

Brady, A.J., and C.H. Dubkin. 1964. Coronary circulation of the turtle ventricle. Comp. Biochem. Physiol. 13:119-128.

Brett, J. R. 1965. The relation of size to rate of oxygen consumption and sustained swimming speed of sockeye salmon (*Onchorhynchus nerka*). J. Fish. Res. Bd. Canada 22:1491-1501.

Brett, J. R., and T. D. D. Groves. 1979. Physiological energetics. In W. S. Hoar and D. J. Randall, eds., *Fish Physiology, Vol. VIII, Bioenergetics and Growth*. New York: Academic Press. pp. 279-352.

Bretz, W. L., and K. Schmidt-Nielsen. 1971. Bird respiration: flow patterns in the duck lung. J. Exp. Biol. 54:103-118.

Bridges, C. R., and A. R. Brand. 1979. The effect of hypoxia on oxygen consumption and blood lactate levels of some marine crustacea. Comp. Biochem. Physiol. 65A:399-409.

Bridges, C. R., S. Morris, and M. K. Grieshaber. 1984. Modulation of haemocyanin oxygen affinity in the intertidal prawn *Palaemon elegans*. Respir. Physiol. 57:189-200.

Brown, W.I., and J.M. Schick. 1979. Bimodal gas exchange and the regulation of oxygen uptake in holothurians. Biol. Bull. 156:272-288.

Burger, R. E., M. Meyer, W. Graf and P. Scheid. 1979. Gas exchange in the parabronchial lung of birds: experiments in unidirectionally ventilated ducks. Respir. Physiol. 36:19-37.

Burggren, W. W. 1975. A quantitative analysis of ventilation tachycardia and its control in two Chelonians, *Pseudemys scripta* and *Testuda graeca*. J. Exp. Biol. 63:367-380.

Burggren, W. W. 1977. The pulmonary circulation of the Chelonian reptile: morphology, haemodynamics and pharmacology. J. Comp. Physiol. B 116:303-323.

Burggren, W. W. 1979. Bimodal gas exchange during variation in environmental oxygen and carbon dioxide in the air breathing fish Trichogaster trichopterus. J. Exp. Biol. 82:197-214.

Burggren, W. W. 1982a. Pulmonary blood plasma filtration in reptiles: a "wet" vertebrate lung? Science 215:77-78.

Burggren, W. W. 1982b. "Air gulping" improves blood oxygen transport during aquatic hypoxia in the goldfish *Carassius auratus*. Physiol. Zool. 55:327-334.

Burggren, W. W. 1985. Hemodynamics and regulation of central cardiovascular shunts in reptiles. In K. Johansen and W. W. Burggren, eds., *Cardiovascular Shunts*. Copenhagen: Munksgaard. pp. 121-142.

Burggren, W. W. 1987. Form and function in reptilian circulations. Amer. Zool. 27:5-20.

Burggren, W. W. 1988a. Role of the central circulation in regulation of cutaneous gas exchange. Amer. Zool. 28:985-998.

Burggren, W. W. 1988b. Cardiovascular responses to diving and their relation to lung and blood oxygen stores in vertebrates. Can. J. Zool. 66:20-28.

Burggren, W. W., and J. N. Cameron. 1979. Anaerobic metabolism, gas exchange, and acid-base balance during hypoxic exposure in the channel catfish, *Ictalurus punctatus*. J. exp. Zool. 213:405-416.

Burggren, W. W., and K. Johansen. 1982. Ventricular haemodynamics in the monitor lizard *Varanus exanthematicus*: pulmonary and systematic pressure separation. J. Exp. Biol. 96:343-354.

Burggren, W. W., B. R. McMahon, and J. W. Costerton. 1975. Branchial water and blood flow patterns and the structure of the gill of the crayfish *Procambarus clarkii*. Can. J. Zool. 52:1511-1522.

Burggren, W. W., and R. Moalli. 1984. 'Active' regulation of cutaneous gas exchange by capillary recruitment in amphibians: experimental evidence and a revised model for skin respiration. Respir. Physiol. 55:379-392.

Burggren, W. W., and D. J. Randall. 1978. Oxygen uptake and transport during hypoxic exposure in the sturgeon (Acipenser transmontanus). Respir. Physiol. 34:171-183.

Burggren, W.W., and G. Shelton. 1979. Gas exchange and transport during intermittent breathing in chelonian turtles. J. Exp. Biol. 82:75-92.

Burleson, M. L., and N. J. Smatresk. 1988. Evidence for internal and external oxygen-sensitive chemoreceptors in channel catfish, Ictalurus punctatus. Physiol. Zool. (In Press)

Burnett, L. E., P. L. DeFur and D. D. Jorgensen. 1981. Application of the thermal dilution tech-que for measuring cardiac output and assessing stroke volume in crabs. J. exp. Zool. 218:165-173.

Burnett, L. E., Jr., and R. L. Infantino. 1984. The CO_2-specific sensitivity of hemocyanin oxygen affinity in the decapod crustacea. J. exp. Zool. 232:59-65.

Burnett, L. E., and K. Johansen. 1981. The role of ventilation in hemolymph acid-base changes in the shore crab *Carcinus maenas* during hypoxia. J. Comp. Physiol. B 141:489-494.

Burnett, L. E., and B. R. McMahon. 1987. Gas exchange, hemolymph acid-base status, and the role of branchial water stores during air exposure in three littoral crab species. Physiol. Zool. 60:27-36.

Burnett, L. E., P. B. J. Woodson, M. G. Rietow, and V. C. Vilicich. 1981. Crab gill intra-epithelial carbonic anhydrase plays a major role in haemolymph CO_2 and chloride ion regulation. J. Exp. Biol. 92:243-254.

Burnstock, G. 1980. Cholinergic and purinergic regulation of blood vessels. In D. F. Bohr, A. P. Somlyo, and H. V. Sparks, Jr., eds., *Handbook of Physiology, Sect. 2: The Cardiovascular System, Vol. II: Vascular Smooth Muscle*. Bethesda: Amer. Physiol. Soc. pp. 567-612.

Burton, A. C. 1954. Relation of structure to function of the tissues of the wall of blood vessels. Physiol. Rev. 34:619-642.

Burton, A. C. 1965. *Physiology and biophysics of the circulation.* Chicago: Year Book Medical Publishers, Inc. 217pp.

Bushinsky, D. A., and R. J. Lechleider. 1987. Mechanism of proton-induced bone calcium release: calcium carbonate dissolution. Am. J. Physiol. 253:F998-F1005.

Butler, P. J. 1982. Respiratory cardiovascular control during diving in birds and mammals. J. Exp. Biol. 100:195-221.

Butler, P. J., and D. R. Jones. 1968. Onset of and recovery from diving bradycardia in ducks. J. Physiol. (London) 196:255-272.

Butler, P. J., and D. R. Jones. 1971. The effect of variations in heart rate and regional distribution of blood flow on the normal pressor response to diving in ducks. J. Physiol. (London) 214:457-479.

Butler, P. J., and E. W. Taylor. 1973. The effect of hypercapnic hypoxia, accompanied by different levels of lung ventilation, on heart rate in the duck. Respir. Physiol. 19:176-187.

Butler, P. J., E. W. Taylor, and S. Short. 1977. The effect of sectioning cranial nerves V, VII, IX and X on the cardiac response to the dogfish *Scyliorhinus canicula* to environmental hypoxia. J. Exp. Biol. 69:233-245.

Butler, P. J., N. H. West, and D. R. Jones. 1977. Respiratory and cardiovascular responses of the pigeon to sustained, level flight in a wind-tunnel. J. Exp. Biol. 71:7-26.

Butler, P. J., and A. J. Woakes. 1979. Changes in heart rate and respiratory frequency during natural behavious of ducks, with particular reference to diving. J. Exp. Biol. 79:283-300.

Cameron, J. N. 1971a. Oxygen dissociation characteristics of the blood of the rainbow trout, *Salmo gairdneri.* Comp. Biochem. Physiol. 38A:699-704.

Cameron, J. N. 1971b. Methemoglobin in erythrocytes of rainbow trout. Comp. Biochem. Physiol. 40A:743-749.

Cameron, J. N. 1975a. Morphometric and flow indicator studies of the teleost heart. Can. J. Zool. 53:691-698.

Cameron, J. N. 1975b. Aerial gas exchange in the terrestrial Brachyura *Gecarcinus lateralis* and *Cardisoma guanhumi.* Comp. Biochem. Physiol. 50A:129-134.

Cameron, J. N. 1975c. Blood flow distribution as indicated by tracer microspheres in resting and hypoxic Arctic grayling (*Thymallus arcticus*). Comp. Biochem. Physiol. 52A:441-444.

Cameron, J. N. 1976. Branchial ion uptake in Arctic grayling: resting values and effects of acid-base disturbance. J. Exp. Biol. 64:711-725.

Cameron, J. N. 1978a. Chloride shift in fish blood. J. exp. Zool. 206:289-295.

Cameron, J. N. 1978b. Effects of hypercapnia on blood acid-base status, NaCl fluxes, and trans-gill potential in freshwater blue crabs, *Callinectes sapidus.* J. Comp. Physiol. B 123:137-141.

Cameron, J. N. 1978c. Regulation of blood pH in teleost fish. Respir. Physiol. 33:129-144.

Cameron, J. N. 1979. Excretion of CO_2 in water-breathing animals. Mar. Biol. Lett. 1:3-13.

Cameron, J. N. 1980. Body fluid pools, kidney function, and acid-base regulation in the freshwater catfish *Ictalurus punctatus*. J. Exp. Biol. 86:171-185.

Cameron, J. N. 1981a. Brief introduction to the land crabs of the Palau Islands: stages in the transition to air breathing. J. exp. Zool. 218:1-5.

Cameron, J. N. 1981b. Acid-base responses to changes in CO2 in two Pacific crabs: the coconut crab, *Birgus latro*, and a mangrove crab, *Cardisoma carnifex*. J. exp. Zool. 218:65-73.

Cameron, J. N. 1984. The acid-base status of fish at different temperatures. Am. J. Physiol. 246:R452-R459.

Cameron, J.N. 1985a. Compensation of hypercapnic acidosis in the aquatic blue crab, *Callinectes sapidus*: The predominance of external seawater over carapace carbonate as the proton sink. J. Exp. Biol. 114:197-206.

Cameron, J. N. 1985b. The bone compartment in a teleost fish, *Ictalurus punctatus*: size, composition, and acid-base response to hypercapnia. J. Exp. Biol. 117:307-318.

Cameron, J. N. 1985c. Post-moult calcification in the blue crab (*Callinectes sapidus*): relationships between apparent net H^+ excretion, calcium and bicarbonate. J. Exp. Biol. 119:275-286.

Cameron, J. N. 1986. Acid-base equilibria in invertebrates. In N. Heisler, ed., *Acid-base Regulation in Animals*. Amsterdam: Elsevier. pp. 357-394.

Cameron, J. N., and C. V. Batterton. 1978a. Temperature and blood acid-base status in the blue crab, *Callinectes sapidus*. Respir. Physiol. 35:101-110.

Cameron, J. N., and C. V. Batterton. 1978b. Antennal gland function in the fresh water blue crab, *Callinectes sapidus*: Water, electrolyte, acid-base and ammonia excretion. J. Comp. Physiol. B 123:143-148.

Cameron, J. N., and J. J. Cech, Jr. 1970. Notes on the energy cost of gill ventilation in teleosts. Comp. Biochem. Physiol. 34:447-455.

Cameron, J. N., and J. C. Davis. 1970. Gas exchange in rainbow trout with varying blood oxygen capacity. J. Fish. Res. Bd. Canada 27:1069-1085.

Cameron, J. N., and G. K. Iwama. 1987. Compensation of progressive hypercapnia in channel catfish and blue crabs: upper limits, and S.I.D. *vs*. bicarbonate analysis. J. Exp. Biol. 133:183-197.

Cameron, J. N., and Kormanik, G. A. 1982a. Intracellular and extracellular acid-base status as a function of temperature in the freshwater channel catfish, *Ictalurus punctatus*. J. Exp. Biol. 99:127-142.

Cameron, J. N., and G. A. Kormanik. 1982b. The acid-base responses of gills and kidneys to infused acid and base loads in the Channel Catfish, *Ictalurus punctatus*. J. Exp. Biol. 99:143-160.

Cameron, J. N., and T. A. Mecklenburg. 1973. Aerial gas exchange in the coconut crab, *Birgus latro*, with some notes on *Gecarcoidea lalandii*. Respir. Physiol. 19:245-261.

Cameron, J. N., and J. A. Polhemus. 1974. Theory of CO_2 exchange in trout gills. J. Exp. Biol. 60:183-194.

Cameron, J. N., and D. J. Randall. 1972. The effect of increased ambient CO_2 on arterial CO_2 tension, CO_2 content and pH in rainbow trout. J. Exp. Biol. 57:673-680.

Cameron, J. N., and Wood, C. M. 1985. Apparent net H^+ excretion and CO_2 dynamics accompanying carapace mineralization in the blue crab, (*Callinectes sapidus*) following moulting. J. Exp. Biol. 114:181-196.

Campbell, G. 1970. Autonomic nervous systems. In W .S. Hoar and D. J. Randall, eds., *Fish Physiology*, Vol. IV. New York: Academic Press. pp. 109-132.

Carey, F. G., and J. M. Teal. 1969. Regulation of body temperatureby the bluefin tuna. Comp. Biochem. Physiol. 28:205-213.

Carey, F. G., J. M. Teal, J. W. Kanwisher, K. D. Lawson, and J. S. Beckett. 1971. Warm-bodied fish. Amer. Zool. 11:137-145.

Carr, A. 1952. *Handbook of Turtles. The Turtles of the United States, Canada and Baja California.* Ithaca: Cornell Univ. Press. 542pp.

Carter, G. S. 1957. Air breathing. In M. E. Brown, ed., *The Physiology of Fishes,* Vol. I. New York: Academic Press. pp. 65-79.

Cech, J. J., Jr., D. W. Bridges, D. M. Rowell and P. J. Balzer. 1976. Cardiovascular responses of winter flounder *Pseudopleuronectes americanus* (Walbaum) to acute temperature increase. Can. Zool. 54:1383-1388.

Cech, J. J., Jr., R. Laurs and J. B. Graham. 1984. Temperature-induced changes in blood gas equilibria in the albacore, *Thunnus alalunga*, a warm-bodied tuna. J. Exp. Biol. 109:21-34.

Chanutin, A., and R. R. Curnish. 1967. Effect of organic and inorganic phosphates on the oxygen equilibrium of human erythrocytes. Arch. Biochem. Biophys. 121:96-102.

Claiborne, J. B., and N. Heisler. 1984. Acid-base regulation and ion transfers in the carp (*Cyprinus carpio*) during and after exposure to environmental hypercapnia. J. Exp. Biol. 108:25-44.

Claiborne, J. B., and N. Heisler. 1986. Acid-base regulation and ion transfers in the carp (*Cyprinus carpio*): pH compensation during graded long- and short-term environmental hypercapnia, and the effect of bicarbonate infusion. J. Exp. Biol. 126:41-62.

Clements, J. A., R. F. Hufstead, R. P. Johnson, and I. Bribetz. 1961. Pulmonary surface tension and alveolar stability. J. Appl. Physiol. 16:444-450.

Cooper, K. E., and W. L. Veale. 1979. Effects of temperature on breathing. In A. P. Fishman *et al.*, eds., *Handbook of Physiology, Sect. 3, Vol. II, Control of Breathing, Part 2*. Bethesda: Amer. Physiol. Soc. pp. 691-702.

Copeland, D. E., and A. T. Fitzjarrell. 1968. The salt absorbing cells in the gills of the blue crab (*Callinectes sapidus* Rathbun) with notes on modified mitochondria. Z. Zellforsch. 9:1-22.

Corliss, J. B., J. Dymond, L. I. Gordon, J. M. Edmond, R. P. von Herzen, R. D. Ballard, K. Green, D. Williams, A. Bainbridge, K. Crane, and T. H. van Andel. 1979. Submarine thermal springs on the Galapagos rift. Science 203:1073-1083.

Crabtree, R. L., and C. H. Page. 1974. Oxygen-sensitive elements in the book gills of *Limulus polyphemus*. J. Exp. Biol. 60:631-640.

Crandall, E. D., and A. Bidani. 1981. Effects of red blood cell HCO_3^-/Cl^- exchange kinetics on lung CO_2 transfer: theory. J. Appl. Physiol. 50:265-271.

Crawford, E. C., Jr., R. N. Gatz, H. Magnussen, S. F. Perry and J. Piiper. 1976. Lung volumes, pulmonary blood flow and carbon monoxide diffusing capacity of turtles. J. Comp. Physiol. B 107:169-178.

Cross, E. C., B. S. Packer, J. M. Linta, H. V. Murdaugh and E. D. Robin. 1969. H^+ buffering and excretion in response to acute hypercapnia in the dogfish *Squalus acanthias*. Am. J. Physiol. 216:440-451.

Cunningham, D. J. C., P. A. Robbins, and C. B. Wolff. 1986. Integration of respiratory responses to changes in alveolar partial pressures of CO_2 and O_2 and in arterial pH. In A. P. Fishman, N. S. Cherniack and J. G. Widdicombe, eds., *Handbook of Physiology, Sect. 3: The Respiratory System, Vol. II, Control of Breathing, Part 2.* Bethesda: Amer. Physiol. Soc. pp. 475-528.

Cuthbert, A. W., and J. Maetz. 1972. Amiloride and sodium fluxes across fish gills in fresh water and in sea water. Comp. Biochem. Physiol. 43A:227-232.

Datta Munshi, J. S. 1976. Gross and fine structure of the respiratory organs of air-breathing fishes. In G. M. Hughes, ed., *Respiration of Amphibious Vertebrates*. New York: Academic Press. pp. 73-104.

Davenport, H. W. 1974. *The ABC of Acid-Base Chemistry*. 6th Ed. Chacago: Univ. of Chicago Press. 296 pp.

Davis, J. C. 1972. An infrared photographic technique useful for studying vascularization of fish gills. J. Fish. Res. Bd. Canada. 29:109-111.

Davis, J. C., and J. N. Cameron. 1970. Water flow and gas exchange at the gills of rainbow trout, *Salmo gairdneri*. J. Exp. Biol. 54:1-18.

Davis, J. C., and K. Watters. 1970. The evaluation of opercular catheterization as a method for sampling expired water in fish. J. Fish Res. Bd. Canada. 27:1627-1635.

Daxboeck, C., and G. F. Holeton. 1978. Oxygen receptors in the rainbow trout, *Salmo gairdneri*. Can. J. Zool. 56:1254-1259.

DeFur, P. L., B. R. McMahon, and C. E. Booth. 1983. Analysis of hemolymph oxygen levels and acid-base status during emersion "in situ" in the red rock crab, *Cancer productus*. Biol. Bull. 165:582-590.

DeFur, P.L., P.R.H. Wilkes, and B.R. McMahon. 1980. Non-equilibrium acid-base status in *Cancer productus*: role of exoskeletal carbonate buffers. Respir. Physiol. 42:247-261.

de Graaf, P. J. F., and C. M. Ballantijn. 1987. Mechanoreceptor activity in the gills of the carp. II. Gill arch proprioceptors. Respir. Physiol. 69:183-194.

de Graaf, P. J. F., C. M. Ballantijn, and F. W. Maes. 1987. Mechanoreceptor activity in the gills of the carp. I. Gill filament and gill raker mechanoreceptors. Respir. Physiol. 69:173-182.

Dejours, P. 1975. *Principles of Comparative Respiratory Physiology*. New York: Elsevier North-Holland. 253pp.

Dejours, P., and H. Beekenkamp. 1977. Crayfish respiration as a function of water oxygenation. Respir. Physiol. 30:241-251.

Dejours, P., A. Toulmond and J.P. Truchot. 1977. The effect of hyperoxia on the breathing of marine fishes. Comp. Biochem. Physiol. 58A:409-412.

DeLaney, R. G., P. Laurent, R. Galante, A. I. Pack, and A. P. Fishman. 1983. Pulmonary mechanoreceptors in the dipnoi lungfish *Protopterus* and *Lepidosiren*. Am. J. Physiol. 244:R418-R428.

Dempsey, J. A., E. H. Vidruk and G. S. Mitchell. 1985. Pulmonary control systems in exercise: update. Fed. Proc. 44:2260-2270.

DeRenzis, G., and Maetz. 1973. Studies on the mechanism of chloride absorption by the goldfish gill: relation with acid-base regulation. J. Exp. Biol. 59:339-358.

Diaz, H., and G. Rodriguez. 1977. The branchial chamber in terrestrial crabs: a comparative study. Biol. Bull. 153:485-504.

Diefenbach, C. O. da C., and C. P. Mangum. 1983. The effects of inorganic ions and acclimation salinity on oxygen binding of the hemocyanin of the horseshoe crab, *Limulus polyphemus*. Mol. Physiol. 4:197-206.

Dodd, G. A. A., and W. K. Milsom. 1987. Effects of H^+ versus CO_2 on ventilation in the Pekin duck. Respir. Physiol. 68:189-201.

Dotterweich, H. 1936. Die Atmung der Vögel. Z. Vergl. Physiol. 23:744-770.

Douse, M. A., and G. S. Mitchell. 1987. Effects of temperature on CO_2-sensitive chemoreceptors (IPC) in the Tegu lizard (*Tupinambis nigropunctatus*). Fed. Proc. 46:Abstr. 162.

Dugal, L.-P. 1939. The use of calcareous shell to buffer the products of anaerobic glycolysis in *Venus mercenaria*. J. Cell. Comp. Physiol. 13:235-251.

Duncker, H.-R. 1972. Structure of avian lungs. Respir. Physiol. 14:44-63.

Duncker, H.-R. 1974. Structure of the avian respiratory tract. Respir. Physiol. 22:1-19.

Dunel, S., and P. Laurent. 1980. Functional organization of the gill vasculature in different classes of fish. In B. Lahlou, ed., *Epithelial Transport in the Lower Vertebrates*. New York: Cambridge Univ. Press. pp. 37-58.

Dunel-Erb, S., Y. Bailly, and P. Laurent. 1982. Neuroepithelial cells in fish gill primary lamellae. J. Appl. Physiol. 53:1342-1353.

Dunnill, M. S. 1962. Post-natal growth of the lung. Thorax 17:329-333.

Eaton, J. W., G. J. Brewer, and R. F. Grover. 1969. Role of red cells 2,3-diphosphoglycerate in the adaptation of man to altitude. Clin. Med. 73:603-609.

Eaton, J. W., G. J. Brewer, J. S. Schultz, and C. F. Sing. 1970. Variation in 2,3-diphosphoglycerate and ATP levels in human erythrocytes and effects on oxygen transport. In G. J. Brewer, ed., *Red Cell Metabolism and Function*. New York: Plenum Press. pp. 21-38.

Eddy, F.B. 1971. Blood gas relationships in the rainbow trout, *Salmo gairdneri*. J. Exp. Biol. 55:695-712.

Edsall, J. T. 1969. Carbon dioxide, carbonic acid and bicarbonate ion: physical properties and kinetics of interconversion. In R. E. Forster, ed., *CO_2:Chemical, Biochemical and Physical Aspects*. Washington, D.C.: NASA. pp. 15-27.

Edsall, J. T., and J. Wyman. 1958. Biophysical chemistry. Academic Press, N.Y. 699 pp.

Edwards, M. J., and R. J. Martin. 1966. Mixing technique for the oxygen-hemoglobin equilibrium and Bohr effect. J. Appl. Physiol. 21:1898-1902.

Ege, R., and A. Krogh 1914. On the relation between the temperature and the respiratory exchange in fishes. Int. Rev. Ges. Hydrobiol. Hydrog. 1:48-55.

Ehrenfeld, J., and F. Garcia-Romeu. 1977. Active hydrogen excretion and sodium absorption through isolated frog skin. Am. J. Physiol. 233:F46-F54.

Ekelund, L. G., and A. Holmgren. 1964. Circulatory and respiratory adaptation during long-term, non-steady state exercise, in the sitting position. Acta Physiol. Scand. 62:240-255.

Eldridge, F. L., D. E. Millhorn, J. P. Kiley and T. G. Waldrop. 1985. Stimulation by central command of locomotion, respiration and circulation during exercise. Respir. Physiol. 59:313-337.

Ellington, W. R. 1983. Phosphorous nuclear magnetic resonance studies of energy metabolism in molluscan tissues: effect of anoxia and ischemia on the intracellular pH and high energy phosphates in the ventricle of the whelk, *Busycon contrarium*. J. Comp. Physiol. B 153:159-166.

Elsner, R. 1965. Heart rate response in forced versus trained experimental dives in pinnipeds. Hvalrad. Skrift. 48:24-29.

Elsner, R. 1969. Cardiovascular adjustments to diving. In H. T. Andersen, ed., *Biology of Marine Mammals*. New York: Academic Press. pp. 117-145.

Epstein, F.H., J. Maetz, and G. DeRenzis. 1973. Active transport of chloride by the teleost gill: inhibition by thiocyanate. Am. J. Physiol. 224:1295-1299.

Evans, D. H. 1975. Ionic exchange mechanisms in fish gills. Comp. Biochem. Physiol. 51A:491-496.

Evans, D. H. 1984. Gill Na^+/H^+ and Cl^-/HCO_3^- exchange systems evolved before the vertebrates entered fresh water. J. Exp. Biol. 113:465-469.

Farhi, L. E., and H. Rahn. 1960. Dynamics of change in carbon dioxide stores. Anesthesiology 21:604-614.

Farrell, A. P. 1978. Cardiovascular events associated with air breathing in two teleosts, *Hoplerythrinus unitaeniatus* and *Arapaima gigas*. Can. J. Zool. 56:953-958.

Farrell, A. P., K. MacLeod, and W. R. Driedzic. 1982. The effects of preload, afterload and epinephrine on cardiac performance in the sea raven, *Hemitripterus americanus*. Can. J. Zool. 60:3165-3171.

Farrell, A. P., and D. J. Randall. 1978. Air-breathing mechanics in two Amazonian teleosts, *Arapaima gigas* and *Hoplerythrinus unitaeniatus*. Can. J. Zool. 56:939-945.

Fedde, M. R. 1986. Avian respiratory system. In P. D. Sturkie, ed., *Avian Physiology,* 4th edition. New York: Springer. pp. 191-220.

Feder, M. E., and W. W. Burggren. 1985. Skin breathing in vertebrates. Sci. Am. 253(5):126-142.

Feder, M. E., and A. W. Pinder. 1988. Ventilation and its effect on "infinite pool" exchangers. Am. Zool. 28:973-983.

Fenn, W. O., H. Rahn and A. B. Otis. 1946. A theoretical study of the composition of alveolar air at altitude. Am. J. Physiol. 146:637-653.

Fidone, S. J., and N. Gonzalez. 1986. Initiation and control of chemoreceptor activity in the carotid body. In A. P. Fishman, N. S. Cherniack and J. G. Widdicombe, eds., *Handbook of Physiology, Sect. 3: The Respiratory System, Vol. II, Control of Breathing, Part 1.* Bethesda: Amer. Physiol. Soc. pp. 247-312.

Fishman, A. P., N. S. Cherniack, and J. G. Widdicombe (eds.) 1986. *Handbook of Physiology*, Sect. 3: *The Respiratory System, Vol. II, Control of Breathing*, Part 2. Amer. Physiol. Soc.:Bethesda.

Fitzgerald, R. S., and S. Lahiri. 1986. Reflex responses to chemoreceptor stimulation. In A. P. Fishman, N. S. Cherniack and J. G. Widdicombe, eds., *Handbook of Physiology, Sect. 3: The Respiratory System, Vol. II, Control of Breathing, Part 1.* Bethesda: Amer. Physiol. Soc. pp. 313-362.

Florey, E. 1966. *An Introduction to General and Comparative Physiology*. Philadelphia: W. B. Saunders. 713pp.

Folkow, B., K. Fuxe and R. R. Sonnenschein. 1966. Response of skeletal musculature and vasculature during "diving" in the duck: peculiarities of the adrenergic vasoconstrictor innervation. Acta Physiol. Scand. 67:327-342.

Forster, R. E. 1964. Diffusion of gases. In W. O. Fenn and H. Rahn, eds., *Handbook of Physiology, Section 3, Vol. I.* Bethesda: Amer. Physiol. Soc. pp. 839-872.

Forster, R. E., and E. D. Crandall. 1976. Pulmonary gas exchange. Ann. Rev. Physiol. 38:69-93.

Forster, R. E., and J. B. Steen. 1969. The rate of Root shift of eel red cells and hemoglobin solution. J. Physiol. (London) 204:259-282.

Fredericq, L. 1978. Recherches sur la physiologie du poulpe commun. Arch. Zool. Exp. 7:535-583.

Fry, F. E. J. 1957. Aquatic respiration of fish. In M. E. Brown, ed., *The Physiology of Fishes*, Vol. I. New York: Academic Press. pp. 1-63.

Fry, F. E. J. 1971. The effect of environmental factors on the physiology of fish. In W. S. Hoar and D. J. Randall, eds. *Fish Physiology*, Vol. VI. New York: Academic Press. pp. 1-98.

Funk, G. D., and W. K. Milsom. 1987. Changes in ventilation and breathing pattern produced by changing body temperature and inspired CO_2 concentration in turtles. Respir. Physiol. 67:37-52.

Furilla, R. A., and D. A. Bartlett, Jr. 1987. Intrapulmonary receptors in the garter snake (*Thamnophis sirtalis*). Fed. Proc. 46:793 (Abstr.).

Furilla, R. A., and D. R. Jones. 1987. The relationship between dive and pre-dive heart rates in restrained and free dives by diving ducks. J. Exp. Biol. 127:333-348.

Gaillard, S., and A. Malan. 1985. Intracellular pH-temperature relationships in a water breather, the crayfish. Mol. Physiol. 7:1-16.

Gannon, B. J., and G. Burnstock. 1969. Excitatory adrenergic innervation of the fish heart. Comp. Biochem. Physiol. 29:765-774.

Gans, C. 1970. Strategy and sequence in the evolution of the external gas exchanges of ectothermal vertebrates. Forma Functio 3:61-104.

Gatten, R. E., Jr. 1987. Cardiovascular and other physiological correlates of hibernation in aquatic and terrestrial turtles. Am. Zool. 27:59-68.

Gaumer, A. E. H., and C. J. Goodnight. 1957. Some aspects of the hematology of turtles as related to their activity. Am. Midl. Nat. 58:332-340.

Gehr, P., D. K. Mwangi, A. Ammann, G. M. O. Maloiy, C. R. Taylor and E. R. Weibel. 1981. Design of the mammalian respiratory system. V. Scaling morphometric pulmonary diffusing capacity to body mass: wild and domestic animals. Respir. Physiol. 44:61-86.

Geiser, F. 1988. Reduction of metabolism during hibernation and daily torpor in mammals and birds: temperature effect or physiological inhibition? J. Comp. Physiol. B 158:25-37.

Gerald, J. W., and J. J. Cech, Jr. 1970. Respiratory responses of juvenile catfish (*Ictalurus punctatus*) to hypoxic conditions. Physiol. Zool. 43:47-54.

Ghidalia, W. 1985. Structural and biological aspects of pigments. In D. E. Bliss and L. H. Mantel, eds., *The Biology of Crustacea. Vol 9: Integument, Pigments and Hormonal Processes*. Orlando: Academic Press. pp. 301-394.

Girard, J.-P., and P. Payan. 1977. Kinetic analysis and partitioning of sodium and chloride influxes across gills of sea water adapted trout. J. Physiol. (London) 267:519-536.

Glass, M. L., R. G. Boutilier, and N. Heisler. 1983. Ventilatory control of arterial P_{O_2} in the turtle *Chrysemys picta bellii*: Effects of temperature and hypoxia. J. Comp. Physiol. B 151:145-153.

Glass, M. L., R. G. Boutilier, and N. Heisler. 1985. Effects of body temperature on respiration, blood gases and acid-base status in the turtle *Chrysemys picta bellii*. J. Exp. Biol. 114:37-51.

Glass, M. L., W. W. Burggren, and K. Johansen. 1978. Ventilation in an aquatic and in a terrestrial chelonian reptile. J. Exp. Biol. 72:165-179.

Glass, M. L., A. Ishimatsu, and K. Johansen. 1986. Responses of aerial ventilation to hypoxia and hypercapnia in *Channa argus*, an air-breathing fish. J. Comp. Physiol. B 156:425-430.

Gleeson, R. A., and P. L. Zubkoff. 1977. The determination of hemolymph volume in the blue crab, *Callinectes sapidus*, using ^{14}C-thiocyanate. Comp. Biochem. Physiol. 56A:411-414.

Goodrich, C. A. 1973. Acid-base balance in euthermic and hibernating marmots. Am. J. Physiol. 224:1185-1189.

Goodrich, E. S. 1930. *Studies on the Structure and Development of Vertebrates*. Dover:New York (1958 reprint of 1930 edition). 837pp.

Gordon, M. S. 1968. Oxygen consumption of red and white muscles from tuna fishes. Science 159:87-90.

Graham, R. A., C. P. Mangum, R. C. Terwilliger, and N. B. Terwilliger. 1983. The effect of organic acids on oxygen binding of hemocyanin from the crab *Cancer magister*. Comp. Biochem. Physiol. 74A:45-50.

Grant, R. T., and M. Regnier. 1926. The comparative anatomy of the cardiac coronary vessels. Heart 12:285-317.

Gray, I. E. 1953. A comparative study of the gill area of crabs. Biol. Bull. 112:34-42.

Grigg, G. C., and K. Johansen. 1987. Cardiovascular dynamics in *Crocodylus porosus* breathing air and during voluntary aerobic dives. J. Comp. Physiol. B 157:381-392.

Gross, W. J. 1964. Water balance in anomuran land crabs on a dry atoll. Biol. Bull. 126:54-68.

Guimond, R. W., and V. H. Hutchison. 1976. Gas exchange of the giant salamanders of North America. In G. M. Hughes, ed., *Respiration of Amphibious Vertebrates*. New York: Academic Press. pp. 313-338.

Haab, P. E., J. Piiper and H. Rahn. 1960. Simple method for rapid determination of an oxygen dissociation curve of the blood. J. Appl. Physiol. 15:1148-1149.

Hamburger, H. J. 1918. Anionenwanderungen in serum und Blut unter dem Einfluss von CO_2, Säure und Alkali. Biochem. Z. 86:309-324.

Hand, S. C., and G. N. Somero. 1983a. Energy metabolism pathways of hydrothermal vent animals: adaptations to a food-rich and sulfide-rich deep-sea environment. Biol. Bull. 165:167-181.

Hand, S. C., and G. N. Somero. 1983b. Phosphofructokinase of the hibernator *Citellus beeche*. Temperature and pH regulation of activity via influence on the tetramer-dimer equilibrium. Physiol. Zool. 56:380-388.

Harms, J. W. 1932. Die Realisation von Genen und die consecutive Adaptation. II. *Birgus latro* L., als Landkrebs und seine Beziehungen zu den Coenobiten. Z. Wiss. Zool. 140:167-290.

Hart, J. S., and O. Z. Roy. 1966. Respiratory and cardiac responses to flight in pigeons. Physiol. Zool. 39:291-306.

Hasselbalch, K. A. 1917. Die Berechnung der Wasserstoffzahl des Blutes auf der freien und gebundenen Kohlensäure desselben, und die Sauerstoffbindung des Blutes als Funktion der Wasserstoffzahl. Biochem. Z. 78:112-144.

Heath, A. G., and G. M. Hughes. 1971. Cardiovascular changes in trout during heat stress. Amer. Zool. 11:664.

Heatwole, H., and R. Seymour. 1975. Pulmonary and cutaneous oxygen uptake in sea snakes and a file snake. Comp. Biochem. Physiol. 51A:399-06.

Heieis, M. R. A., and D. R. Jones. 1988. Blood flow and volume distribution during forced submergence in Pekin ducks (*Anas platyrhynchos*). Can. J. Zool. 66:1589-1596.

Heisler, N. 1978. Bicarbonate exchange between body compartments after changes of temperature in the larger spotted dogfish (*Scyliorhinus stellaris*). Respir. Physiol. 33:145-160.

Heisler, N. 1980. Regulation of the acid-base status in fishes. In M. A. Ali, ed., *Environmental Physiology of Fishes*. NATO Adv. Study Inst., Series A, Vol. 35. New York: Plenum Press. pp. 123-162.

Heisler, N. 1984. Acid-base regulation in fishes. In W. S. Hoar and D. J. Randall, eds., *Fish Physiology*, Vol. XA. New York: Academic Press. pp. 315-401.

Heisler, N. 1986a. Acid-base regulation in fishes. In N. Heisler, ed., *Acid-Base Regulation in Animals*. Amsterdam: Elsevier. pp. 309-356.

Heisler, N. 1986b. Comparative aspects of acid-base regulation. In N. Heisler, ed., *Acid-Base Regulation in Animals*. Amsterdam: Elsevier. pp. 397-450.

Heisler, N., H. Weitz, and A. M. Weitz. 1976. Extracellular and intracellular pH with changes of temperature in the dogfish *Scyliorhinus stellaris*. Respir. Physiol. 26:249-263.

Heller, H. C. 1988. Sleep and hypometabolism. Can. J. Zool. 66:61-69.

Henderson, L. J. 1909. Das Gleichgewicht zwischen Basen und Säuren im tierischen Organismus. Ergeb. Physiol. 8:254-325.

Henry, J. D., and M. R. Fedde. 1970. Pulmonary circulation time in the chicken. Poult. Sci. 49:1286-1290.

Henry, R. P. 1987. Membrane-associated carbonic anhydrase in gills of the blue crab, *Callinectes sapidus*. Am. J. Physiol. 252:R966-R971.

Henry, R. P. 1988. Multiple functions of carbonic anhydrase in the crustacean gill. J. exp. Zool. 248:19-24.

Henry, R. P., and J. N. Cameron. 1982a. Acid-base balance in *Callinectes sapidus* during acclimation from high to low salinity. J. Exp. Biol. 101:255-264.

Henry, R. P., and J. N. Cameron. 1982b. The distribution and partial characterization of carbonic anhydrase in selected aquatic and terrestrial decapod crustaceans. J. Exp. Zool. 221:309-321.

Henry, R. P., and J. N. Cameron. 1983. The role of carbonic anhydrase in respiration, ion regulation and acid-base balance in the aquatic crab *Callinectes sapidus* and the terrestrial crab *Gecarcinus lateralis*. J. Exp. Biol. 103:205-223.

Henry, R. P., G. A. Kormanik, N. J. Smatresk, and J. N. Cameron. 1981. The role of $CaCO_3$ dissolution as a source of HCO_3^- for buffering hypercapnic acidosis in aquatic and terrestrial decapod crustaceans. J. Exp. Biol. 94:269-274.

Henry, R. P., N. J. Smatresk and J. N. Cameron. 1988. The distribution of branchial carbonic anhydrase, and the effects of gill and erythrocyte carbonic anhydrase inhibition in the channel catfish, *Ictalurus punctatus*. J. Exp. Biol. 134:201-218.

Hickman, C. P, Jr., and B. F. Trump. 1969. The kidney. In W. S. Hoar and D.J. Randall, Eds., *Fish Physiology*, Vol. I. New York: Academic Press. pp. 91-239.

Hills, A. G. 1973. *Acid-Base Balance: Chemistry, Physiology, Pathophysiology*. Baltimore: Williams and Wilkins. 381pp.

Hitzig, B. M., J. C. Allen and D. C. Jackson. 1985. Central chemical control of ventilation and response of turtles to inspired CO_2. Am. J. Physiol. 249:R323-R328.

Hochachka, P. W. 1973. Comparative intermediary metabolism. In C. L. Prosser, ed., *Comparative Animal Physiology*, 3rd edition. Philadelphia: W. B. Saunders. pp. 212-278.

Hochachka, P. W., and T. P. Mommsen. 1983. Protons and anaerobiosis. Science 219:1391-1398.

Hodler, J. E., H. O. Heinemann, A. P. Fishman, and H. W. Smith. 1955. Urine pH and carbonic anhydrase activity in the marine dogfish. Am. J. Physiol. 183:155-162.

Hoffman, R. A. 1964. Terrestrial animal in cold: hibernators. In *Handbook of Physiology; Adaptations to the environment*, ed. D. B. Dill. Baltimore:Williams & Wilkins. pp. 541-550.

Hoffman, R. J., and C. P. Mangum. 1970. The function of coelomic cell hemoglobin in the polychaete *Glycera dibranchiata*. Comp. Biochem. Physiol. 36:211-228.

Høglund, L. B. 1961. *The Reactions of Fish in Concentration Gradients*. Reports of the Inst. Freshwater Research, Drottningholm. No. 43. 147 pp.

Holeton, G. F. 1973. Respiration of Arctic char (*Salvelinus alpinus*) from a high Arctic lake. J. Fish. Res. Bd. Canada 30:717-723.

Holeton, G. F. 1974. Metabolic cold adaptation of polar fish: fact or artefact? Physiol. Zool. 47:137-152.

Holeton, G. F., and N. Heisler. 1983. Contribution of net ion transfer mechanisms to acid-base regulation after exhausting activity in the larger spotted dogfish (*Scyliorhinus stellaris*). J.

Exp. Biol. 103:31-46.

Holeton, G. F., and D. R. Jones. 1975. Water flow dynamics in the respiratory tract of the carp (*Cyprinus carpio* L.). J. Exp. Biol. 63:537-549.

Holeton, G. F., P. Neumann, and N. Heisler. 1983. Branchial ion exchange and acid-base regulation after strenuous exercise in rainbow trout (*Salmo gairdneri*). Respir. Physiol. 51:303-318.

Holeton, G. F., and D. J. Randall. 1967. The effect of hypoxia upon the partial pressure of gases in the blood and water afferent and efferent to the gills of rainbow trout. J. Exp. Biol. 46:317-327.

Holle, J. P., M. Meyer, and P. Scheid. 1977. Oxygen affinity of duck blood determined by in vivo and in vitro technique. Respir. Physiol. 29:355-361.

Holman, J. P. 1981. *Heat Transfer*. New York: McGraw-Hill. 570pp.

Holmes, W. N., and E. M. Donaldson. 1969. The body compartments and distribution of electrolytes. In W.S. Hoar and D.J. Randall, eds., *Fish Physiology,* Vol. I. New York: Academic Press. pp. 1-89.

Houk, J. C. 1988. Control strategies in physiological systems. Faseb J. 2:97-107.

Houlihan, D. F., C. K. Govind, and A. El Haj. 1985. Energetics of swimming in *Callinectes sapidus* and walking in *Homarus americanus*. Comp. Biochem. Physiol. 82A:267-280.

Houlihan, D. F., A. J. Innes, M. J. Wells, and J. Wells. 1982. Oxygen consumption and blood gases of *Octopus vulgaris* in hypoxic conditions. J. Comp. Physiol. B 148:35-40.

Houston, A. H., and M. A. DeWilde. 1968. Hematological correlations in the rainbow trout, *Salmo gairdneri*. J. Fish. Res. Bd. Canada 25:173-176.

Hoyt, D. F. 1980. Adaptation of avian eggs to incubation period: variability around allometric regressions is correlated with time. Am. Zool. 20:417-425.

Hudson, D. M., and D. R. Jones. 1986. The influence of body mass on the endurance to restrained submergence in the Pekin duck. J. Exp. Biol. 120:351-367.

Hufner, G. 1897. Ueber die verschiedenen Geschwindigkeiten mit denen sich die atmosphärischen Gase in Wasser verbreiten. Arch. Anal. Physiol., Physiol. Abt. pp. 112-131.

Hughes, G. M. 1964. Fish respiratory homeostasis. Soc. Exp. Biol. Symp. 28:81-107.

Hughes, G. M. 1966. The dimensions of fish gills in relation to their function. J. Exp. Biol. 18:81-107.

Hughes, G. M. 1976. Fish respiratory physiology. In P. Spencer-Davies, ed., *Perspectives in Experimental Biology*, Vol. 1. New York: Pergamon Press. pp. 235-245.

Hughes, G. M. 1980. Functional morphology of fish gills. In B. Lahlou, ed., *Epithelial Transport in the Lower Vertebrates*. New York: Cambridge Univ. Press. pp. 15-36.

Hughes, G. M. 1984. General anatomy of the gills. In W. S. Hoar and D. J. Randall, eds., *Fish Physiology*, Vol. XA. Orlando: Academic Press. pp. 1-72.

Hughes, G. M., B. Knights, and C. A. Scammel. 1969. The distribution of PO_2 and hydrostatic pressure changes within the branchial chambers in relation to gill ventilation of the shore crab *Carcinus maenas* L. J. Exp. Biol. 51:203-220.

Hughes, G. M., and M. Morgan. 1973. The structure of fish gills in relation to their respiratory function. Biol. Rev. 48:419-475.

Hughes, G. M., and R. L. Saunders. 1970. Responses of the respiratory pumps to hypoxia in the rainbow trout (*Salmo gairdneri*). J. Exp. Biol. 53:529-545.

Hughes, G. M., and G. Shelton. 1962. Respiratory mechanisms and their nervous control in fish. Adv. Comp. Physiol. Biochem. 1:275-364.

Hughes, G. M., and E. R. Weibel. 1976. Morphometry of fish lungs. In G. M. Hughes, ed., *Respiration of Amphibious Vertebrates*. New York: Academic Press. pp. 213-232.

Hyman, L. H. 1951. *The Invertebrates: Platyhelminthes and Rhynchocoela. The Acoelomate Bilateria*. Vol. II. McGraw-Hill:New York. 550pp.

Hyman, L. H. 1955. *The invertebrates: Echinodermata. The Coelomate Bilateria*. Vol. IV. McGraw-Hill:New York. 763pp.

Hyman, L. H. 1967. *The Invertebrates: Mollusca I.* Vol. VI. McGraw-Hill:New York. 792pp.

Innes, A. J., E. W. Taylor, and A. J. El Haj. 1987. Air-breathing in the Trinidad mountain crab: a quantum leap in the evolution of the invertebrate lung? Comp. Biochem. Physiol. 87A:1-8.

Isaacks, R. E., and D. R. Harkness. 1980. Erythrocyte organic phosphates and hemoglobin function in birds, reptiles and fishes. Am. Zool. 20:115-129.

Ishii, K., K. Honda, and K. Ishii. 1966. The function of the carotid labyrinth of the toad. Tohoku J. exp. Med. 91:119-128.

Ishii, K., K. Ishii, and T. Kusakabe. 1985a. Chemo- and baroreceptor innervation of the aortic trunk of the toad *Bufo vulgaris*. Respir. Physiol. 60:365-375.

Ishii, K., K. Ishii, and T. Kusakabe. 1985b. Electrophysiological aspects of reflexogenic area in the Chelonian, *Geoclemmys reevesii*. Respir. Physiol. 59:45-54.

Ishii, K., K. Ishii and P. Dejours. 1986. Activity of afferent vagal fibers innervating carbon dioxide sensitive receptors in the tortoise, *Testudo hermanni*. Jpn. J. Physiol. 36:1015-1026.

Ishimatsu, A., and Y. Itazawa. 1983. Difference in blood oxygen levels in the outflow vessels of the heart of an air-breathing fish, *Channa argus*: Do separate blood streams exist in a teleostean heart? J. Comp. Physiol. B 149:435-440.

Ishimatsu, A., G. K. Iwama and N. Heisler. 1988. In vivo analysis of partitioning of cardiac output between systemic and central venous sinus circuits in rainbow trout: a new approach using chronic cannulation of the branchial vein. J. Exp. Biol. 137:75-88.

Jackson, D. C. 1971. The effect of temperature on ventilation in the turtle *Pseudemys scripta* elegans. Respir. Physiol. 12:131-140.

Jackson, D. C. 1987. Cardiovascular function in turtles during anoxia and acidosis: In vivo and in vitro studies. Am. Zool. 27:49-58.

Jackson, D.C., and B.A. Braun. 1979. Respiratory control in bullfrogs: cutaneous versus pulmonary response to selective CO_2 exposure. J. Comp. Physiol. B. 129:339-342.

Jackson, D. C., and N. Heisler. 1982. Plasma ion balance of submerged anoxic turtles at 3°C: the role of calcium lactate formation. Respir. Physiol. 49:159-174.

Jackson, D. C., and N. Heisler. 1983. Intracellular and extracellular acid-base and electrolyte status of submerged anoxic turtles at 3°C. Respir. Physiol. 53:187-202. Jackson, D. C., C. V. Herbert and G. R. Ultsch. 1984. The comparative physiology of diving in North American freshwater turtles. II. Plasma ion balance during prolonged anoxia. Physiol. Zool. 57:632-640.

Jackson, D. C., S. E. Palmer and W. L. Meadow. 1974. The effects of temperature and carbon dioxide breathing on acid-base status of turtles. Respir. Physiol. 20:131-146.

Jackson, D. C., and K. Schmidt-Nielsen. 1964. Counter-current heat exchangers in the respiratory passages. Proc. Natl. Acad. Sci. U.S.A. 51:1192-1197.

Jackson, D. C., and H. Silverblatt. 1974. Respiration and acid-base status of turtles following experimental dives. Am. J. Physiol. 226:903-909.

Jackson, D. C., and G. R. Ultsch. 1982. Long-term submergence at 3°C of the turtle, *Chrysemys picta bellii*, in normoxic and severely hypoxic water. II. Extracellular ionic responses to extreme lactic acidosis. J. Exp. Biol. 96:29-43.

Jaffe, E. R. 1964. Metabolic processes involved in the formation and reduction of methemoglobin in human erythrocytes. In C. Bishop and D. M. Surgeoner, eds., *The Red Blood Cell*. New York: Academic Press. pp. 397-422.

Johansen, K. 1959. Circulation in the three-chambered snake heart. Circ. Res. 7:828-832.

Johansen, K. 1963. Cardiovascular dynamics in the amphibian, *Amphiuma tridactylum*. Acta Med. Scand. Suppl. 402, 82 pp.

Johansen, K. 1964. Regional distribution of circulating blood during submersion asphyxia in the duck. Acta Physiol. Scand. 62:1-9.

Johansen, K. 1968. Air-breathing fishes. Sci. Amer. 219(4):102-111.

Johansen, K. 1970. Air breathing in fishes. In W. S. Hoar and D. J. Randall, eds., *Fish Physiology*, Vol. IV. New York: Academic Press. pp. 361-411.

Johansen, K. 1972. Heart and circulation in gill, skin and lung breathing. Respir. Physiol. 14:193-210.

Johansen, K., D. Hanson, and C. Lenfant. 1970. Respiration in a primitive air breather, *Amia calva*. Respir. Physiol. 62-174.

Johansen, K., and C. Lenfant. 1966. Gas exchange in the cephalopod, *Octopus dofleini*. Am. J. Physiol. 210:910-918.

Johansen, K., C. Lenfant, and D. Hanson. 1968. Cardiovascular dynamics in the lungfishes. Z. Vergl. Physiol. 61:137-163.

Johansen, K., and A. W. Martin. 1965. Circulation in a giant earthworm, *Glossoscolex giganteus*. I. Contractile processes and pressure gradients in the large blood vessels. J. Exp. Biol. 43:333-347.

Johansen K., and J. A. Peterson. 1971. Gas exchange and active ventilation in a starfish, *Pteraster tesselatus*. Z. Vergl. Physiol. 71:365-381.

Johnson, P. C. 1980. The myogenic response. In D. F. Bohr, A. P. Somlyo and H. V. Sparks, Jr., eds., *Handbook of Physiology, Sect. 2: The Cardiovascular System, Vol. II: Vascular Smooth Muscle*. Bethesda: Amer. Physiol. Soc. pp. 409-442.

Johnson, P. T. 1980. *Histology of the Blue Crab, Callinectes sapidus: A Model for the Decapoda*. New York: Praeger. 440pp.

Jones, D. R. 1970. Experiments on amphibian respiratory and circulatory systems. Exp. in Physiol. and Biochem. 3:233-293.

Jones, D. R. 1971. The effect of hypoxia and anemia on the swimming performance of rainbow trout (*Salmo gairdneri*). J. Exp. Biol. 55:541-.

Jones, D. R., R. W. Brill, and D. C. Mense. 1986. The influence of blood gas properties on gas tensions and pH of ventral and dorsal aortic blood in free-swimming tuna, *Euthynnus affinis. J. Exp. Biol. 120:201-213.*

Jones, D. R. & C. Chu. 1988. Effect of denervation of carotid labyrinths on breathing in unrestrained Xenopus laevis. Respir. Physiol. 73:243-256.

Jones, D. R., R. A. Furilla, M. R. A. Heieis, G. R. J. Gabbott, and F. M. Smith. 1988. Forced and voluntary diving in ducks: cardiovascular adjustments and their control. Can. J. Zool. 66:75-83.

Jones, D. R., and G. F. Holeton. 1972a. Cardiovascular and respiratory responses of ducks to progressive hypocapnic hypoxia. J. Exp. Biol. 56:657-666.

Jones, D. R., and G. F. Holeton. 1972b. Cardiac output of ducks during diving. Comp. Biochem. Physiol. 41:639-646.

Jones, D. R., and K. Johansen. 1972. The blood vascular system of birds. In *Avian Biology*, Vol. II. New York: Academic Press. pp. 157-285.

Jones, D. R., and W. K. Milsom. 1979. Functional characteristics of slowly adapting pulmonary stretch receptors in the turtle (*Chrysemys picta*). J. Physiol. (London) 291:37-49.

Jones, D. R., and W. K. Milsom. 1982. Peripheral receptors affecting breathing and cardiovascular function in non-mammalian vertebrates. J. Exp. Biol. 100:59-91.

Jones, D. R., and M. J. Purves. 1970a. The effect of carotid body denervation upon the respiratory response to hypoxia and hypercapnia in the duck. J. Physiol. (London) 211:295-309.

Jones, D. R., and M. J. Purves. 1970b. The carotid body in the duck and the consequences of its denervation upon the cardiac responses to immersion. J. Physiol. (London) 211:279-294.

Jones, D. R., and T. Schwarzfeld. 1974. The oxygen cost to the metabolism and efficiency of breathing in the trout (*Salmo gairdneri*). Respir. Physiol. 21:241-254.

Jones, H. D. 1983. Circulatory systems of gastropods and bivalves. In A. S. M. Saleuddin and K. M. Wilbur, eds., *The Mollusca, Vol. 5, Physiology, Part 2*. New York: Academic Press. pp. 189-238.

Jones, J. E. 1952. The reactions of fish to water of low oxygen content. J. Exp. Biol. 29:403-415.

Juhasz-Nagy, A., M. Szentivanyi, M. Szabo, and S. Vamosi. 1963. Coronary circulation of the tortoise heart. Acta Physiol. Hung. 23:33-48.

Kawashiro, T., and P. Scheid. 1975. Arterial blood gases in undisturbed resting birds: measurements in chicken and duck. Respir. Physiol. 23:337-342.

Kays, W. M., and A. L. London. 1958. *Compact Heat Exchangers*. New York: McGraw-Hill.

Keuthe, D. O. 1988. Fluid mechanical valving of air flow in bird lungs. J. Exp. Biol. 136:1-12.

Khouri, E. M., D. E. Gregg, and C. R. Raymond. 1965. Effects of exercise on cardiac output, left coronary flow, and myocardial metabolism in the unanesthetized dog. Circ. Res. 17:427-437.

Kiceniuk, J. W., and D. R. Jones. 1977. The oxygen transport system in trout (*Salmo gairdneri*) during sustained exercise. J. Exp. Biol. 69:247-260.

King, A. S. 1966. Structural and functional aspects of the avian lungs and air sacs. Int. Rev. Gen. Exp. Zool. 2:171-267.

Kinney, J. L., D. T. Matsuura and F. N. White. 1977. Cardiorespiratory effects of temperature in the turtle. Respir. Physiol. 31:309-325.

Kinney, J. L., and F. N. White. 1977. Oxidative cost of ventilation in a turtle *Pseudemys floridana*. Respir. Physiol. 31:327-332.

Kirschner, L. B., L. Greenwald, and T. H. Kerstetter. 1973. Effect of amiloride on sodium transfer across body surfaces of fresh water animals. Am. J. Physiol. 224:832-837.

Kisch, B. 1949. Observations on the hematology of fishes and birds. Exp. Med. and Surgery. 7:318-326.

Kleiber, M. 1961. *The Fire of Life*. New York: J. Wiley and Sons. 453pp.

Klippenstein, G. L. 1980. Structural aspects of hemerythrin and myohemerythrin. Amer. Zool. 20:39-51.

Kooyman, G. L. 1988. Pressure and the diver. Can. J. Zool. 66:84-88.

Kooyman, G. L., D. D. Hammond, and J. P. Schroeder. 1970. Bronchograms and tracheograms of seals under pressure. Science 169:82-84.

Kooyman, G. L., D. H. Kerem, W. B. Campbell, and J. J. Wright. 1971. Pulmonary function in freely diving Weddell seals, *Leptonychotes weddelli*. Respir. Physiol. 12:271-282.

Krahl, V. E. 1964. Anatomy of the mammalian lung. In W. O. Fenn and H. Rahn, eds., *Handbook of Physiology, Sect. 3, Vol. I*. Bethesda: Amer. Physiol. Soc. pp. 213-284.

Kramer, D. L., and J. B. Graham. 1976. Synchronous air breathing, a social component of respiration in fishes. Copeia (1976):689-697.

Kreuzer, F. 1970. Facilitated diffusion of oxygen and its possible significance: a review. Respir. Physiol. 9:1-30.

Krogh, A. 1904. On the cutaneous and pulmonary respiration of the frog. Skand. Arch. Physiol. 15:328-419.

Krogh, A. 1919a. The rate of diffusion of gases through animal tissues, with some remarks on the coefficient of invasion. J. Physiol. (London) 52:391-408.

Krogh, A. 1919b. The number and distribution of capillaries in muscle with calculation of the oxygen pressure head necessary for supplying the tissue. J. Physiol., London 52:409-415.

Krogh, A. 1937. Osmotic regulation in fresh water fishes by active absorption of chloride ions. Z. Vergl. Physiol. 24:656-666.

Krogh, A. 1938. The active absorption of ions in some freshwater animals. Z. Vergl. Physiol. 25:335-350.

Krogh, A. 1941. *Comparative Physiology of Respiratory Mechanisms*. University Park: Univ. of Pennsylvania Press. 172pp.

Kuhn, W. & H. J. Kuhn. 1961. Multiplikation von Aussalz- und Einzel-effekten für die Bereitung hoher Gasdrucke in der Schwimmblase. Zeits. Elektrochem. 65:426-439.

Lahiri, S., J. P. Szidon, and A. P. Fishman. 1970. Potential respiratory and circulatory adjustments to hypoxia in African lungfish. Fed. Proc. 29:1141-1148.

Laurent, P. 1967. La pseudobranchie des teleosteens: preuves electrophysiologiques de ses fonctions chemoreceptrice et baroreceptrice. C. R. Acad. Sci. Paris 264:1879-1882.

Laurent, P. 1984. Gill internal morphology. In W. S. Hoar and D. J. Randall, eds., *Fish Physiology*, Vol. XA. New York: Academic Press. pp. 73-183.

Laurent, P., and S. Dunel. 1980. Morphology of gill epithelia in fish. Am. J. Physiol. 238:R147-R159.

Laurent, P., and J.D. Rouzeau. 1972. Afferent neural activity from pseudobranch of teleosts. Effects of Po_2, pH, osmotic pressure and Na^+ ions. Respir. Physiol. 14:307-331.

Lawson, H. C. 1962. The volume of blood – a critical examination of methods for its measurement. In W. F. Hamilton and P. Dow, eds., *Handbook of Physiology, Sect. 2: Circulation, Vol. I.* Bethesda: Amer. Physiol. Soc. pp. 23-49.

Lehninger, A. L. 1975. *Biochemistry*. 2nd ed. New York: Worth Publishers. 1104pp.

Lenfant, C., and K. Johansen. 1966. Respiratory function in the elasmobranch *Squalus suckleyi*. Respir. Physiol. 1:13-29.

Lenfant, C., and K. Johansen. 1972. Gas exchange in gill, skin and lung breathing. Respir. Physiol. 211-218.

Lenfant, C., K. Johansen, J. A. Peterson, and K. Schmidt-Neilsen. 1970. Respiration in the fresh water turtle *Chelys fimbriata*. Respir. Physiol. 8:261-275.

Lenfant, C., J. E. Torrance and C. Reynafarje. 1971. Shift of the O_2-Hb dissociation curve at altitude: mechanism and effect. J. Appl. Physiol. 30:625-631.

Liem, K. F. 1984. The muscular basis of aquatic and aerial ventilation in the air-breathing teleost fish *Channa*. J. Exp. Biol. 113:1-18.

Lillo, R. S. 1980. Localisation of chemoreceptors which may cause diving bradycardia in bullfrogs. Can. J. Zool. 52:931-936.

Lindsay, F. E. 1967. The cardiac veins of *Gallus domesticus*. J. Anat. 101:555.

Lomholt, J. P., and K. Johansen. 1979. Hypoxia acclimation in carp – how it affects O_2 uptake, ventilation and O_2 extraction from water. Physiol. Zool. 52:38-49.

Lückner, H. 1939. Ueber die Geschwindigkeit des Austausches der Atemgase im Blut. Pflügers Arch. Ges. Physiol. 241:753-758.

Lutz, P. 1980. On the oxygen affinity of bird blood. Am. Zool. 20:187-198.

Lykkeboe, G., O. Brix and K. Johansen. 1980. Oxygen-linked CO_2 binding independent of pH in cephalopod blood. Nature 287:330-331.

Mackinnon, M. R., and H. Heatwole. 1981. Comparative cardiac anatomy of the reptilia. IV: The coronary arterial circulation. J. Morphol. 170:1-27.

Maetz, J., and F. Garcia-Romeu. 1964. The mechanism of sodium and chloride uptake by the gills of a fresh water fish, *Carassius auratus*. I. Evidence for a independent uptake of sodium and chloride ions. J. Gen. Physiol. 47:1195-1207.

Maginnis, L. A., Y. K. Sang, and R. B. Reeves. 1980. Oxygen equilibria of ectotherm blood containing multiple hemoglobins. Respir. Physiol. 42:329-343.

Malan, A. 1973. Ventilation measured by plethysmography in hibernating mammals and in poikilotherms. Respir. Physiol. 17:32-44.

Malan, A. 1988. pH and hypometabolism in mammalian hibernation. Can. J. Zool. 66:95-98.

Malan, A., H. Arens, and A. Waechter. 1973. Pulmonary respiration and acid-base state in hibernating marmots and hamsters. Respir. Physiol. 17:45-61.

Malan, A., J. L. Rodeau, and F. Daull. 1985. Intracellular pH in hibernation and respiratory acidosis in the European hamster. J. Comp. Physiol. B 156:251-258.

Malan, A., T.L. Wilson, and R.B. Reeves. 1976. Intracellular pH in cold-blooded vertebrates as a function of body temperature. Respir. Physiol. 28:29-48.

Malte, H., and R. E. Weber. 1985. A mathematical model for gas exchange in the fish gill based on non-linear blood gas equilibrium curves. Respir. Physiol. 62:359-374.

Malvin, G. M. 1988. Microvascular regulation of cutaneous gas exchange in amphibians. Amer. Zool. 28:999-1007.

Malvin, G. M., and M. P. Hlastala. 1986. Regulation of cutaneous gas exchange by environmental O_2 and CO_2 in the frog. Respir. Physiol. 65:99-111.

Mangum, C. P. 1973. Evaluation of the functional propertiesof invertebrate hemoglobins. Neth. J. Sea. Res. 7:303-315.

Mangum, C. P. 1976. Primitive respiratory adaptations. In R. C. Newell, ed.,*Adaptation to Environment*. London: Butterworth. pp. 191-278.

Mangum, C. P. 1980. Respiratory function of the hemocyanins. Am. Zool. 20:19-38.

Mangum, C. P. 1983. On the distribution of lactate sensitivity among the hemocyanins. Mar. Biol. Lett. 4:139-150.

Mangum, C. P. 1985. Oxygen transport in invertebrates. Am. J. Physiol. 248:R505-R514.

Mangum, C. P., and L. E. Burnett, Jr. 1986. The CO_2 sensitivity of the hemocyanins and its relationship to Cl^- sensitivity. Biol. Bull. 171:248-263.

Mangum, C. P., J. M. Colacino, and T. L. Vandergon. 1989. Oxygen binding of single red blood cells of the annelid bloodworm *Glycera dibranchiata*. J. Exp. Zool. 249:144-149.

Mangum, C. P., and K. Johansen. 1975. The colloid osmotic pressures of the body fluids of invertebrates. J. Exp. Biol. 63:661-671.

Mangum, C. P., and G. Lykkeboe. 1979. The influence of inorganic ions and pH on the oxygenation properties of the blood in the gastropod mollusc *Busycon canaliculatum*. J. exp. Zool. 207:417-430.

Mangum, C. P., and N. A. Mauro. 1985. Metabolism of invertebrate red cells: a vacuum in our knowledge. In R. Gilles, ed., *Circulation, Respiration and Metabolism: Current Comparative Approaches*. Berlin: Springer-Verlag. pp. 280-289.

Mangum, C. P., B. R. McMahon, P. L. DeFur, and M. G. Wheatly. 1985. Gas exchange, acid-base balance, and the oxygen supply to the tissues during a molt of the blue crab *Callinectes sapidus*. J. Crust. Biol. 5:188-206.

Mangum, C. P., K. I. Miller, J. L. Scott, K. E. van Holde, and M. P. Morse. 1987. Bivalve hemocyanin: structural, functional, and phylogenetic relationships. Biol. Bull. 173:205-221.

Mangum, C. P., and J. M. Schick. 1972. The pH of body fluids of marine invertebrates. Comp. Biochem. Physiol. 42A:693-697.

Mangum, C. P., S. U. Silverthorne, J. L. Harris, D. W. Towle, and A. R. Krall. 1976. The relationship between blood pH, ammonia excretion and adaptation to low salinity in the blue crab *Callinectes sapidus*. J. Exp. Zool. 195:129-136.

Mangum, C. P., and D. W. Towle. 1977. Physiological adaptation to unstable environments. Amer. Scientist. 65:67-75.

Mangum, C. P., and W. Van Winkle. 1973. Responses of aquatic invertebrates to declining oxygen conditions. Am. Zool. 13:529-542.

Mangum, C. P., and A. L. Weiland. 1975. The function of hemocyanin in respiration of the blue crab *Callinectes sapidus*. J. exp. Zool. 193:257-264.

Manwell, C. 1977. Superoxide dismutase and NADH diaphorase in haemerythrocytes of sipunculans. Comp. Biochem. Physiol. 58A:331-338.

Maren, T. H. 1967. Carbonic anhydrase in the animal kingdom. Fed. Proc. 26:1097-1103.

Martin-Body, R. L., G. J. Robson and J. D. Sinclair. 1985. Respiratory effects of sectioning the carotid sinus glossopharyngeal and abdominal vagal nerves in the awake rat. J. Physiol. (London) 361:353-45.

Marvin, D. E., and A. G. Heath. 1968. Cardiac and respiratory responses to gradual hypoxia in three ecologically distinct species of fresh-water. Comp. Biochem. Physiol. 27:349-355.

Mason, R. P., C. P. Mangum and G. Godette. 1983. The influence of inorganic ions and acclimation salinity on hemocyanin-oxygen binding in the blue crab *Callinectes sapidus*. Biol. Bull. 164:104-123.

Matthew, J. B., G. I. H. Hanania, and F. R. N. Gurd. 1979. Electrostatic effects in hemoglobin: hydrogen ion equilibria in human deoxy- and oxyhemoglobin A. Biochemistry 18:1919-1928.

Mauer, F. 1974. Seasonal changes in organic phosphates in the erythrocytes of some Alaskan fish. M.Sc. Thesis, University of Alaska, Fairbanks, May, 1974.

Maynard, D. M. 1960. Circulation and heart function. In T. H. Waterman, ed., *The Physiology of Crustacea*, Vol. I. New York: Academic Press. pp. 161-226.

McDonald, D. G., B. R. McMahon, and C. M. Wood. 1979. An analysis of acid-base disturbances in the haemolymph following strenuous activity in the Dungeness crab, *Cancer magister*. J. Exp. Biol. 79:47-58.

McDonald, D. G., R. L. Walker, P. R. H. Wilkes, and C. M. Wood. 1982. H^+ excretion in the marine teleost *Parophrys vetulus*. J. Exp. Biol. 98:403-414.

McLaughlin, P. A. 1982. Comparative morphology of appendages. In L. G. Abele, ed., *The Biology of Crustacea, Vol. 2: Embryology, Morphology and Genetics*. New York: Academic Press. pp. 197-256.

McMahon, B. R. 1969. A functional analysis of the aquatic and aerial respiratory physiology of an African lungfish *Protopterus aethiopicus* with reference to the evolution of vertebrate lung ventilation mechanisms. J. Exp. Biol. 51:407-430.

McMahon, B. R., and W. W. Burggren. 1981. Acid-base balance following temperature acclimation in land crabs. J. exp. Zool. 218:45-52.

McMahon, B. R., and J. L. Wilkens. 1975. Respiratory and circulatory responses to hypoxia in the lobster *Homarus americanus*. J. Exp. Biol. 62:637-656.

McMahon, B. R., and J. L. Wilkens. 1983. Ventilation, perfusion and oxygen uptake. In L. H. Mantel, ed., *The Biology of Crustacea, Vol. 5: Internal Anatomy and Physiological Regulation*. New York: Academic Press. pp. 289-372.

Metcalfe, J. D., and P. J. Butler. 1984a. Changes in activity and ventilation in response to hypoxia in unrestrained, unoperated dogfish (*Scyliorhinus canicula* L.). J. Exp. Biol. 108:411:418.

Metcalfe, J. D., and P. J. Butler. 1984b. On the nervous regulation of gill blood flow in the dogfish (*Scyliorhinus canicula*). J. Exp. Biol. 113:253-267.

Milhorn, H. T. 1966. *The Application of Control Theory to Physiological Systems*. Philadelphia: W. B. Saunders. 396pp.

Milhorn, H.T., Jr., R. Benton, R. Ross, and A.C.Guyton. 1965. A mathematical model of the human respiratory control system. Biophys. J. 5:27-46.

Milligan, C. L., and C. M. Wood. 1986. Intracellular and extracellular acid-base status and H^+ exchange with the environment after exhaustive exercise in the rainbow trout. J. Exp. Biol. 123:93-122.

Milsom, W. K., and R. W. Brill. 1986. Oxygen sensitive afferent information arising from the first gill arch of yellowfin tuna. Respir. Physiol. 66:193-203.

Milsom, W. K., and D. R. Jones. 1977. Carbon dioxide sensitivity of pulmonary receptors in the frog. Experientia 33:1167-1168.

Milsom, W. K., and D. R. Jones. 1980. The role of vagal afferent information and hypercapnia in control of the breathing pattern in chelonia. J. Exp. Biol. 87:53-63.

Milsom, W. K., and D. R. Jones. 1985. Characteristics of mechanoreceptors in the air-breathing organ of the holostean fish, *Amia calva*. J. Exp. Biol. 117:389-400.

Milsom, W. K., D. R. Jones, and G. R. J. Gabbott. 1981. On chemoreceptor control of ventilatory responses ito CO_2 in unanesthetized ducks. J. Appl. Physiol. 50:1121-1128.

Milsom, W. K., F. L. Powell, and G. S. Mitchell. 1987. CO_2-sensitive pulmonary receptors in *Alligator mississipiensis*. Fed. Proc. 46:793 (Abstr.)

Milsum, J. H. 1966. *Biological Control Systems Analysis*. New York: McGraw-Hill. 466pp.

Mission, B. H., and B. M. Freeman. 1972. Organic phosphates and oxygen affinity of chick blood before and after hatching. Respir. Physiol. 14:343-352.

Moalli, R., R.S. Meyers, G.R. Ultsch, and D.C. Jackson. 1981. Acid-base balance and temperature in a predominantly skin-breathing salamander, *Cryptobranchus alleganiensis*. Respir. Physiol. 43:1-12.

Morris, S., C. R. Bridges, and M. K. Grieshaber. 1985. A new role for uric acid: modulator of haemocyanin oxygen affinity. J. exp. Zool. 235:135-140.

Morrison, P. R., and W. Galster. 1975. Patterns of hibernation in the arctic ground squirrel. Can. J. Zool. 1345-1355.

Morrison, P. R., M. Rosenmann, and J. A. Estes. 1977. Sea otter metabolism and heat economy. In M. L. Merritt and R. G. Fuller, eds., *The Environment of Amchitka Island, Alaska*. Tech. Information Ctr., Energy Res. and Dev. Admin., NTIS TID-26712. pp. 569-577.

Neumann, P., G. F. Holeton, and N. Heisler. 1983. Cardiac output and regional blood flow in gills and muscles after exhausting exercise in rainbow trout (*Salmo gairdneri*). J. Exp. Biol. 105:1-14.

Nicol, J. A. C. 1952. Autonomic nervous systems in lower chardates. Biol. Rev. 27:1-49.

Nicol, S. C., M. L. Glass, and N. Heisler. 1983. Comparison of directly determined and calculated plasma bicarbonate concentration in the turtle *Chrysemys picta bellii* at different temperatures. J. Exp. Biol. 107:521-

Obaid, A. L., A. M. Critz, and E. D. Crandall. 1979. Kinetics of bicarbonate/chloride exchange in dogfish erythrocytes. Am. J. Physiol. 237:R132-R138.

Olson, K. R. 1984. Distribution of flow and plasma skimming in isolated perfused gills of three teleosts. J. Exp. Biol. 109:97-108.

Pappenheimer, J. R. 1950. Standardization of definitions an symbols in respiratory physiology. Fed. Proc. 9:602-603.

Parmley, W. M., and L. E. Talbot. 1979. Heart as a pump. In R. M. Berne, N. Sperelakis and S. R. Geiger, eds., *Handbook of Physiology, Sect. 2: The Cardiovascular System, Vol. I., The Heart*. Bethesda: Amer. Physiol. Soc. pp. 429-460.

Pasztor, V. M., and H. Kleerkoper. 1962. The role of the gill filament musculature in teleost. Can. J. Zool. 49:785-802.

Pattle, R. E. 1955. Properties, function, and origin of the alveolar lining layer. Nature 175:1125-1126.

Pattle, R. E. 1976. The lung surfactant in the evolutionary tree. In G. M. Hughes, ed., *Respiration of Amphibious Vertebrates*. New York: Academic Press. pp. 233-255.

Paul, R., T. Fincke, and B. Linzen. 1987. Respiration in the tarantula *Eurypelma californicum*: evidence for diffusion lungs. J. Comp. Physiol. B 157:209-217.

Payan, P., and J.-P. Girard. 1977. Adrenergic receptors regulating the pattern of blood flow through the gill of trout. Am. J. Physiol. 232:H18-H23.

Pearse, A. S. 1929. Observations on certain littoral and terrestrial animals at Tortugas, Florida with special reference to migrations from marine to terrestrial habitats. Papers Tortugas Lab. 26:205-223.

Pennycuick, C. J. 1975. Mechanics of flight. In D. S. Farner and J. R. King, eds., *Avian Biology*, Vol. V. New York: Academic Press. pp. 1-75.

Perkins, J. F., Jr. 1964. Historical development of respiratory physiology. In W. O. Fenn and H. Rahn, eds., *Handbook of Physiology, Sect. 3: Respiration, Vol. 1.* Washington, D. C.: Amer. Physiol. Soc. pp. 1-62.

Piiper, J. 1964. Geschwindigkeit des CO_2 Austausches zwischen Erythrocyten und Plasma. Archiv. Gesam. Physiol. 278:500-512.

Piiper, J. 1969. Rates of chloride-bicarbonate exchange between red cells and plasma. In R. E. Forster, A. B. Otis, and F. J. W. Roughton, eds., *CO_2:Chemical, Biochemical and Physical Aspects*. Washington: NASA. pp. 267-273.

Piiper, J. 1982. Respiratory gas exchange at lungs, gills and tissues: mechanisms and adjustments. J. Exp. Biol. 100:5-22.

Piiper, J., P. Dejours, P. Haab, and H. Rahn. 1971. Concepts and basic quantities in gas exchange physiology. Respir. Physiol. 13:292-304.

Piiper, J., R. N. Gatz, and E. C. Crawford, Jr. 1976. Gas transport characteristics in an exclusively skin-breathing salamander, *Desmognathus fuscus* (Plethodontidae). In G. M. Hughes, ed., *Respiration of Amphibious Vertebrates*. New York: Academic Press. pp. 339-356.

Piiper, J., and P. Scheid. 1972. Maximum gas transfer efficacy of models for fish gills, avian lungs and mammalian lungs. Respir. Physiol. 23:209-221.

Piiper, J., and P. Scheid. 1975. Gas transfer efficacy of gills, lungs and skin: theory and experimental data. Respir. Physiol. 14:115-124.

Piiper, J., and P. Scheid. 1984. Gas transfer in gills. In W. S. Hoar and D. J. Randall, eds., *Fish Physiology*, Vol. XA. New York: Academic Press. pp. 229-262.

Pitts, R. F. 1973. Production and excretion of ammonia in relation to acid-base regulation. In J. Orloff and R. W. Berliner, eds., *Handbook of Physiology, Sect. 8: Renal Physiology*. Bethesda: Amer. Physiol. Soc. pp. 445-496.

Poole, C. A., and G. H. Satchell. 1979. Nociceptors in the gills of the dogfish *Squalus acanthias*. J. Comp. Physiol. A 130:1-7.

Pörtner, H.-O. 1987. Contributions of anaerobic metabolism to pH regulation in animal tissues: theory. J. Exp. Biol. 131:69-88.

Pörtner, H.-O., M. K. Grieshaber, and N. Heisler. 1984a. Anaerobiosis and acid-base status in marine invertebrates: effect of environmental hypoxia on extracellular and intracellular pH in *Sipunculus nudus* L. J. Comp. Physiol. B 155:13-20.

Pörtner, H.-O., N. Heisler, and M. K. Grieshaber. 1984b. Anaerobiosis and acid-base status in marine invertebrates: a theoretical analysis of proton generation by anaerobic metabolism. J. Comp. Physiol. B 155:1-12.

Pörtner, H.-O., U. Kreutzer, B. Siegmund, N. Heisler, and M. K. Grieshaber. 1984c. Metabolic adaptation of the intertidal worm *Sipunculus nudus* to functional and environmental hypoxia. Marine Biol. 79:237-247.

Potter, G. E. 1927. Respiratory function of the swim bladder of *Lepisosteus*. J. exp. Zool. 49:45-67.

Pough, F. H. 1980. Blood oxygen transport and delivery in reptiles. Am. Zool. 20:173-185.

Powell, F. L., and R. W. Mazzone. 1983. Morphometrics of rapidly frozen goose lungs. Respir. Physiol. 51:319-332.

Poyart, C.E., E. Bursaux, and A. Freminet. 1975a. The bone CO_2 compartment: evidence for a bicarbonate pool. Respir. Physiol. 25:84-99.

Poyart, C. E., A. Freminet, and E. Bursaux. 1975b. The exchange of bone CO_2 in vivo. Respir. Physiol. 25:101-107.

Price, J. W. 1931. Growth and gill development in the small-mouth black bass. Franz Theodore Stone Lab. Contribution, 4:1-46.

Prosser, C. L. 1973. *Comparative Animal Physiology*, 3rd ed. Philadelphia: W. B. Saunders. pp. 190-191.

Pyle, R. W., and L. E. Cronin. 1950. The general anatomy of the blue crab *Callinectes sapidus* Rathbun. Solomons, Md.: Chesapeake Biol. Lab., Publ. No. 87, 40pp.

Rabalais, N. N., and R. H. Gore. 1985. Abbreviated development in decapods. In A. M. Wenner, ed., *Crustacean Growth: Larval Growth*. Rotterdam: A. A. Balkema. pp. 67-126.

Radford, E. P. 1964. The physics of gases. In W. O. Fenn and H. Rahn, eds., *Handbook of Physiology, Sect. 3, Vol. I. Respiration*. Washington, D. C.: Amer. Physiol. Soc. pp. 125-152.

Rahim, S. M., J.-P. Delaunoy, and P. Laurent. 1988. Identification and immunocytochemical localization of two different carbonic anhydrase isoenzymes in teleostean fish erythrocytes and gill epithelia. Histochemistry 89:451-459.

Rahn, H. 1966. Aquatic gas exchange: theory. Respir. Physiol. 1:1-12.

Rahn, H. 1967. Gas transport from the external environment to the cell. In A. V. S. deReuck and R. Porter, eds., *Development of the Lung*. CIBA Foundation Symposium. London: Churchill. pp. 3-23.

Rahn, H., K. B. Rahn, B. J. Howell, C. Gans, and S. M. Tenney. 1971. Air breathing of the gar fish (*Lepisosteus osseus*). Respir. Physiol. 11:285-307.

Ramirez, J. M., and M. H. Bernstein. 1976. Compound ventilation during therm panting in pigeons: a possible mechanism for minimizing hypocapnic alkalosis. Fed. Proc. 35:2562-2565.

Randall, D. J. 1968. Functional morphology of the heart in fishes. Amer. Zool. 8:179-189.

Randall, D. J. 1970a. The circulatory system. In W.S. Hoar and D.J. Randall, eds., *Fish Physiology*, Vol. IV. New York: Academic Press. pp. 133-172.

Randall, D. J. 1970b. Gas exchange in fish. In W. S. Hoar and D. J. Randall, eds., *Fish Physiology*, Vol. IV. New York: Academic Press. pp. 253-292.

Randall, D. J., and J. N. Cameron. 1973. Respiratory control of arterial pH as temperature changes in rainbow trout *Salmo gairdneri*. Am. J. Physiol. 225:997-1002.

Randall, D. J., J. N. Cameron, C. S. Daxboeck, and N. J. Smatresk. 1981. Aspects of bimodal gas

exchange in the bowfin, *Amia calva* L. (Actinopterygii; Amiiformes). Respir. Physiol. 43:339-348.
Randall, D. J., A. P. Farrell, and M. S. Haswell. 1978. Carbon dioxide excretion in the pirarucu (*Arapaima gigas*), an obligate air-breathing fish. Can. J. Zool. 56:977-982.
Randall, D. J., N. Heisler, and F. Drees. 1976. Ventilatory response to hypercapnia in the larger spotted dogfish *Scyliorhinus stellaris*. Am. J. Physiol. 230:590-594.
Randall, D. J., and D. R. Jones. 1973. The effect of deafferentation of the pseudobranch on the respiratory response to hypoxia and hyperoxia in the trout (*Salmo gairdneri*). Respir. Physiol. 17:291-301.
Randall, D. J., and G. Shelton. 1963. The effects of changes in environmental gas concentrations on the breathing and heart rate of a teleost fish. Comp. Biochem. Physiol. 9:229-239.
Rashevsky, N. 1960. *Mathematical Biophysics: Physico-mathematical Foundations of Biology*. New York: Dover. 488pp.
Redfield, A. C. 1934. The haemocyanins. Biol. Rev. Cambr. Phil. Soc. 9:175-212.
Redmond, J. R. 1955. The respiratory function of haemocyanin in crustacea. J. Cell. Comp. Physiol. 46:209-247.
Reeves, R. B. 1972. An imidazole alphastat hypothesis for vertebrate acid-base regulation: tissue carbon dioxide content and body temperature in bull frogs. Respir. Physiol. 14:219-236.
Reeves, R. B. 1976. Temperature-induced changes in blood acid-base status: pH and pCO_2 in a binary buffer. J. Appl. Physiol. 40:752-761.
Rhodin, J. A. G. 1980. Architecture of the vessel wall. In D. F. Bohr, A. P. Somlyo and H. V. Sparks, Jr., eds., *Handbook of Physiology, Sect. 2: The Cardiovascular System, Vol. 2, Vascular Smooth Muscle*. Bethesda: Amer. Physiol. Soc. pp. 1-31.
Roberts, J. L. 1975. Active branchial and ram gill ventilation in fishes. Biol. Bull. 148:85-105.
Roberts, G. C. K., D. C. Meadows and O. Jardetzky. 1969. Solvent and temperature effects on the ionization of histidine residues of ribonuclease. Biochemistry 8:2053-2056.
Robin, E. D. 1962. Relationship between temperature and plasma pH and carbon dioxide tension in the turtle. Nature 195:249-251.
Robin, E. D., J. W. Vester, H. V. Murdaugh, and J. E. Millen. 1963. Prolonged anaerobiosis in a vertebrate: anaerobic metabolism in the freshwater turtle. J. cell. Comp. Physiol. 63:287-297.
Romanoff, A. L. 1960. *The Avian Embryo*. New York: Macmillan.
Romer, A. S. 1972. Skin breathing – primary or secondary? Respir. Physiol. 14:183-192.
Roos, A., and W. Boron. 1981. Intracellular pH. Physiol. Rev. 61:296-435.
Root, R. W. 1931. Respiratory function of the blood of marine fishes. Biol. Bull. 61:426-456.
Root, R. W., L. Irving and E. C. Black. 1939. The effects of hemolysis upon the combination of oxygen with the blood of some marine fishes. J. cell. comp. Physiol. 13:303-313.
Rosenthal, T.B. 1948. The effect of temperature on the pH of blood and plasma in vitro. J. Biol. Chem. 173:25-30.
Roughton, F. J. W. 1964. Transport of oxygen and carbon dioxide. In W. O. Fenn and H. Rahn, eds., *Handbook of Physiology, Sect. 3: Respiration, Vol. 1*. Bethesda: Amer. Physiol. Soc. pp. 767-825.
Ruch, T. C., and H. D. Patton. 1965. *Physiology and Biophysics*, 19th ed. Philadelphia: W. B. Saunders. 1242pp.
Rushmer, R. F. 1961. *Cardiovascular dynamics*, 2nd ed. Philadelphia: W. B. Saunders.
Rushmer, R. F. 1965. General characteristics of the cardiovascular system. In T. C. Ruch and H. D. Patton, eds., *Physiology and Biophysics*, 19th ed. Philadelphia: W. B. Saunders. pp. 543-549.
Santer, R. M., and M. Greer-Walker. 1980. Morphological studies on the ventricle of teleost and elasmobranch hearts. J. Zool. (London). 190:259-272.
Satchell, G. H. 1971. *Circulation in fishes*. Cambridge: Cambridge Univ. Press. 131 pp.
Satchell, G. H. 1976. The circulatory system of air-breathing fish. In G. M. Hughes, ed., *Respira-*

tion of Amphibious Vertebrates. New York: Academic Press. pp. 105-123.
Satchell, G. H., and D. J. Maddalena. 1972. The cough or expulsion reflex in the Port Jackson shark, *Heterodontus portusjacksoni*. Comp. Biochem. Physiol. 41A:49-62.
Saunders, R. L. 1961. The irrigation of the gills in fishes. I. Studies of the mechanism of branchial irrigation. Can. J. Zool. 39:637-653.
Scheid, P. 1979. Mechanisms of gas exchange in bird lungs. Rev. Physiol. Biochem. Pharmacol. 86:137-186.
Scheid, P., C. Hook, and J. Piiper. 1986. Model for analysis of counter-current gas transfer in fish gills. Respir. Physiol. 64:365-374.
Scheid, P., and J. Piiper. 1986. Control of breathing in birds. In A. P. Fishman, N. S. Cherniack, J. G. Widdicombe, and S. Geiger, eds., *Handbook of Physiology, Sect. 3: The Respiratory System, Vol. II, Control of Breathing, Part 2*. Bethesda: Amer. Physiol. Soc. pp. 815-832.
Scheid, P., H. Slama, and J. Piiper. 1972. Mechanisms of unidirectional flow in parabronchi of avian lungs: measurements in duck lung preparations. Respir. Physiol. 14:83-95.
Scheid, P., H. Slama, and H. Willmer. 1974. Volume and ventilation of air sacs in ducks studied by inert gas wash-out. Respir. Physiol. 21:19-36.
Scheipers, G., T. Kawashiro, and P. Scheid. 1975. Oxygen and carbon dioxide dissociation of duck blood. Respir. Physiol. 24:1-13.
Scher, A. M., and M. S. Spach. 1979. Cardiac depolarization and repolarization and the electrocardiogram. In R. M. Berne, N. Sperelakis and S. R. Geiger, eds., *Handbook of Physiology, Sect. 2: The Cardiovascular System, Vol. I: The Heart*. Bethesda: Amer. Physiol. Soc. pp. 357-392.
Schlue, W. R., and R. C. Thomas. 1985. A dual mechanism for intracellular pH regulation by leech neurones. J. Physiol. (London) 364:327-338.
Schmidt-Nielsen, K. 1964. *Desert Animals, Physiological Problems of Heat and Water*. Oxford: Clarendon Press. 277 pp.
Schmidt-Nielsen, K., and J. L. Larimer. 1958. Oxygen dissociation curves of mammalian blood in relation to body size. Am. J. Physiol. 195:424-428.
Scholander, P. F. 1940. Experimental investigations in diving mammals and birds. Hvalradets Skrifter 22:1-131.
Scholander, P. F. 1954. Secretion of gases against high pressures in the swimbladder of deep sea fishes. II. The rete mirabile. Biol. Bull. 107:260-277.
Scholander, P. F. 1958. Counter current exchange. A principle in biology. Hvalrad. Skrifter 44:1-24.
Scholander, P. F., R. Hock, V. Walters, F. Johnson, and L. Irving. 1950. Heat regulation in some Arctic and tropical mammals and birds. Biol. Bull. 99:237:258.
Scholander, P. F., W. Flagg, V. Walters, and L. Irving. 1953. Climatic adaptation in arctic and tropical poikilotherms. Physiol. Zool. 26:67-92.
Scholander, P. F., and L. van Dam. 1954. Secretion of gases against high pressures in the swimbladder of deep sea fishes. I. Oxygen dissociation in blood. Biol. Bull. 107:247-259.
Schumann, D., and J. Piiper. 1966. Der Sauerstoffbedarf der Atmung bei Fischen nach Messungen an der narkotisierten Schleie (*Tinca tinca*). Pflügers Arch. Ges. Physiol. 288:15-26.
Semper, C. 1878. Ueber die Lunge von *Birgus latro*. Z. Wiss. Zool. 30:282-287.
Shadwick, R. E., and J. M. Gosline. 1981. Elastic arteries in invertebrates: mechanics of the octopus aorta. Science 213:759-761.
Shafie, S. M., S. N. Vinogradov, L. Larson, and J. J. McCormick. 1976. RNA and protein synthesis in the nucleated erythrocytes of *Glycera dibranchiata*. Comp. Biochem. Physiol. 53A:85-88.
Shaw, J. 1964. The control of the salt balance in the Crustacea. Symp. Soc. Exp. Biol. 18:237-254.
Shelton, G. 1970. The regulation of breathing. In W. S. Hoar and D. J. Randall, eds., *Fish Physiology*, Vol. IV. New York: Academic Press. pp. 293-359.
Shelton, G., and W. W. Burggren. 1976. Cardiovascular dynamics of the Chelonia during apnoea and lung ventilation. J. Exp. Biol. 64:323-343.
Shelton, G., D. R. Jones, and W. K. Milsom. 1986. Control of breathing in ectothermic vertebrates.

In A. P. Fishman, N. S. Cherniack, and J. G. Widdicombe, eds., *Handbook of Physiology, Sect. 3: The Respiratory System, Vol. II: Control of Breathing, Part 2.* Bethesda: Amer. Physiol. Soc. pp. 857-909.

Siesjö, B. K. 1971. Quantification of pH regulation in hypercapnia and hypocapnia. Scand. J. Clin. Lab. Invest. 28:113-119.

Sigaard-Andersen, O. 1974. *The Acid-Base Status of the Blood.* 4th ed., Copenhagen: Munksgaard.

Silver, R. B., and D. C. Jackson. 1985. Ventilatory and acid-base responses to long-term hypercapnia in the freshwater turtle, *Chrysemys picta bellii*. J. Exp. Biol. 114:661-672.

Silver, R. B., and D. C. Jackson. 1986. Ionic compensation with no renal response to chronic hypercapnia in *Chrysemys picta bellii*. Am. J. Physiol. 251:R1228-R1234.

Smatresk, N. J. 1986. Ventilatory and cardiac reflex responses to hypoxia and NaCN in *Lepisosteus osseus*, an air–breathing fish. Physiol. Zool. 59:385-397.

Smatresk, N. J. 1988a. Control of the respiratory mode in air-breathing fishes. Can. J. Zool. 66:144-151.

Smatresk, N. J. 1988b. Chemoreflex control of bimodal breathing in gar (*Lepisosteus*). Proc. Comroe Symp., New York: Oxford Univ. Press. (In Press).

Smatresk, N. J., and S. Q. Azizi. 1987. Characteristics of lung mechanoreceptors in spotted gar, *Lepisosteus oculatus*. Am. J.Physiol. 252:R1066-R1072.

Smatresk, N. J., M. L. Burleson and S. Q. Azizi. 1986. Chemoreflexive responses to hypoxia and NaCN in longnose gar: evidence for two chemoreceptor loci. Am. J. Physiol. 251:R116-R125.

Smatresk, N. J., and J. N. Cameron. 1981. Post-exercise acid-base balance and ventilatory control in *Birgus latro*, the coconut crab. J. exp. Zool. 218:75-82.

Smatresk, N. J., and J. N. Cameron. 1982a. Respiration and acid-base physiology of the spotted gar, a bimodal breather. I. Normal values, and the response to severe hypoxia. J. Exp. Biol. 96:263-280.

Smatresk, N. J., and J. N. Cameron. 1982b. Respiration and acid-base physiology of the spotted gar, a bimodal breather. II. Responses to temperature change and hypercapnia. J. Exp. Biol. 96:281-293.

Smatresk, N. J., and J. N. Cameron. 1983. Respiration and acid-base physiology of the spotted gar, a bimodal breather. III. Response to transfer from fresh water to 50% sea water, and control of ventilation. J. Exp. Biol. 96:295-306.

Smatresk, N. J., A. J. Preslar, and J. N. Cameron. 1979. Post-exercise acid-base disturbances in *Gecarcinus lateralis*, a terrestrial crab. J. exp. Zool. 210:205-210.

Smith, D. G., and D. W. Johnson. 1977. Oxygen exchange in a simulated trout gill secondary lamella. Am. J. Physiol. 233:R145-R161.

Smith, F. M., and P. S. Davie. 1984. Effects of sectioning cranial nerves IX and X on the cardiac response to hypoxia in the coho salmon, *Onchorhynchus kisutch*. Can. J. Zool. 62:766-768.

Smith, F. M., and D. R. Jones. 1978. Localization of receptors causing hypoxic bradycardia in trout (*Salmo gairdneri*). Can. J. Zool. 56:1260-1265.

Smith, F. M., and D. R. Jones. 1982. The effect of changes in blood oxygen-carrying capacity on ventilation volume in the rainbow trout (*Salmo gairdneri*). J. Exp. Biol. 97:325-334.

Smith, H. W. 1961. *From Fish to Philosopher*. New York: Doubleday.

Smith, L. S. and G. R. Bell. 1975. *A Practical Guide to the Anatomy and Physiology of Pacific Salmon*. Misc. Spec. Publ., 27, Dept. of Envir., Fish. and Marine Serv., Ottawa, Canada. 14pp.

Snyder, G. K. 1973. Erythrocyte evolution: the significance of the Fahraeus-Lindqvist phenomenon. Respir. Physiol. 19:271-278.

Somero, G. N. 1981. pH-temperature interactions on proteins: principles of optimal pH and buffer system design. Mar. Biol. Lett. 2:163-178.

Somero, G. N. 1986. Protons, osmolytes, and fitness of internal milieu for protein function. Am. J. Physiol. 251:R197-R213.

Stadie, W. C., J. H. Austin, and H. W. Robinson. 1925. The effect of temperature on the acid-base-protein equilibrium and its influence on the CO_2 absorption curve of whole blood, true and separated serum. J. Biol. Chem. 66:901-920.

Standaert, T., and K. Johansen. 1974. Cutaneous gas exchange in snakes. J. Comp. Physiol. 89:313-320.

Steen, J. B. 1963. The physiology of the swimbladder in the eel *Anguilla vulgaris*. III. The mechanism of gas secretion. Acta Physiol. Scand. 59:221-241.

Steen, J. B., and A. Kruysse. 1964. The resiratory function of teleostean gills. Comp. Biochem. Physiol. 12:127-142.

Steen, I., and J. B. Steen. 1965. The importance of the legs in the thermoregulation of birds. Acta Physiol Scand. 63:285-291.

Stevens, E. D. 1968. The effect of exercise on the distribution of blood to various organs in rainbow trout. Comp. Biochem. Physiol. 25:615-625.

Stevens, E. D., and D. J. Randall. 1967a. Changes in blood pressure, heart rate and breathing rate during moderate swimming activity in rainbow trout. J. Exp. Biol. 46:307-315.

Stevens, E. D., and D. J. Randall. 1967b. Changes of gas concentrations in blood and water during moderate swimming activity in rainbow trout. J. Exp. Biol. 46:329-337.

Stewart, P. A. 1978. Independent and dependent variables of acid-base control. Respir. Physiol. 33:9-26.

Stewart, P. A. 1981. *How to Understand Acid-Base*. New York: Elsevier. 144 pp.

Strazny, F., and S. F. Perry. 1984. Morphometric diffusing capacity and functional anatomy of the book lungs in the spider *Tegenaria* spp. (Agelenidae). J. Morphol. 182:339-354.

Sund, T. 1977. A mathematical model for counter-current multiplication in the swim-bladder. J. Physiol. (London) 267:679-696.

Sutterlin, A. M., and R. L. Saunders. 1969. Proprioceptors in the gills of teleosts. Can. J. Zool. 47:1209-1212.

Swezey, R. R., and G. N. Somero. 1982. Polymerization thermodynamics and structural stabilities of skeletal muscle actins from vertebrates adapted to different temperatures and hydrostatic pressures. Biochemistry 21:4496-4503.

Takeda, R., J. E. Remmers, J. P. Baker, K. P. Madden, and J. P. Farber. 1986. Postsynaptic potentials of bulbar respiratory neurons of the turtle. Respir. Physiol. 64:149-160.

Talor, Z., W.-C. Yang, J. Shuffield, E. Sack, and J. A. Arruda. 1987. Chronic hypercapnia enhances V_{max} of Na-H antiporter of renal brush-border membranes. Am. J. Physiol. 253:F394-F400.

Taylor, C. R., G. M. O. Maloiy, E. R. Weibel, V. A. Langman, J. M. Z. Kamau, H. J. Seeherman, and N. C. Heglund. 1981. Design of the mammalian respiratory system. III. Scaling maximum aerobic capacity to body mass: wild and domestic animals. Respir. Physiol. 44:25-38.

Taylor, E. W., and D. J. Barrett. 1985. Evidence of a respiratory role for the hypoxic bradycardia in the dogfish *Scyliorhinus canicula* L. Comp. Biochem. Physiol. 80A:99-102.

Taylor, E. W., P. J. Butler, and A. Alwassia. 1977. Some responses of shore crab, *Carcinus maenas* (L.) to progressive hypoxia at different acclimation temperatures and salinities. J. Comp. Physiol. B 122:391-402.

Taylor, E. W., and P. Greenaway. 1979. The structure of the gills and lungs of the air-zone crab, *Holthuisana (Austrothelphusa) transversa* (Brachyura: Sundathelphusidae) including observations on arterial vessels within the gills. J. Zool. (London) 189:359-384.

Taylor, E. W., and P. Greenaway. 1984. The role of the gills and branchiostegites in gas exchange in a bimodally breathing crab, *Holthuisana transversa*: evidence for a facultative change in the distribution of the respiratory circulation. J. Exp. Biol. 111:103-121.

Taylor, E. W., S. Short, and P. J. Butler. 1977. The role of the cardiac vagus in the response of the dogfish *Scyliorhinus canicula* to hypoxia. J. Exp. Biol. 70:57-75.

Taylor, H. H., and E. W. Taylor. 1986. Observations of valve-like structures and evidence for rectification of flow within gill lamellae of the crab *Carcinus maenas* (Crustacea, Decapoda). Zoomorphology 106:1-11.

Tazawa, H. 1971. Measurement of respiratory parameters in blood of chicken embryo. J. Appl.

Physiol. 30:17-20.

Tazawa, H. 1980. Oxygen and CO_2 exchange and acid-base regulation in the avian embryo. Amer. Zool. 20:395-404.

Tazawa, H., A. H. J. Visschedijk, and J. Piiper. 1983. Blood gases and acid-base status in chicken embryos with naturally varying egg shell conductance. Respir. Physiol. 54:137-144.

Tenney, S. M., and D. F. Boggs. 1986. Comparative mammalian respiratory control. In A. P. Fishman, N. S. Cherniack, and J. G. Widdicombe, eds., *Handbook of Physiology, Sect. 3: The respiratory System, Vol. II: Control of Breathing, Part 2*. Bethesda: Amer. Physiol. Soc. pp. 833-855.

Tenney, S. M., and J. B. Tenney. 1970. Quantitative morphology of cold-blooded lungs: Amphibia and Reptilia. Respir. Physiol. 9:197-215.

Terwilliger, R. C. 1978. The respiratory pigment of the serpulid polychaete, *Serpula vermicularis*. I. Structure of its chlorocruorin and hemoglobin (erythrocruorin). Comp. Biochem. Physiol. 61B:463-470.

Terwilliger, R. C. 1980. Structures of invertebrate hemoglobins. Amer. Zool. 20:53-67.

Terwilliger, R. C., N. B. Terwilliger and E. Schabtach. 1978. Extracellular hemoglobin of the clam *Cardita borealis* (Conrad): an unusual polymeric hemoglobin. Comp. Biochem. Physiol. 59B:9-14.

Thomas, R. C. 1977. The role of bicarbonate, chloride and sodium ion in the regulation of intracellular pH in snail neurones. J. Physiol. (London) 273:317-388.

Thomas, R. C. 1988. Changes in the surface pH of voltage-clamped snail neurones apparently caused by H^+ fluxes through a channel. J. Physiol. (London) 398:313-328.

Thomas, S., and G. M. Hughes. 1982. A study of the effects of hypoxia on acid-base status of rainbow trout blood using an extra-corporeal blood circulation. Respir. Physiol. 49:371-382.

Thomas, S., B. Fievet, and R. Motais. 1986. Effect of deep hypoxia on acid-base balance in trout: role of ion transfer processes. Am. J. Physiol. 250:R319-R327.

Thorson, T. B. 1968. Body fluid partitioning in reptilia. Copeia (1968, No.3):592-601.

Toews, D., G. Shelton, and R. G. Boutilier. 1982. The amphibian carotid labyrinth: some anatomical and physiological relationships. Can. J. Zool. 60:1153-1160.

Torrance, J. D., C. Lenfant, J. Cruz, and E. Marticorena. 1970. Oxygen transport mechanisms in residents at high altitude. Respir. Physiol. 11:1-15.

Tota, B. 1983. Vascular and metabolic zonation in the ventricular myocardium of mammals and fishes. Comp. Biochem. Physiol. 76A:423-427.

Toulmond, A. 1977. Temperature-induced variations of blood acid-base status in the lugworm, *Arenicola marina* (L.): II. In vivo study. Respir. Physiol. 31:151-160.

Toulmond, A., P. Dejours, and J.-P. Truchot. 1982. Cutaneous O_2 and CO_2 exchanges in the dogfish, *Scyliorhinus canicula*. Respir. Physiol. 48:169-181.

Towle, D. W., and C. P. Mangum. 1985. Ionic regulation and transport ATPase activities during the molt cycle in the blue crab *Callinectes sapidus*. J. Crust. Biol. 5:216-222.

Towle, D. W., C. P. Mangum, B. A. Johnson, and N. A. Mauro. 1982. The role of the coxal gland in ionic, osmotic and pH regulation in the horseshoe crab, *Limulus polyphemus*. In J. Bonaventura, C. Bonaventura, and S. Tesh, eds., *Physiology and Biology of Horseshoe Crabs*. New York: Alan R. Liss. pp. 147-172.

Truchot, J.-P. 1973a. Action spécifique du dioxyde de carbone sur l'affinité pour l'oxygène de l'hémocyanine de *Carcinus maenas* (L.). C.R. Acad. Sci. Paris, [D] 276:2965-2968.

Truchot, J.-P. 1973b. Temperature and acid-base regulation in the shore crab *Carcinus maenas*. Respir. Physiol. 17:11-20.

Truchot, J.-P. 1975a. Factors controlling the in vivo and in vitro oxygen affinity of the hemocyanin in the crab *Carcinus maenas* (L.). Respir. Physiol. 24:173-179.

Truchot, J.-P. 1975b. Blood acid-base changes during experimental emersion and reimmersion of the intertidal crab *Carcinus maenas* (L.). Respir. Physiol. 23:351-360.

Truchot, J.-P. 1979. Mechanisms of the compensation of blood respiratory acid-base disturbances in the shore crab, *Carcinus maenas* (L.). J. exp. Zool. 210:407-416.

Truchot, J.-P. 1980. Lactate increases the oxygen affinity of crab hemocyanin. J. exp. Zool. 214:205-208.

Truchot, J.-P. 1983. Regulation of acid-base balance, In L. H. Mantel, ed., *Biology of the Crustacea*, Vol. 5. New York: Academic Press. pp. 431-457.

Truchot, J.-P. 1987. *Comparative Aspects of Extracellular Acid-Base Balance*. Berlin: Springer-Verlag. 248pp.

Tucker, V. A. 1966. Oxygen consumption of a flying bird. Science 154:150-151.

Tucker, V. A. 1968. Respiratory exchange and evaporative water loss in the flying budgerigar. J. Exp. Biol. 48:67-87.

Tucker, V. A. 1971. Flight energetics in birds. Am. Zool. 11:115-124.

Tucker, V. A. 1972. Respiration during flight in birds. Respir. Physiol. 14:75-82.

Turner, J. D., C. M. Wood, and D. Clark. 1983. Lactate and proton dynamics in the rainbow trout (*Salmo gairdneri*). J. Exp. Biol. 104:247-268.

Tyler, C. 1964. A study of the eggshells of the Anatidae. Proc. Zool. Soc. Lond. 142:547-583.

Ultsch, G. R., and D. C. Jackson. 1982. Long-term submergence at 3°C of the turtle, *Chrysemys picta bellii*, in normoxic and severely hypoxic water. I. Survival, gas exchange and acid-base status. J. Exp. Biol. 96:11-28.

Ultsch, G. R., C. V. Herbert, and D. C. Jackson. 1984. The comparative physiology of diving in North American freshwater turtles. I. Submergence tolerance, gas exchange, and acid-base balance. Physiol. Zool. 57:620-631.

Van Bruggen, E. F. 1983. An electron microscopist's view of the quaternary structure of arthropodan and molluscan hemocyanins. Life Che. Rep. Suppl. 1:1-14.

Van Mierop, L. H. S. 1979. Morphological development of the heart. In R. M. Berne, ed., *Handbook of Physiology, Sect. 2: The Cardiovascular System, Vol. I: The Heart*. Bethesda: Amer. Physiol. Soc. pp. 1-28.

Vuillemin, S. 1963. Système artériel de *Bottia madagascariensis reticulata* G. Pretzman, 1961 (Crustacé, Brachyoure). Bull. Soc. Zool. Fr. 88:603-607.

Vogel, W. O. P. 1978. Arteriovenous anastomoses in the afferent region of trout gill filaments (*Salmo gairdneri* Richardson, Teleostei). Zoomorphologie 90:205-212.

Von Raben, K. 1934. Veranderungen im Kiemendeckel und in den Kiemen einiger Brachyuren (Decapoden) im Verlauf der Anpassung an die Feuchtluftatmung. Z. Wiss. Zoologie. 145:425-461.

Waddell, W. J., and R. G. Bates. 1969. Intracellular pH. Physiol. Rev. 49:285-329.

Walsh, P. J., and T. W. Moon. 1982. The influence of temperature on extracellular and intracellular pH in the American eel, *Anguilla rostrata* (Le Sueur). Respir. Physiol. 50:129-140.

Wang, N., R. B. Banzett, J. P. Butler, and J. J. Fredberg. 1988. Bird lung models show that convective inertia effects inspiratory aerodynamic valving. Respir. Physiol. 73:111-124.

Wangensteen, O. D. 1972. Gas exchange by a bird's embryo. Respir. Physiol. 14:64-74.

Wasser, J. S., and D. C. Jackson. 1988. Acid-base balance and the control of respiration during anoxic and anoxic-hypercapnic gas breathing in turtles. Respir. Physiol. 71:213-226.

Wasserman, K., B. J. Whipp and R. Casaburi. 1979. Respiratory control during exercise. In A. P. Fishman et al., eds., *Handbook of Physiology, Sect. 3, Vol. 2, Part 2*. Bethesda: Amer. Physiol. Soc. pp. 595-619.

Weast, A. C. 1971. *Handbook of Chemistry and Physics*. Cleveland: CRC Press.

Webb, P. W. 1971a. The swimming energetics of trout. I. Thrust and power output at cruising speeds. J. Exp. Biol. 55:489-520.

Webb, P. W. 1971b. The swimming energetics of trout. II. Oxygen consumption and swimming speed. J. Exp. Biol. 55:521-529.

Webb, P. W. 1978. Hydrodynamics: nonscombroid fish. In W. S. Hoar and D. J. Randall, eds., *Fish Physiology*, Vol. VII. New York: Academic Press. pp. 190-237.

Weibel, E. R. 1963. Morphometry of the human lung. Heidelberg: Springer-Verlag.

Weibel, E. R. 1964. Morphometrics of the lung. In W. O. Fenn and H. Rahn, eds., *Handbook of Physiology, Sect. 3, Vol. I.* Bethesda: Amer. Physiol. Soc. pp. 285-307.

Weibel, E. R. 1972. Morphometric estimation of pulmonary diffusion capacity. V. Comparative morphometry of alveolar lungs. Respir. Physiol. 14:26-43.

Weibel, E. R. 1973. Morphological basis of alveolar-capillary gas exchange. Physiol. Rev. 53:419-495.

Weibel, E. R. 1984. *The Pathway for Oxygen*. Harvard Univ. Press, Cambrdge, Mass. 425 pp.

Weinstein, Y., R. A. Ackerman, and F. N. White. 1986. Influence of temperature on the CO_2 dissociation curve of the turtle *Pseudemys scripta*. Respir. Physiol. 63:53-63.

Weis-Fogh, T. 1964. Diffusion in insect wing muscle, the most active tissue known. J. Exp. Biol. 41:229-256.

Weis-Fogh, T. 1967. Respiration and tracheal ventilation in locusts and other flying insects. J. Exp. Biol. 47:561-587.

Welch, B. 1988. Human powered flight takes a great leap upward. Private Pilot 23(3):18-27.

Wells, G. P. 1949. Respiratory movements of *Arenicola marina* L.: intermittent irrigation of the tube and intermittent aerial respiration. J. Mar. Biol. Assoc. U.K. 28:447-464.

Wells, M. J. 1979. The heartbeat of *Octopus vulgaris*. J. Exp. Biol. 78:87-104.

Wells, M. J. 1983. Circulation in cephalopods. In A. S. M. Saleuddin and K. M. Wilbur, eds., *The Mollusca, Vol. 5: Physiology, Part 2*. New York: Academic Press. pp. 239-290..

West, J. 1962. Regional differences in gas exchange in the lung of erect man. J. Appl. Physiol. 17:893-898.

West, N. H., and D. R. Jones. 1976. The initiation of diving apnoea in the frog, *Rana pipiens*. J. Exp. Biol. 64:25-38.

White, F. N., and P. E. Bickler. 1987. Cardiopulmonary gas exchange in the turtle: a model analysis. Am. Zool. 27:31-40.

White, F. N. & G. Ross. 1966. Circulatory changes during experimental diving in the turtle. Am. J. Physiol. 211:15-18.

Wigglesworth, V. B. 1984. *Insect Physiology*, 8th ed. London: Chapman & Hall, 186pp.

Willmer, E. N. 1934. Some observations on the respiration of certain tropical fresh-water fishes. J. Exp. Biol. 11:283-306.

Wilson, T. L. 1977. Theoretical analysis of the effects of two pH regulation patterns on the temperature sensitivities of biological systems in nonhomeothermic animals. Arch. Biochem. Biophys. 182:409-419.

Winterstein, H. 1909. Über die Atmung der Holothurien. Arch. Fisiol. 7:87-93.

Withers, P. C. 1981. The effects of ambient air pressure on oxygen consumption of resting and hovering honeybees. J. Comp. Physiol. B 141:433-437.

Wittenberg, J. B., and B. A. Wittenberg. 1962. Active secretion of oxygen into the eye of fish. Nature 194:106.

Wohlschlag, D. E. 1964. Respiratory metabolism and ecological characteristics of some fishes in McMurdo Sound, Antarctica. In H. W. Wells, ed., *Biology of the Antarctic Seas*. Research Series. Vol. I. Amer. Geophysical Union, Washington, D. C. pp. 33-62.

Wood, C. M. 1974. A critical examination of the physical and adrenergic factors affecting blood flow through the gills of the rainbow trout. J. Exp. Biol. 60:241-265.

Wood, C. M., and F. H. Caldwell. 1978. Renal regulation of acid-base balance in a freshwater fish. J. exp. Zool. 205:301-307.

Wood, C. M., and Cameron, J. N. 1985. Temperature and the physiology of intracellular and extracellular acid-base regulation in the blue crab *Callinectes sapidus*. J. Exp. Biol. 114:151-179.

Wood, C. M., and E. B. Jackson. 1980. Blood acid-base regulation during environmental hyperoxia in the rainbow trout (*Salmo gairdneri*). Respir. Physiol. 42:351-372.

Wood, C. M., D. G. McDonald, and B. R. McMahon. 1982. The influence of experimental anaemia on blood acid-base regulation in vivo and in vitro in the starry flounder (*Platichthys stellatus*) and the rainbow trout (*Salmo gairdneri*). J. Exp. Biol. 96:221-237.

Wood, C. M., B. R. McMahon, and D. G. McDonald. 1977. An analysis of changes in blood pH following exhausting activity in the starry flounder, *Platichthys stellatus*. J. Exp. Biol. 69:173-186.

Wood, C. M., and D. J. Randall. 1981a. Oxygen and carbon dioxide exchange during exercise in the land crab (*Cardisoma carnifex*). J. exp. Zool. 218:7-22.

Wood, C. M., and D. J. Randall. 1981b. Haemolymph gas transport, acid-base regulation, and anaerobic metabolism during exercise in the land crab (*Cardisoma carnifex*). J. exp. Zool. 218:23-35.

Wood, C. M., M. G. Wheatly, and H. Hobe. 1984. The mechanisms of acid-base and ionoregulation in the freshwater rainbow trout during environmental hyperoxia and subsequent normoxia. III. Branchial exchanges. Respir. Physiol. 55:175-192.

Wood, S. C. 1984. Invited opinion: Cardiovascular shunts and oxygen transport in lower vertebrates. Am. J. Physiol. 247:R3-R14.

Wood, S. C., J. W. Hicks, and R. K. Dupre. 1987. Hypoxic reptiles: blood gases, temperature regulation and control of breathing. Amer. Zool. 27:21-30.

Wood, S. C., and K. Johansen. 1972. Adaptation to hypoxia by increased HbO_2 affinity and decreased red cell ATP concentration. Nature New Biology. 237:278-279.

Wunnenberg, W. and D. Baltruschat. 1982. Temperature regulation of golden hamsters during acute hypercapnia. J. Therm. Biol. 7:83-86.

Yancey, P. H., and G. N. Somero. 1978. Temperature dependence of intracelular pH: its role in the conservation of pyruvate apparent Km values of vertebrate. J. Comp. Physiol. B 125:129-134.

Yee, H. F., Jr. and D. C. Jackson. 1984. The effects of different types of acidosis and extracellular calcium on the mechanical activity of turtle atria. J. Comp. Physiol. B 154:385-392.

Young, R. E. 1972a. The physiological ecology of haemocyanin in some selected crabs. I. The characteristics of haemocyaninin a tropical population of the blue crab *Callinectes sapidus* Rathbun. J. exp. mar. Biol. Ecol. 10:183-192.

Young, R. E. 1972b. The physiological ecology of haemocyanin in some selected crabs. II. The characteristics of haemocyanin in relation to terrestrialness. J. exp. mar.Biol. Ecol. 10:193-206.

Young, R. E. 1975. Neuromuscular control of ventilation in the crab *Carcinus maenas*. J. Comp. Physiol. B 101:1-37.

Zandee, D. J., D. A. Holwerda and A. de Zwaan. 1980. Energy metabolism in bivalves and cephalopods. In R. Gilles, ed., *Animals and Environmental Fitness*, Vol. 1. Oxford: Pergamon Press. pp. 185-206.

Zapol, W. M., G. C. Liggins, R. C. Schneider, J. Qvist, M. T. Snider, R. K. Creasy, and P. W. Hochachka. 1979. Regional blood flow during simulated diving in the conscious Weddell seal. J. Appl. Physiol. 47:968-973.

Zeuthen, E. 1942. The ventilation of the respiratory tract in birds. II. Respiration and heat regulation in the bird at rest at high temperature and in the flying bird. K. Danske. Vidensk. Selskab. Biol. Medd. 17:41-51.

Zeuthen, E. 1953. Oxygen uptake as related to body size in organisms. Quart. Rev. Biol. 28:1-12.

Zotterman, Y. 1949. The response of the frog's taste fibers to the application of pure water. Acta Physiol. Scand. 18:181-189.

Zwaan, A. de, A. M. T. de Bont and A. Verhoeven. 1982. Anaerobic energy metabolism in isolated adductor muscle of the sea mussel *Mytilus edulis* L. J. Comp. Physiol. 149:137-143.

Author Index

L

M

N

O

P

V

W

Y

Z

Subject Index

L

M

N

O